SMS and MMS Interworking in Mobile Networks

For a listing of recent titles in the *Artech House Mobile Communications Series,* turn to the back of this book.

SMS and MMS Interworking in Mobile Networks

Arnaud Henry-Labordère
Vincent Jonack

Artech House, Inc.
Boston • London
www.artechhouse.com

Library of Congress Cataloging-in-Publication Data
Henry-Labordère, A.
     SMS and MMS interworking in mobile networks/Arnaud Henry-Labordère, Vincent Jonack.
        p. cm.—(Artech House mobile communications series)
        Includes bibliographical references and index.
        ISBN 1-58053-890-8 (alk. paper)
        1. Mobile communication systems. 2. Internetworking (Telecommunications).
I. Jonack, Vincent. II. Title. III. Series.

        TK6570.M6H42 2004
        621.3845'6—dc22   2004053825

British Library Cataloguing in Publication Data
Henry-Labordère, Arnaud
SMS and MMS interworking in mobile networks. —(Artech House mobile communication series).
    1. Global system for mobile communications   2. Text messages (Telephone systems)
3. Multimedia systems
    I. Title II. Jonack, Vincent
    658.3'845

    ISBN 1-58053-890-8

Cover design by Igor Valdman

International Standard Book Number: 1-58053-890-8

10 9 8 7 6 5 4 3 2 1

Contents

Introduction   xiii

CHAPTER 1
Standard Procedures for SMS in GSM Networks   1

1.1 GSM Network Architecture and Principle of the SMS Procedure   1
1.2 Implementation of SMS Services   3
    1.2.1 SMS-MO Implementation   3
    1.2.2 The SMS-MT Implementation   6
    1.2.3 Sending Commands to the SMSC   14
    1.2.4 Addressing the Foreign Network HLRs for SMS-MT   15
    1.2.5 Summary of the Network Equipment Model for SMS   16
1.3 MAP Dialogue Models at the Application Level   16
    1.3.1 Request and CNF (Simple) Dialogue   17
    1.3.2 Concatenated SMS Dialogue: More Message to Send   17
    1.3.3 Update Location Dialogue   17
    1.3.4 Send Routing Information for SM Dialogue   18
1.4 SCCP Addresses: The Tool for Flexible International Roaming   18
1.5 Mobility Procedures   19
    1.5.1 Update Location Procedure   20
    1.5.2 Making a Telephone Call to a Mobile   22
1.6 GPRS Procedures: The Gc Interface   23
1.7 SMS Billing Records and Methods   23
    1.7.1 SMS-MO CDRs   25
    1.7.2 SMS-MT CDRs   26
1.8 Load Test of an SMSC   26
    1.8.1 SMS-MT Test Configuration   26
    1.8.2 Results and Performance Model   26
    Exercises   28
    References   28

CHAPTER 2
SS7 Network and Protocol Layers   29

2.1 History   29
2.2 Efficient and Secure Worldwide Telecommunications   29
2.3 MTP Protocol (OSI Layers 1–3)   30
    2.3.1 MTP Layer 1: Signaling Data Link Level   31

2.3.2	MTP Layer 2: Signaling Link Functions	31
2.3.3	MTP Layer 3: Signaling Network Functions	34
2.4	Signaling Connection Control Part	37
2.4.1	SCCP Message Format	38
2.4.2	SCCP Layer Architecture	38
2.4.3	SCCP Routing	39
2.5	Transaction Capability Application Part (TCAP)	42
2.5.1	Main Features of TCAP	43
2.5.2	TCAP Architecture	43
2.5.3	TCAP Operation Invocation Example	44
2.6	User-Level Application Parts: MAP, INAP, CAMEL	45
2.6.1	User Part Mapping onto TCAP: MAP Example	45
2.6.2	Routing Design	48
2.6.3	Service-Oriented Design: Application to an SS7-Based Fault-Tolerant System	50
2.7	SS7 and VoIP Interworking Overview SIGTRAN	51
2.7.1	SCTP	51
2.7.2	Interworking with SS7	52
2.7.3	M3UA Layer	52
2.7.4	M2UA Layer	52
2.7.5	SUA Layer	52
2.7.6	TUA Layer	52
2.8	Conclusions	52
2.8.1	Powerful, Efficient Network Architecture	52
2.8.2	Application to a Worldwide SMS Service Network	53
	References	54

CHAPTER 3

Standard Procedures for SMS in IS-41 Networks   57

3.1	Introduction	57
3.1.1	IS-41 Networks	57
3.1.2	Inefficient Handover Chain Procedure	57
3.1.3	MIN and IMSI for IS-41 Networks	59
3.2	Implementation of SMS Services	61
3.2.1	SMS-MO Implementation	61
3.2.2	SMS-MT Implementation	63
3.3	IS-41 Procedure for SMS	63
3.3.1	Functional Description of IS-41 SMS Services	64
3.3.2	IS-41 SMS Protocol Description	68
3.3.3	Specification of the SMS Interworking Network IS-41 SMS Router	70
3.4	Interworking Between IS-41 and GSM	75
3.4.1	GSM Specifications of User Information	75
3.4.2	Mapping GSM to IS-637	76
3.4.3	Mapping GSM to IS-136-710	78
3.4.4	SMS Delivery from IS-41 SME to MAP SME	78
3.4.5	SMS Delivery from MAP SME to IS-41 SME	82

3.4.6   IS-41 Numbering for SMS Delivery   83
3.5   Addressing HLRs in TDMA and CDMA Networks for SMS
Interworking: Updating Point Code–Based Addressing Information   83
References   84

CHAPTER 4
Implementation of Mobile Number Portability and GSM-to-IS-41
Conversion   85

4.1   Business Model   85
4.2   Basics of Roaming Agreement Implementation   85
4.3   Implementations of Number Portability   86
      4.3.1   MNP Handled by Each Individual Operator (Level N)   87
      4.3.2   MNP Handled by the Entry International SCCP Gateway
      (Level $N-1$)   90
      4.3.3   Unregulated Countries' MNP Process Must Be Handled
      by the SMS Interworking Network   91
4.4   SMS Routing Strategies for an SMS Interworking Operator
to a Regulated MNP Country   91
4.5   MNP for SMS in Countries That Have Both GSM and IS-41 Operators   92
      4.5.1   SMS-MT GSM to an IS-41 Destination   92
      4.5.2   SMS-MT from an IS-41 Network to a GSM Destination   95
4.6   Identification of the Destination Network   96
      4.6.1   MMS Interconnection   96
      4.6.2   Fixed-Line SMS Interconnection   96
      4.6.3   MMS and Fixed-Line SMS Interconnection Business   97
References   99

CHAPTER 5
Barring Inbound SMS-MT   101

5.1   Barring Inbound SMS-MT: An Important Business Issue   101
      5.1.1   Filtering Service Offered by IGPs at the SCCP Level   101
      5.1.2   Selective E164 Translation Facility Barring of the SMS-MT
      at the GMSCs SCCP Level   102
      5.1.3   HLR Barring   103
      5.1.4   Origin Address Type Barring at the MSC Level   103
      5.1.5   MAP Barring by the GMSC   103
5.2   Barring or Restricting the SMS-MO of One's Own Subscribers   104
5.3   Intelligent Barring of SMS-MT   104
      5.3.1   Origin Address-Based Barring   104
      5.3.2   Filtering Based on Content of Incoming SMS-MT   105

CHAPTER 6
Virtual SMSC Implementation and Transit Agreements   109

6.1   Business Model   109
6.2   Principle of the Virtual SMSC: Architecture and Billing of SMS-MO   109
      6.2.1   Architecture   109
      6.2.2   Payment Issues   110

      6.2.3   Billing Coherence: Dynamic Originating SMSC GT   111
      6.2.4   Use of a Local Virtual SMSC GT in the SIM Card   111
  6.3   Detailed Implementation of the Virtual SMSC   112
      6.3.1   Half-SCCP Roaming for SMS-MO   112
      6.3.2   Failure of Half-SCCP Roaming for SMS-MO   113
      6.3.3   Solving This Failure Case   113
  6.4   Implementation of Transit Agreements (SMS-MT)   114
      6.4.1   Cases When a Virtual SMSC Has All Roaming Agreements
      of the Operator   114
      6.4.2   Optimization of the Implementation of a Transit Agreement   118
      6.4.3   Use of an International Point Code: The Solution in Difficult
      Setup Cases   118
  6.5   Super-Routing Gateway and Multiple Virtual SMSCs in the Same
  Equipment   120
      Reference   121

CHAPTER 7
Connecting Mobile Operators for SMS-MO   123

  7.1   Business Need for an SMS Interworking Operator to Connect
  Multiple Mobile Operators   123
  7.2   Principle of the Virtual HLR/MSC Approach   123
      7.2.1   Relay Mode   123
      7.2.2   Transparent Mode   125
      7.2.3   Direct Interrogation of the HLR by the Client Operator   126
      7.2.4   SMS Interworking Network and the Status Report   127
  7.3   Configuration the SMSC or GMSC to Route to the Third Party   127
      7.3.1   GT Address Translation in the GMSC   127
      7.3.2   Doing the Address Translation in the SMSC   130
      7.3.3   Use of a Private Conversion Unit   131
      7.3.4   Intelligent SCCP Routing by Your IGP   133
  7.4   Creating Third-Party SCCP Routing When a GT Translation Is
  Unavailable   134
      7.4.1   Case in Which Connected Operator Acts as Its Own SCCP
      Gateway   134
      7.4.2   Case in Which Connected Operator Uses an International
      SCCP Gateway Service: No Solution   135
      7.4.3   Case in Which GT Translation Is Not Possible and the
      Operator Is Not Its Own SCCP Provider: Use a Conversion Unit   135
      7.4.4   Transmission of Signaling Between a GSM and an IS-41
      Network   136
  7.5   Conclusion   136
      Reference   136

CHAPTER 8
Connecting ASPs and ISPs with SMPP   137

  8.1   Introduction   137
  8.2   SMPP Sessions   137

8.3	SMPP Commands	138
8.4	Example of SMPP Sessions	138
8.5	Example of Message Operations	138
	8.5.1 Session Management: Transceiver PDUs	138
	8.5.2 Message Submission Operation	139
	8.5.3 Other SMPP Operations	143
8.6	GSM IS-41 Interworking Through SMPP	143
	Reference	144

CHAPTER 9

MMS Interworking   145

9.1	Introduction	145
9.2	Standard Model for MMS Sending and Receiving	145
	9.2.1 MMS Relay/Server	145
	9.2.2 MMS User Databases	145
	9.2.3 MMS User Agent	146
	9.2.4 MMS VAS Applications	146
9.3	Standard Protocols for MMS	147
	9.3.1 MM1 Protocol over WAP	147
	9.3.2 MM1 over M-IMAP	149
	9.3.3 MM4 Protocol	150
	9.3.4 MM7 Protocol	151
9.4	MMS Interworking Architectures Using a Third Party	151
9.5	Setting Up the MMS Profile in the Cell Phone	156
	9.5.1 Data Access Profile	157
	9.5.2 MMSC Profile	159
	References	160

CHAPTER 10

Optimal Routing Algorithms for an SMS Interworking Network   161

10.1	Maximizing the Margin of an SMS Interworking Network	161
10.2	Enumerating All Loopless Paths with the Latin Multiplication Algorithms	161
10.3	Shortest Path: Djsktra Algorithm	165
10.4	Least Cost Path	165
10.5	Least Trouble Path	165
10.6	The Best Flow Problem—Not a Classical Graph Problem	165
	10.6.1 Income Model for Customer Charges and Notations	166
	10.6.2 Noncontinuous Price Function Paid to the Interworking Network for an Unsatisfied Demand	166
	10.6.3 Continuous Concave Price Function	167
	10.6.4 Network Model	167
	10.6.5 Mathematical Model for Optimization	168
	10.6.6 Algorithm to Find the Global Optimum	171
	10.6.7 Centralized Network Traffic Regulation Principle	171
10.7	Example: Detailed Modeling of a Real SMS Interworking Network	172
	10.7.1 Modeling a Simple SS7 Router or a Relay	172

10.7.2   Modeling Traffic to Subscribers of a Network Hosting an SS7
Router   173
10.7.3   Modeling a Virtual SS7 Router with Several IGPs and Transit
Agreements   173
10.7.4   Connection of Hosting Partners   176
10.7.5   Path Valuations   176
References   176

CHAPTER 11
INAP and CAMEL Overview and Other Solutions for Prepaid SMS   177

11.1   Use of CAMEL for SMS Prepaid Services   178
        11.1.1   SMS Payment from Prepaid Customers   178
        11.1.2   Credit Reloading for Prepaid Customers   179
11.2   Useful Subset of CAMEL Services for Prepaid Customers   179
        11.2.1   Example 1: Prepaid SMS   179
        11.2.2   Example 2: Simple Prepaid Voice Call   179
        11.2.3   Example 3: Voice Call Rerouted to an Announcement Machine   181
        11.2.4   Details of Applicable CAMEL Services   182
        11.2.5   Specificity of the CAMEL Services   183
11.3   Implementation: Multiple-Protocol Services-Oriented Platform:
CAMEL Gateways   184
11.4   Example of Analyzer Traces of a CAMEL Transaction   185
11.5   Other Solutions for Prepaid SMS   187
        11.5.1   Prepaid SMS with Service Nodes   187
        11.5.2   Prepaid SMS with AoC-Enabled Networks   188
        References   189

CHAPTER 12
USSD: A Still-Relevant Conversational Application Service   191

12.1   USSD Advantages over SMS   191
12.2   How Does Mobile-Initiated USSD Service Work?   191
12.3   Example of USSD Service   194
12.4   USSD Is Free: A Call-Back Application   195

CHAPTER 13
Location-Based Services   197

13.1   Location-Based Services: Examples and Revenue Possibilities   197
13.2   Mobile-Originated LBS   197
13.3   Methods   198
        13.3.1   MSC Location Method   198
        13.3.2   Cell ID Method   198
        13.3.3   Extended Cell ID Method   200
        13.3.4   Mobile Location Units and BSSAP-LE   200
13.4   Other Methods: Mobile Measured Power Level   201
13.5   3G UMTS Networks   202
13.6   Best Estimate of a Location Using Hyperbolic n-Triangulation   203
        13.6.1   Algebraic Equation of a Hyperbola   203

13.6.2   Finding the Best Localization Estimate   204
13.6.3   Exact Solution (True Optimum)   205
13.7   Main Results in the Theory of Resultants and Sturm's Theorem   206
13.7.1   Purpose of the Theory of Resultants   206
13.7.2   Main Result for Two Algebraic Equations   206
13.7.3   Sturm's Theorem   208
13.7.4   Bounds on the Value of Roots   210
13.7.5   Application: Recursive Algorithm to Find All the Real Roots   211
References   213

CHAPTER 14

SMS-MO Premium Number Services and Architectures   215

14.1   The Premium SMS-MO Number Business   215
14.1.1   Use of a GSM Modem: Small Throughput   215
14.1.2   Use of a Direct IP Connection to an SMSC: Negotiation and
Setup Tasks   216
14.2   Virtual Roaming Subscriber Architecture   216
14.2.1   Case 1: Omnitel and Third-Party Operator   216
14.2.2   Case 2: Mobile Operator Has a Virtual MSC   217
14.3   SMS-MO with a Real SIM Card   218
14.4   Short Code: A Costly and Time-Consuming Setup   218
14.5   FSG Architecture   219
References   220

CHAPTER 15

Numbering Plan Creation and Maintenance Algorithms   221

15.1   Purpose of Computing Numbering Plans for an SMS Interworking
Network   221
15.2   Entropy of a Numbering Plan as a Quality Indicator   222
15.2.1   Avoiding the Multiple Spanning of HLRs   222
15.2.2   Average Entropy of the Numbering Plan   222
15.2.3   Resulting Global Entropy   223
15.3   "Little Prince" Algorithm to Compute an HLR Numbering Plan   223
15.3.1   Numbering Plan After One Try   224
15.3.2   Numbering Plan After Two Tries   224
15.3.3   Numbering Plan After Three Tries   224
15.4   MSC Search Problem   224
15.4.1   Problem 1   225
15.4.2   Problem 2   225
15.5   Definitions and Properties   225
15.6   Problem 1: Average Number of Searches for a Known N 228
15.6.1   Case $N = 2$ MSCs   228
15.6.2   Case $N = 3$ MSCs   229
15.6.3   Asymptotic Bound of M_N 230
15.7   Problem 2: Estimate of the Probability That the Number of
MSCs $N = j$ 231
References   232

CHAPTER 16

Worked-Out Examples   233

16.1   Example 1   233
16.2   Example 2   250
16.3   Example 3   268
16.4   Example 4   268
16.5   Example 5   269
16.6   Example 6   270
16.7   Example 7: Connection of a GSM to a Third-Party SMS Network   280
16.8   Example 8: SMS Interworking Between CDMA Networks   294

Abbreviations and Acronyms   301
About the Authors   319
Index   321

Introduction

In December 2002, an invitation to lecture at the Belgacom Corporate University (Brussels) offered the opportunity to put together several years of research on the subject of short message service. The *Global System for Mobile Communications* (GSM) includes a large number of beautifully specified standards, with visionary ideas. Comparatively, IS-41 (the U.S.-originated mobile standard specifications for CDMA and TDMA) is much less developed. The GSM specifications are a wonderful reference but they give only a canonical description of the subject.

Short message service (SMS) is a very clever and economical resource that was designed back in the 1980s when GSM specifications were taken from CNET (the research center of France Telecom) and redeveloped as a worldwide standard. These services have been tremendously successful and *multimedia messaging* (MMS) will have the same success. For the same network resources as a telephone call, SMS services provide about 100 times more revenue to the operators.

I have been interested in the subject of SMS and now MMS for several years. Currently, when a telephone call is made to any number in the world, the called party is reached. For SMS, this is not yet the case by far because of a lack of connections, lack of commercial agreements, and differences in standards among GSM, IS-41 (CDMA, TDMA), and others, including Japanese standards. While developing solutions to interwork SMS and later MMS, several nonstandard (noncanonical) procedures were implemented to provide termination and two-way SMS, such as the dynamic reply path procedure.

When I was working at FERMA, a French voice mail manufacturer, we received in 1995 our first order from Telkomsel (Indonesia) for a distributed *voice mail system* (VMS) covering 22 provinces. It was also to be equipped with 22 simplified *SMS centers* (SMSCs) for SMS notification: Each time a message was left on one of the VMSs, its associated SMSC would send the called party the information that they had one new message. It was very basic and we did not anticipate the future interest in sending SMS to other networks. I left FERMA at the end of 1998; after 15 years as chairman-founder-CEO, I thought that I had given everything to voice processing, and wanted to do something completely different.

In early 1999, at NILCOM (Paris), I started working on SMS interworking, that is, the ability to use SMS with anyone in the mobile world. We conducted several experiments with cell phones that could be controlled (with a V24 data cable) by an outside PC, which could change the service center address. We found that many SMSCs accepted the SMSs (with a French operator's originating address) and sent the SMS to their own subscribers along with the true originating addresses. We naively thought that if this was made general, we would have a "worldwide" system. We then attempted to build a "worldwide numbering plan" so that from a destination number, we would get the service center number. This prompted our interest in building an accurate numbering plan with *mobile number portability* (MNP, whose first appearance was in Hong Kong in 1999), but this was just part of the problem. We then created a start-up company to commercialize this PC software.

At the beginning of 1999, we found that when sending an SMS (for a French operator's number) to a Luxembourg SMSC, the SMS reached this French operator with the real originating number! We therefore got the wrong idea that the Luxembourg SMSC was copying the SMS to the French operator's SMSC, using a SMS-MO procedure, to an open SMSC.

By mid-1999, we had created our first SMS interworking gateway, based on the above method. This first gateway was installed in Taipei for a Taiwan operator with my friend Alain Lardenois. The routing strategy was to relay the SMS to the destination SMSC (over an SS7 roaming agreement), the *SMS mobile originated* (SMS-MO) method. It worked well because (although we did not know this at the time) most of the SMSCs were open and SMS-MOs were not charged. From these Taiwanese roaming agreements, we could reach about 70 networks by July 1999, 470 by September 2000, 530 by December 2002, and more than 600 by the end of July 2003.

We had started using the standard *SMS message terminated* (SMS-MT) procedure immediately after the Taiwan experiment: Address the *home location register* (HLR) to get the routing information and send the SMS-MT. We thought, however, that we needed the real HLR address for each addressed mobile number. Hence we began to research the process of HLR finding. Understanding came in September 1999 from a discussion with an expert colleague at France Telecom about *mobile station international ISDN number* (MSISDN) addressing. However, many mobile networks had not implemented a translation facility in the gateway MSC of their network, so direct HLR addressing was kept as backup.

Our full understanding of the subject came in April 2000, after we had installed another SMS interworking gateway in Singapore and conducted extensive tests with our partner. That was when we discovered all of the billing issues. Our backup SMS-MO procedure (we were forcing the originating address to equal the destination address whenever it was rejected) was creating complaints because the recipient was charged! This was because the call detail records of the SMSC were being used for billing in certain Asian countries. From then on, we knew that only SMS-MT standard (or derived) procedures could be used. As a result, we discovered all the procedures associated with the *"network destination code* (NDC) not opened" situation so that our backup direct HLR addressing procedure had to be used and was very useful.

At the beginning of 2000, we also installed an SMS gateway in Indonesia and one of our partners asked us if we could provide two-way SMS; that is, provide a way to receive SMS from another network. This was the demand behind the automatic reply path technique, which we first implemented in March 2000. With these other nodes, we could materialize the relaying techniques, where one node sends to another node, which roams with the destination network. In an important effort to make the software stable, one can say that by the end of 2000, all the ideas concerning the SMS interworking had been identified.

In mid-2000, we also discovered HLR barring with my friends at Chungwha Telecom (Taiwan) who pointed out the effect in SMS-MT *called detail records* (CDRs) at the receiving *mobile switching center* (MSC) of the wrong originating SC addresses.

We had signed a SMS interworking contract with Scancom (Ghana), which had a talented group of engineers, and they asked us if we could provide an *unstructured supplementary service data* (USSD) platform for value-added services, such as registration on a portal for breaking news or demands for weather forecast. USSD allows conversational services, using text, with a server. It exists in GSM 2G, 2.5G, and 3G and is somewhat ignored. USSD is a better alternative to SMS in conversational application cases such as chat, service registration, and customer provisioning.

With Scancom, we also implemented the notion of a private SS7 network with a *signaling transfer point* (STP) in Monaco and the virtual SMSC concept that we had patented. The use of private leased lines was justified by the high traffic and the savings on the volume-dependent charges of the international SS7 carrier. We also acquired a lot of practical experience on the configuration of STPs and SCCP address translation. As a result of teaching operations research for more than 30 years, it was inevitable that many ideas on routing SMS were drawn from this experience. Routing algorithms are instrumental to these nonstandard SMS interworking methods. To implement them easily, one must rely on the SS7 network. The description of the network layer (SCCP and MTP) of SS7 is therefore necessary and is found in Chapter 2.

Working on the standard SMS-MT procedure, we wanted to provide receiving time zone accuracy; that is, when an SMS is received, the time stamp should be that of the receiving region. So we developed a huge MSC database, which today has more than 4,000 entries. The idea of developing a *mobile location center* (MLC) arose very early. The coming of *general packet radio service* (GPRS) and *customized application for mobile network* (CAMEL) made it easy as more and more networks implemented the additional software and hardware facilities that could provide localization data. We delivered our first MLC early in 2002. MLCs use the SMS procedure in the initial phase (gross resolution location). The subject has been included, as well as the original work on the best approximation of the position.

MMS is a very promising topic for future business because 2 years from now, there will not be any costs associated with an MMS-enabled phone. With MNP, it is a key interworking issue. With our work on MNP, we thought that we had a several-years edge for providing MMS interworking as well as MM5 network name resolution.

Also we met in 2002 with the problem of interworking in the fast growing fixed-line SMS market. Our approach with a domain resolution server (which uses the same idea for SMS as a DNS for the Internet) proved well worth pursuing for MMS interworking (described in Chapter 9). You will see that the interworking issue is quite different from the mobile GSM interworking.

Maximizing the operating margin of SMS interworking is a key issue: Optimal routing algorithms will be required to decide the best paths over which to forward the SMS traffic. The algorithm uses the multicommodity flow model described in Chapter 10 and requires linear and nonlinear programming. My goal is for the modeling to be understandable by any reader, although the solution algorithms are for specialists.

Many of the acronyms used in this book are known to people who work in the subject. Those who are unfamiliar with the acronyms will find them defined at the end of the book and detailed explanations in the main text.

This book is aimed at marketing, sales, technical, and operations professionals working with mobile operators or *application service providers* (ASPs) as well as consultants. They will understand the business, billing, and very detailed technical issues associated with SMS and MMS interworking, that is, with the exchange of SMS and MMS among all of the world's mobile operators as well as with content providers. The book will also be useful to SMSC and *multimedia messaging center* (MMSC) manufacturers, because it provides a global view of interworking that they can integrate into their designs.

The text of this book has been used for teaching nonengineer, business graduates and engineering students without difficulty, including its worked-out examples and exercises.

Those who do not have a second-year college level in mathematics can skip the appendixes on the location-based methods (Chapter 13) and the second part of Chapter 15, which is on computation of numbering plans. To understand Chapter 10 (optimal routing algorithms in SMS/MMS interworking networks), you must have studied graph theory and linear/nonlinear programming.

Throughout the book, we assume that the reader has a basic knowledge of graph theory vocabulary, which is common now in many courses. The rest is self-contained including the necessary explanations of the SS7 network (the packet-based network used to exchange data between mobile operators).

## CHAPTER 1

Standard Procedures for SMS in GSM Networks

To Kunlun now I say,
Neither all your height
Nor all your snow is needed.
Could I but draw my sword overtopping heaven,
I'd cleave you in three:
One piece for Europe,
One for America,
One to keep in the East.
Now a world in peace, sharing together
The same warmth and cold throughout the globe.

—Mao Zedong, *Poems*, 1935

1.1 GSM Network Architecture and Principle of the SMS Procedure

A GSM network uses two separate networks: one for telephone calls (using the ISUP protocol, which is the foundation of the ISDN protocol, for call setup and release) and one for signaling, mainly for mobility-related messages (otherwise it would not be a mobile network) and for SMS. Each mobile operator has its own two private separate such networks that it builds with equipment and lines. To communicate with each other (in particular, to be able to offer roaming services), providers use international carriers (France Telecom, Belgacom, Teleglobe, Swisscom, British Telecom, and so on), which also provide them with an international telephone network (ISUP) and a signaling network for the exchange of signaling messages (using the SCCP protocol; see Chapter 2).

In a given mobile network, such as that shown in Figure 1.1, the gateway between the private networks and the international carrier is a piece of equipment called a *gateway mobile switching center* (GMSC), which is both a switch for telephone calls and a router for the SCCP signaling messages. The SMS service involves only the SCCP data signaling network (a packet datagram network reserved for mobile operators) and is covered in Chapter 2. An SMS uses only about 1 Kb and, because it brings about 0.1 euros of revenue for the operator, it provides a high margin compared to a voice call of 30 seconds.

A cell phone (1) in Figure 1.1 sends an SMS with a text to a destination number (MSISDN). The *subscriber identity module* (SIM) card, a smart card with a processor and a memory that carries all subscription information if someone is using

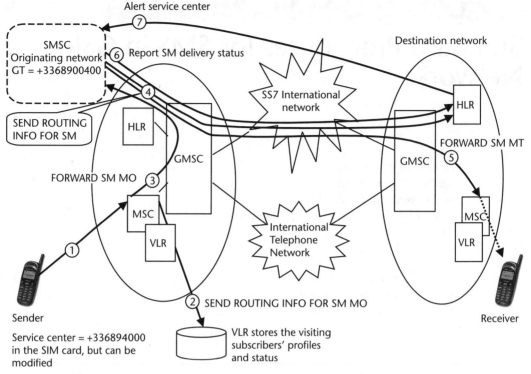

Figure 1.1 Implementation of SMS service in GSM networks.

another cell phone, stores the SMSC number. The SMS-MO reaches this SMSC through the *Signaling System No. 7* (SS7) signaling network (3); this is the SMS-MO procedure Then the SMSC sends it to the destination cell phone, either within the same network or in another network (as in Figure 1.1) using the SMS-MT procedure.

To send the SMS, the SMSC interrogates the HLR of the destination cell phone (4) to get the address of the equipment (the MSC/VLR which can be in another country) where this cell phone is currently visiting. It can then send the SMS-MT signal to this MSC [see (5) in Figure 1.1].

If the signal fails for any of various reasons (e.g., the mobile is out of the coverage area, the memory of the cell phone is full), the SMSC has a retry scheme, including triggering (6) by automatic alerts whenever the conditions required to deliver the SMS are met again (7). This is the purpose of the retry and alert mechanisms. The role of the *visitor location register* (VLR) will be explained in the SMS-MO implementation discussion and in the mobility procedure.

At this level, there is no fundamental difference between the SMS procedure in GSM, which uses the *Mobility Application Part* (MAP) protocol, and IS-41, the ANSI standard used for *code division multiple access* (CDMA) and *time division multiple access* (TDMA) cell phones (more precisely TIA/EIA-41-D; see Chapter 3).

In GSM the SMSC can easily address the HLR of its roaming partners because an international MSISDN is used to interrogate the HLRs. In IS-41 this is not the case, as explained in the later chapter on IS-41. This is why there is a major difference in CDMA or TDMA when it comes to sending a SMS to a subscriber in another network (CDMA or TDMA). The SMS is sent from the originating SMSC to the SMSC of the destination network. In GSM, there is no difference between sending a SMS to a subscriber of the same network or another network.

We must also consider *multimedia messaging service* (MMS) for still and animated pictures, quality sound, and so on. MMS uses the well-known Internet MIME format. The delivery mechanism from one network to the other is very different from the SMS procedure. It is based on the standard Internet *Simplified Mail Transfer Protocol* (SMTP), the same protocol used in PCs to send e-mail.

1.2   Implementation of SMS Services

Referring to Figure 1.1, note that it shows the exchanges that occur when a cell phone (on the left) sends an SMS to a cell phone in another network (on the right). We now detail these exchanges one by one. The detailed contents of the various messages or primitives of the MAP protocol are given and it is necessary to fully understand them. The mandatory parameters are referred to as M, conditional as C (they depend on the context), and optional as O.

1.2.1   SMS-MO Implementation

The SIM card has a service center address (a normal international number) that is your operator's SMSC, a piece of equipment designed for receiving SMS (SMS-MO) and sending it to other phones (SMS-MT). It has a store-and-forward capability so that if the destination is not reachable, it can retry later until the end of a validity period.

To send an SMS, a user writes the text on his or her mobile, enters the destination number, and presses the SEND button. The SMS then goes to the SMSC, which replies back SENT. In many cases (even recently), the SMSC will reply SENT, and some operators charge the customer even if it has no roaming agreement with the destination area and is completely unable to perform the service. If you have a GSM, try to send to +821112345678 (a CDMA number in Korea). If you receive the message SENT, this is not normal because it is very likely that the destination number has not received your SMS. This is the objective of the implementation of SMS interworking—to remedy this situation and to provide a worldwide SMS termination service, just like that which exists for voice telephone calls.

In Figure 1.1, we see various MAP primitives. MAP is the high-level protocol used in GSM mobile networks to implement mobility, SMS, and other functions [1]. Various evolutions of this protocol are planned: V1, V2, V3, and V4. This is shown in the context of the various primitives.

Let's look in detail at the content of the various messages used in the implementation. We start with FWD_SM_MO [(3) in Figure 1.1] (see Table 1.1), which is sent by the MSC that you are currently visiting to your SMSC. (Note that you are registered in a base station that is connected to this MSC by landlines belonging to your operator.) We use the normalized name of the parameters as defined in [2]. SM-RP stands for *Short Message Relay Protocol,* the upper layer of the MAP protocol layer (see Chapter 2).

Table 1.1 FORWARD_SHORT_MESSAGE_REQ and CNF
Message Details

FORWARD_SHORT_MESSAGE_REQ (MO)

Parameter	Class	Context
Primitive type octet	M	V1,V2
Timeout	O	V1,V2
Invoke ID	M	V1,V2
SM-RP-DA (the service center GT)	M	V1,V2
SM-RP-OA (the sending cell phone number)	M	V1,V2 *(MSISDN)*
SM-RP-UI (a SM_SUBMIT type)	M	V1,V2 *Text mss*

FORWARD_SHORT_MESSAGE_CNF

Parameter	Class	Context
Primitive type octet	M	V1,V2
Invoke ID	M	V1,V2
User error	O	V1,V2
Network resource	O	V1,V2
SM delivery failure cause	C	V1,V2

The SM-RP-DA (destination address) is the address (called global title or abbreviated GT) of the SMSC in your SIM card for a SMS-MO. You can modify this address.

The SM-RP-OA (originating address) is your ordinary number (MSISDN) for a SMS-MO. It is inserted by the VLR, the database integrated with the MSC. If follows the sending of a SMS over the radio channel [(1) in Figure 1.1] to the MSC, the mobile being identified by the *International Mobile Subscriber Identity* (IMSI) of the SIM card. Then the VLR interrogates the MSC [SEND_INFO_FOR_SMS-MO, which is (2) in Figure 1.1], which returns the MSISDN. It has obtained this MSISDN in the mobility procedure (see the later section).

The SM-RP-UI is the text of the message. It has a type called SUBMIT and contains many other parameters that you have specified; for instance, whether you want a status receipt when the SMS actually reaches the destination, whether the SMS that you send is text, but also logos or ring tones. The details of the content of a SM_SUBMIT (the SMS-MO) are shown in Table 1.2.

The cell phone is also able to specify (1) if it wants a status report (when the SMS is received by the cell phone), (2) its own validity period (the SMS is discarded after this), and (3) if it wants to set a Reply-Path for the reply of the destination cell phone.

1.2.1.1 Purpose of the Reply-Path Function

When this function was specified about 1990, the underlying idea was that SMS sending party A would offer to its receiving party B the possibility of a free (not charged) reply to the SMS that was received. To do this, sending cell phone A sets up an option, sometimes called free reply (in the cell phone menu), which will set the TP-Reply-Path parameter. So the SMSC of A will know that is has to set the Reply-Path as explained later in the SMS-MT procedure.

1.2.1.2 Roaming Case

If the sender is roaming in another network, nothing is changed. The SMS-MO will reach the home SMSC through the international SS7 network. The visited network

Table 1.2   SM_SUBMIT Message Details

Abbreviation	Reference	Class	Length	Description
TP-MTI	TP-Message-Type-Indicator	M	2b	Parameter describing the message type.
TP-RD	TP-Reject-Duplicates	M	B	Parameter indicating whether or not the SC shall accept an SMS-SUBMIT for an SM still held in the SC, which has the same TP-MR and the same TP-DA as a previously submitted SM from the same OA.
TP-VPF	TP-Validity-Period-Format	M	2b	Parameter indicating whether or not the TP-VP field is present.
TP-RP	TP-Reply-Path	M	B	Parameter indicating the request for a Reply-Path.
TP-UDHI	TP-User-Data-Header-Indicator	O	B	Parameter indicating that the TP-UD field contains a header.
TP-SRR	TP-Status-Report-Request	O	B	Parameter indicating if the MS is requesting a status report.
TP-MR	TP-Message-Reference	M	I	Parameter identifying the SMS-SUBMIT.
TP-DA	TP-Destination-Address	M	2-12o	Address of the destination SME.
TP-PID	TP-Protocol-Identifier	M	O	Parameter identifying the top layer protocol, if any.
TP-DCS	TP-Data-Coding-Scheme	M	O	Parameter identifying the coding scheme within the TP–User–Data.
TP-VP	TP-Validity-Period	O	o/7o	Parameter identifying the time from where the message is no longer valid.
TP-UDL	TP-User-Data-Length	M	I	Parameter indicating the length of the TP-User-Data field to follow.
TP-UD	TP-User-Data	O		

(VPLMN) will create a SMS-MO CDR in the MSC that was used. It may claim an agreed-on charge with the *home public lands mobile network* (HPLMN). Depending on the SMS billing implementation, the HPLMN may use this roaming CDR to increase the charges to their subscriber or with their SMSC's own SMS-MO records.

1.2.1.3   Changing the Service Center

If one changes the service center address and sends an SMS-MO to another service center, then the following holds:

- If there is no MSC barring (a rare feature implemented by Orange in the United Kingdom, SFR, and others), and if there is roaming with the target SMSC, the SMS-MO will reach the SMSC. This control is implemented with the SEND_INFO_FOR_MO_SMS_REQ sent by the MSC to the VLR.
- In 2002, 99% of the world's SMSCs controlled the originating address. If the originator is not one of their subscribers, they will reject the SMS-MO (you see the message "REFUSED" within a few seconds on your cell phone). If after about 30 seconds, you see REFUSED, it indicates that there is no roaming.

1.2.1.4   Controlling the Originating Address in the SMSC

The FORWARD_SHORT_MESSAGE (FWD_SM_MO) sent by the visited MSC to the SMSC includes a SM_RP_OA message, that is, the originating cell phone number (MSISDN). The owner of the SMSC wants to reserve the usage of the

MNP

SMSCs to its own subscribers. Two methods are used to control this: (1) without MNP, in which the SMSC has a table of its range of numbers, and (2) with MNP, in which the SMSC must interrogate its HLRs in order to verify that the sender is one of its subscribers.

1.2.1.5   SMS-MO Teleservice Provisioning Control in the Visited MSC

The later discussion on mobility explains that the VLR associated with the MSC has the profile of all cell phones registered in the area. In particular, it knows the list of the teleservices they can use. So if an operator wishes to restrict the SMS-MO for its roaming subscribers because the users are charged for the service, it may suppress the SMS-MO teleservice, in which case the cell phone immediately gets a REJECT message. This is because the MSC interrogates the VLR with a SEND_INFO_FOR_MO_SMS containing the IMSI of the cell phone and receives the MSISDN, as well as a possible reject cause (see Table 1.3).

1.2.2   The SMS-MT Implementation

The procedure to send an SMS-MT is much more complex than the SMS-MO case and is now explained in detail.

1.2.2.1   Interrogation of the HLR

Once the SMS-MO has reached the SMSC, this equipment will address the HLR of the destination phone in order to get another number, the IMSI, and the address of the visited MSC. It does this with a SEND_ROUTING_INFO_FOR_SM [(4) in Figure 1.1; see Table 1.4]. The key for the interrogation is the MSISDN (ordinary number). Then it uses a FORWARD_SHORT_MESSAGE [(5) in Figure 1.1].

Various important parameters include the following:

- The MSISDN;
- The SM-RP-PRI, which designates the priority of the request;

Table 1.3   SEND_INFO_FOR_MO_SMS_REQ and CNF Message Details

Interrogate VLR for MSISDN of originator

SEND_INFO_FOR_MO_SMS_REQ		
Parameter	*Class*	*Context*
Primitive type octet	M	V1,V2
Timeout	O	V1,V2
Invoke ID	M	V1,V2
Service center address*	M	V1,V2
SEND_INFO_FOR_MO_SMS_CNF		
Parameter	*Class*	*Context*
Primitive type octet	M	V1,V2
Invoke ID	M	V1,V2
MSISDN (the MSISDN was stored in the VLR)	C	V1,V2
User error	C	V1,V2
Call barring cause	O	V1,V2

* Used to ensure that the subscribers are using only their operator's service center and not that of somebody else; this disables the possibility of Reply-Paths.

Table 1.4 SEND_ROUTING_INFO_FOR_SM_REQ and CNF Message Details

SEND_ROUTING_INFO_FOR_SM_REQ

Parameter	Class	Context
Primitive type octet	M	V1,V2
Timeout	O	V1,V2
Invoke ID	M	V1,V2
MSISDN	M	V1,V2
SM-RP-PRI	M	V1,V2
Service center address	M	V1,V2
CUG interlock	O	V1,V2
GPRS support indicator	O	V2
Teleservice	O	V1,V2

SEND_ROUTING_INFO_FOR_SM_CNF

Parameter	Class	Context
Primitive type octet	M	V1,V2
Invoke ID	M	V1,V2
Where User Error Not Included:		
IMSI	M	V1,V2
MSC number	M	V1,V2
LMSI	O	V1,V2
MWD set	O	V1,V2
GPRS node indicator	O	V1,V2
Network node number	O	V1,V2
Where User Error Included:		
User error	M	V1,V2
Network resource	O	V1,V2
Call barring cause	O	V1,V2

[Handwritten annotations: "Sent to HLR of destination phone"; "white list of SMSC"; "⟺ Deliver VIA SGSN?"; "IMSI = Intl - Mobile Subscriber Identity of SIM card"]

- The service center address of the requesting SMSC, which is used by the HLR to authorize the request from a white list of authorized SMSCs, if the operator wants to restrict the reception of SMS-MT from other networks;

- The GPRS support indicator, which indicates whether the destination MSISDN supports the delivery of SMS-MT using the GPRS network. The procedure is very much like using the circuit mode when the SMS is sent by the visited MSC. Using GPRS, the SMS is sent using the visited *support GPRS service node* (SGSN). (See Chapter 9 on MMS for a description of the GPRS architecture.)

The key to the possibility of easy SMS interworking between GSM networks is this function. *From the known MSISDN (the ordinary number), you can get the IMSI to actually address the SMS.* This is not the case with the IS-41 networks and makes SMS interworking very difficult to implement and maintain.

If the home network supports the delivery of SMS-MT by GPRS, and if it has been requested with the setting of the GPRS support indicator in the SEND_ROUTING_INFO_FOR_SM_REQ, the HLR also returns the SGSN address (a GT) in the parameter network node number. The SMS may then choose the circuit mode through the visited MSC or the GPRS mode through the visited SGSN.

1.2.2.2 Forwarding the SMS-MT

The SMS-MT is sent to the GT of the visited MSC (or the GT of the SGSN if the GPRS mode is chosen), which has been returned by the HLR interrogation. The key to addressing a particular subscriber is the IMSI. Although they have the same names, the content of the SMS-RP-DA and SM-RP-OA parameters are different from the SMS-MO case. Table 1.5 shows the details for the FORWARD_SHORT_MESSAGE_REQ and CNF.

If the SMSC sets the "more message to send" feature, the MSC will not page between blocks, which is a very important feature for *over the air* (OTA) provisioning; that is, downloading of a large amount of data automatically, such as an address book, to customize a cell phone or a SIM card. The basic elements of the SMS-DELIVER type SMS are shown in Table 1.6.

The User-Error indicates the type of failure:

27: Subscriber has switched off his phone (detached) or is temporarily not reachable (when in a tunnel, for example).

32: The total memory of the SIM card and of the SMS memory extension in the cell phone itself is full and cannot receive more SMS. The SMS must be retried later.

34: Receiving MSC congestion, for example, if two SMSs are sent at the same time to the cell phone.

Depending on the error, various retry algorithms will be used by the SMSC. In Figure 1.1, we have shown a case when the FORWARD_SHORT_ MESSAGE_MT is not successful.

1.2.2.3 Setting Up a Reply-Path

If the SMSC has received an SMS-MO from A requesting a Reply-Path, it will set the field TP-Reply-Path of the SMS-MT sent to B. When receiving cell phone B wants to

Table 1.5 FORWARD_SHORT_MESSAGE_REQ and CNF Message Details

FORWARD_SHORT_MESSAGE_REQ (MT)		
Parameter	*Class*	*Context*
Primitive type octet	M	V1,V2
Timeout	O	V1,V2
Invoke ID	M	V1,V2
SM-RP-DA (the IMSI)	M	V1,V2
SM-RP-OA (the sending service center GT)	M	V1,V2
SM-RP-UI (a SM_DELIVER type)	M	V1,V2
More message to send	O	V2
FORWARD_SHORT_MESSAGE_CNF		
Parameter	*Class*	*Context*
Primitive type octet	M	V1,V2
Invoke ID	M	V1,V2
User error	O	V1,V2
Network resource	O	V1,V2
SM delivery failure cause	C	V1,V2

Table 1.6   Basic Elements of the SMS-DELIVER Type SMS (SMS-MT)

Abbreviation	Reference	Class	Length of Element	Description
TP-MTI	TP-Message-Type-Indicator	M	2b	Parameter describing the message type.
TP-MMS	TP-More-Messages-to-Send	M	b	Parameter indicating whether or not there are more messages to send.
TP-RP	TP-Reply-Path	M	b	Parameter indicating that Reply-Path exists.
TP-UDHI	TP-User-Data-Header-Indicator	O	b	Parameter indicating that the TP-UD field contains a header.
TP-SRI	TP-Status-Report-Indication	O	b	Parameter indicating whether the SME has requested a status report.
TP-OA	TP-Originating-Address	M	2–12o	Address of the originating SME.
TP-PID	TP-Protocol-Identifier	M	o	Parameter identifying the top layer protocol, if any.
TP-DCS	TP-Data-Coding-Scheme	M	o	Parameter identifying the coding scheme within the TP-User-Data.
TP-SCTS	TP-Service-Center-Time-Stamp	M	7o	Parameter identifying time when the SC received the message.
TP-UDL	TP-User-Data-Length	M	I	Parameter indicating the length of the TP-User-Data field to follow.
TP-UD	TP-User-Data	O	Variable	The text or the data of the SMS

b = bit, o = octet, and I = integer.

use the Reply function to reply to this SMS, the SMS-MO will be sent to the SMSC of A instead of its own SMSC B (whose address is in the SIM card). It will jump to its own SMSC B. The developers of the GSM standard thought that, in this way, B would not be charged by its operator for the SMS replied to A. This feature is now almost never used in standard SMSCs because, as explained later in the section on billing procedures, B will still be charged in many cases. The SMSC of A will see an originating address in the SMS-MO sent by B that is not one of its subscribers and will bar this SMS-MO.

Situations in which it is used for third-party SMS interworking networks that have a number of SMSCs and voluntarily set the Reply-Path offer a reply possibility. The possibility to reply is limited by a reply token, which is allocated and contains (1) the origin MSISDN, (2) the destination MSISDN, and (3) the date-time of creation, and (4) is canceled once the reply has been performed by means of the SMSC that originated the SMS-MT.

1.2.2.4   Concatenated Short Messages

This section can be skipped at first reading; however, it is useful for people implementing the sending of SMS for various value-added services.

It is possible to concatenate to form a longer message. Applications are picture messages and OTA provisioning, in which a particular profile (for example, a WAP profile) is downloaded to the cell phone using a number of SMS. The format guarantees that they are resequenced properly by the receiving cell phone, if needed.

A compressed format is available that conforms to specifications. In the case of uncompressed 8-bit data, the maximum length of the short message within the TP-UD field (user data) is 134 (140 – 6) octets. In the case of uncompressed GSM default 7-bit data, the maximum length of the short message within the TP-UD field is 153 (160 – 7) characters. In the case of 16-bit uncompressed USC2 data, the maximum length of the short message within the TP-UD field is 67 [(140 – 6)/2] characters. A UCS2 character must not be split in the middle; if the length of the user data header is odd, the maximum length of the whole TP-UD field is 139 octets.

In the case of compressed GSM default alphabet 7-bit data, 8-bit data, or UCS2, the maximum length of the compressed short message within the TP-UD field is 134 (140 – 6) octets including the compression header and compression footer, both or either of which may be present.

The maximum length of an uncompressed concatenated short message is 39,015 (255×153) default alphabet characters, 34,170 (255×134) octets or 17,085 (255×67) UCS2 characters.

The maximum length of a compressed concatenated message is 34,170 (255×134) octets including the compression header and compression footer (see Figure 1.2).

The Information–Element–Data field contains information set by the application in the SMS_SUBMIT so that the receiving entity is able to reassemble the short messages in the correct order. Each concatenated short message contains a reference number that, together with the originating address and service center address, allows the receiving entity to discriminate between concatenated short messages sent from different originating SMEs and/or SCs. In a network that has multiple SCs, it is possible for different segments of a concatenated SM to be sent via different SCs and so it is recommended that the SC address *not* be checked by the MS unless the application specifically requires such a check.

The TP elements in the SMS_SUBMIT PDU, apart from TP-MR, TP-SRR, TP-UDL, and TP-UD, should remain unchanged for each SM that forms part of a concatenated SM, otherwise this may lead to irrational behavior. TP-MR must be incremented for every segment of a concatenated message. An SC will handle segments of a concatenated message like any other short message. The relationship between segments of a concatenated message is made only at the originator, where the message is segmented, and at the recipient, where the message is reassembled. SMS_COMMANDs identify messages by TP-MR (message reference number) and therefore apply to only one segment of a concatenated message. It is up to the originating SME to issue SMS_COMMANDs for all required segments of a concatenated message.

The Information–Element–Data octets shall be coded as follows:

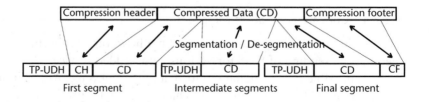

Figure 1.2   Concatenation of a compressed short message.

- *Octet 1: Concatenated short message reference number.* This octet shall contain a modulo 256 counter indicating the reference number for a particular concatenated short message. This reference number shall remain constant for every short message that makes up a particular concatenated short message.
- *Octet 2: Maximum number of short messages in the concatenated short message.* This octet shall contain a value in the range 0 to 255, indicating the total number of short messages within the concatenated short message. The value shall start at 1 and remain constant for every short message that makes up the concatenated short message. If the value is zero then the receiving entity shall ignore the whole information element.
- *Octet 3: Sequence number of the current short message.* This octet shall contain a value in the range 0 to 255, indicating the sequence number of a particular short message within the concatenated short message. The value shall start at 1 and increment by one for every short message sent within the concatenated short message. If the value is zero or the value is greater than the value in octet 2, then the receiving entity shall ignore the whole information element.

1.2.2.5   The Retry and Alert Mechanisms and SMS Delivery Failures

Whenever the cell phone becomes either reachable again or the SMS memory is not full anymore, an exchange with the MSC will occur over the radio interface (the A interface) with the visited MSC. To do this a CM_Service_Request is sent to the MSC, which relays it to its associated VLR. As a result of it the VLR will send to the HLR a MAP_READY_FOR_SM_REQ to inform it of the current situation. The parameters are the IMSI (the HLR can find the MSISDN) and the alert reason (whether it is a change of reachability or cell phone memory state) as shown in Table 1.7 (VLR→HLR message).

The HLR then provides the ALERT function so that an SMSC knows immediately when a subscriber is reachable again or when the memory is not full. To be alerted, the SMSC must have previously sent a REPORT_SM_DELIVERY_ STATUS to the HLR [(6) in Figure 1.1], specifying in the SM delivery outcome parameter the reason for the alert (see Table 1.8). The HLR will send an

Table 1.7   READY_FOR_SM_REQ and CNF Message Details

READY_FOR_SM_REQ		
Parameter	*Class*	*Context*
Primitive type octet	M	V1,V2
Invoke ID	M	V1,V2
IMSI or TMSI	M (one of them)	V1,V2
Alert reason	M	V1,V2
Alert reason indicator	M	V1,V2
READY_FOR_SM_CNF		
Parameter	*Class*	*Context*
Primitive type octet	M	V1,V2
Invoke ID	M	V1,V2
User error	C[1]	V1,V2

Table 1.8 REPORT_SM_DELIVERY_STATUS_REQ and CNF Message Details

REPORT_SM_DELIVERY_STATUS_REQ		
Parameter	*Class*	*Context*
Primitive type octet	M	V1,V2
Timeout	O	V1,V2
Invoke ID	M	V1,V2
MSISDN	M	V1,V2
Service center address	M	V1,V2
SM delivery outcome (not reachable or SIM full)	M	V2
REPORT_SM_DELIVERY_STATUS_CNF		
Parameter	*Class*	*Context*
Primitive type octet	M	V1,V2
Invoke ID	M	V1,V2
MSISDN	C	V2

ALERT_SERVICE_CENTER [(7) in Figure 1.1] to the SMSC (with the MSISDN) so that the SMSC can immediately resend the SMS (see Table 1.9).

In MAP V2, it also sends an INFORM_SERVICE_CENTER_REQ message, indicating the status of the *message waiting data* (MWD) (see Table 1.10). However, the SMSC must implement also a regular retry mechanism for the cases where there is no alert:

- MSC congestion, for example, once every minute during a 10-minute time period;
- Error 27, if it is a transient that is not reachable, there is no way to know, and the SMS must be repeated regularly (for example, every minute, then every hour).

But for the nonreachable reason or when the memory is full, it is sufficient to rely on the alert mechanism, which saves a lot of SS7 traffic.

In the SMSC a validity period is set. After this period has passed, the SMS is deleted. This is one reason (in addition to temporary network congestion or failures) why it is never possible to guarantee a 100% delivery rate. The SS7 link protocol (see

Table 1.9 ALERT_SERVICE_CENTER and CNF Message Details

ALERT_SERVICE_CENTER		
Parameter	*Class*	*Context*
Primitive type octet	M	V2
Timeout	O	V2
Invoke ID	M	V2
MSISDN	M	V2
Service center address	M	V2
ALERT_SERVICE_CENTER_CNF		
Parameter	*Class*	*Context*
Primitive type octet	M	V2
Invoke ID	M	V2
User error	O	V2
Network resource	O	V2

Table 1.10   INFORM_SERVICE_CENTER_REQ
Message Details

INFORM_SERVICE_CENTER_REQ		
Parameter	*Class*	*Context*
Primitive type octet	M	V2
Timeout	O	V2
Invoke ID	M	V2
MSISDN	O	V2
MWD status	O	V2

Chapter 2) will drop the packets (more specifically called MSUs) if the flow of data that is received overflows the input buffer because there is no flow control in the SS7 protocol. In particular, an excessive rate of SMS-MT (spamming or bulk messaging) could even provoke a failure of the addressed network.

Figure 1.3 shows average delivery rate as a growing function of time in the system. Because little progress is gained after 24 or 48 hours, this validity period may be retained to reduce the system load.

1.2.2.6   Status Reporting Back to the Cell Phone

When you set up your cell phone, you can request status reports. That is, when the SMS is effectively received by the destination, you will know it. This can happen several hours after you have sent the SMS-MO to your SMSC.

Before you get this status report back from the SMSC, in your cell phone, you see a PENDING message. The status is sent in a special SMS-MT from the SMSC to the cell phone, using an SMS_STATUS_REPORT type (see Table 1.11). It is an ordinary SMS-MT with retry in case of nonreachability (or alert mechanism). It may also contain text.

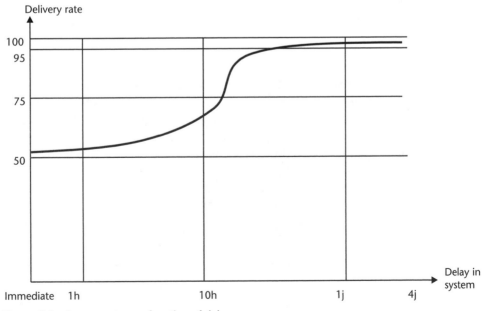

Figure 1.3   Success rate as a function of delay.

Table 1.11   Basic Elements of the SMS_STATUS_REPORT Type *↳ Message delivery receipt*

Abbreviation	Reference	Class	Length	Description
TP–MTI	TP–Message–Type–Indicator	M	2b	Parameter describing the message type.
TP-UDHI	TP–User–Data–Header–Indication	O	b	Parameter indicating that the TP–UD field contains a header.
TP–MMS	TP–More–Messages–to–Send	M	b	Parameter indicating whether or not there are more messages to send.
TP–SRQ	TP–Status–Report–Qualifier	M	b	Parameter indicating whether the previously submitted TPDU was an SMS_SUBMIT or an SMS_COMMAND.
TP–MR	TP–Message–Reference	M	I	Parameter identifying the previously submitted SMS_SUBMIT or SMS_COMMAND.
TP–RA	TP–Recipient–Address	M	2–12o	Address of the recipient of the previously submitted mobile-originated short message.
TP–SCTS	TP–Service–Center–Time–Stamp	M	7o	Parameter identifying time when the SC received the previously sent SMS_SUBMIT.
TP–DT	TP–Discharge-Time	M	7o	Parameter identifying the time associated with a particular TP–ST outcome.
TP–ST	TP–Status	M	o	Parameter identifying the status of the previously sent mobile-originated short message.
TP–PI	TP–Parameter–Indicator	O	o	Parameter indicating the presence of any of the optional parameters that follow.
TP–PID	TP–Protocol–Identifier	O	o	TP–PID of original SMS_SUBMIT.
TP–DCS	TP–Data–Coding–Scheme	O	o	Parameter identifying the coding scheme within the TP–UD.
TP–UDL	TP–User–Data–Length	O	o	Parameter identifying the length of the TP–UD field.
TP–UD	TP–User–Data	O		Variable user data.

The TP–Status field can be a short message received by the destination, an unable-to-confirm-delivery message (at the end of the validity period), and so on. The TP–Message–Reference field contains the number of a SMS sent by the cell phone with its request for a status report, so the cell phone can match the number and show a status received when one looks at the list of status reports for his or her already-sent SMS.

1.2.3   Sending Commands to the SMSC

This is a special SMS-MO sent by the cell phone to the SMSC for uses such as these:

- Inquiring on the delivery status of an SMS-MO (with the reference);
- Deleting a previously submitted short message;
- Canceling status report requests.

In your cell phone, it corresponds typically to the menu Messages, then Service Commands. It is specified in the TP_COMMAND_TYPE field (with the message reference provided). Very few cell phones implement these functions. The basic elements of the SMS_COMMAND are listed in Table 1.12.

1.2.4   Addressing the Foreign Network HLRs for SMS-MT

The SEND_ROUTING_INFO_FOR_SM must be sent by the SMSC to the receiving party HLR to obtain the IMSI and the visited MSC GT. There is no practical way to know which HLR holds the data. Table 1.13 shows the HLRs for several different networks.

The HLR levels are a proprietary table for each operator that gives, for each range of MSISDN, the GT of the HLR and the point code. It may change when the operators reorganize their networks and is not exchanged with their roaming partners (because they do not need to exchange it).

This HLR table (MSISDN E164 → HLR GT) is created in the GMSC of the operators for the *sole utility of receiving SMS-MT from their roaming partners*. For voice roaming (allowing their subscribers to roam in other networks), they need only to install the table (cell phone E214 address → HLR GT). In the section on mobility, we explain the differences between E164 and E214 addresses.

So in the standard SMS-MT procedure, the SMSC creates a SCCP called party address for the HLR which is just the MSISDN with routing on GT of the destination cell phone (E164) and it expects the destination GMSC to do the address translation.

Some networks are not interested in doing anything to receive SMS-MT. In this case they have not created the GT E164 translation table in their GMSC, and the classical procedure fails because the *SEND_ROUTING_INFO_FOR_SM (SRI_SM)* is not routed to the HLR.

From Mobile set SMSC to SMSC   *Not widely used*

Table 1.12   Basic Elements of the SMS_COMMAND Type

Abbreviation	Reference	Class	Length	Description
TP–MTI	TP–Message–Type–Indicator	M	2b	Parameter describing the type.
TP–UDHI	TP–User–Data–Header–Indication	O	b	Parameter indicating that the TP–CD field contains a header.
TP–SRR	TP–Status–Report–Request	O	b	Parameter indicating if the SMS_COMMAND is requesting a status report.
TP–MR	TP–Message Reference	M	I	Parameter identifying the SMS_COMMAND.
TP-PID	TP–Protocol–Identifier	M	o	Parameter identifying the top layer protocol, if any.
TP–CT	TP–Command–Type	M	o	Parameter specifying which operation is to be performed on an SM.
TP–MN	TP–Message–Number	M	o	Parameter indicating on which SM in the SC to operate.
TP–DA	TP–Destination–Address	M	2–12o	Parameter indicating the destination address to which the TP–Command refers.
TP–CDL	TP–Command–Data–Length	M	o	Parameter indicating the length of the TP–CD field in octets.
TP–CD	TP–Command–Data	O	o	Parameter containing user data.

Table 1.13   Number of HLTs for Various Networks

Network	Number of HLRs (in 2002)
Telecel Centrafrique	1
Sabafon	1
Spacetel Yemen	1
Telecom Maroc	5
Orange France	50
Mannesmann	100
China Telecom Mobile	300

1.2.5   Summary of the Network Equipment Model for SMS

Note that for simplification (see the section on mobility and Figure 1.4), we do not show the initial provisioning of the subscriber data in the VLR from the HLR where the cell phone makes an UPDATE_LOCATION_REQ when it registers in the network.

1.3   MAP Dialogue Models at the Application Level

To understand how an SMSC works, one must understand the MAP protocol, which is backed by the *transaction capability application part* (TCAP). This MAP protocol allows sessions to be established and exchange coordinated operations between two network entities. That is, the invoker makes a request (REQ) and someone else (the consumer) receives an indication (IND) so that the various MAP

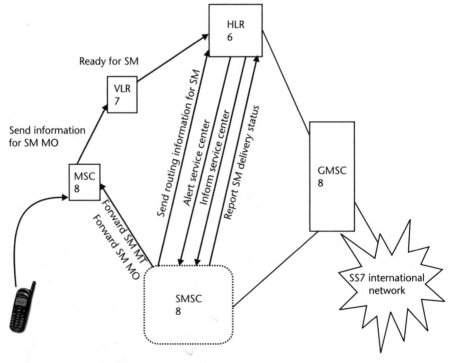

Figure 1.4   Network equipment model for SMS service.

services for SMS presented previously use their protocol with the initiator (invoker) doing an OPEN and the receiver (consumer) closing by means of a CLOSE. Note that the receiver *always* closes the dialogue.

The OPEN includes a number of parameters, discussed in the following sections, that may be included.

1.3.1   Request and CNF (Simple) Dialogue

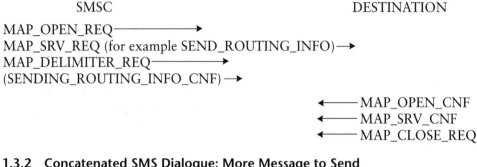

1.3.2   Concatenated SMS Dialogue: More Message to Send

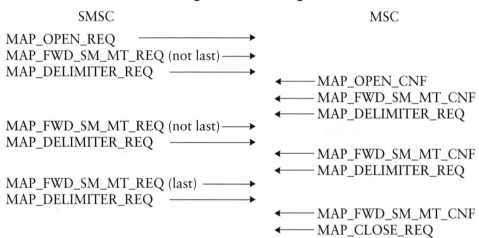

The big advantage is that, if the receiving MSC implements MAP V2+, there is no paging, that is, searching for the cell phone in a given *location area code* (LAC), between the various blocks. So the transmission is quick (paging takes 5–7 seconds).

1.3.3   Update Location Dialogue

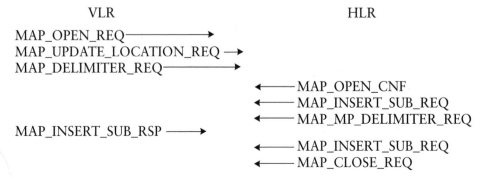

1.3.4   Send Routing Information for SM Dialogue

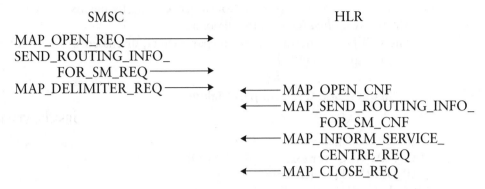

SMSC   HLR

MAP_OPEN_REQ————————▶
SEND_ROUTING_INFO_
        FOR_SM_REQ————————▶
MAP_DELIMITER_REQ————————▶   ◀————MAP_OPEN_CNF
                                  ◀————MAP_SEND_ROUTING_INFO_
                                              FOR_SM_CNF
                                  ◀————MAP_INFORM_SERVICE_
                                              CENTRE_REQ
                                  ◀————MAP_CLOSE_REQ

In the preceding case and in application context V2, the HLR will send an INFORM_ SERVICE_CENTER_REQ to the SMSC in order to give it the message waiting data status.

1.4   SCCP Addresses: The Tool for Flexible International Roaming

Any MAP primitive, such as SEND_ROUTING_INFO_FOR_SM is carried in a TCAP transaction, which itself is carried in an SCCP envelope, which itself is carried in an MTP3 envelope (see Figure 1.5).

The SCCP envelope has (1) a called party address (the destination network equipment)—the HLR, for example; and (2) a calling party address (the equipment originating the message)—the SMSC, for example. The SCCP level network is a datagram network. To answer, the receiving equipment just inverts the called party and the calling party address. The format is the same for destination addresses (called party addresses) and originating addresses (calling party addresses) and is specified in the Q713 recommendations:

• Routing on GT/routing on PC;

• Signaling point code (or absence thereof);

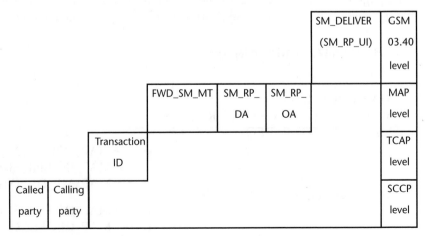

Figure 1.5 SS7 stack for SMS (GSM).

- Subsystem number (or absence thereof);
- GT (or absence thereof) with type of GT: E164 or E214.

Here is the trace of a MAP_OPEN_REQ with the destination address and the originating address :

```
MAPE-E Instance = 0
  MAPE-E Type = MAP_MSG_DLG_REQ (0000C7E2)
  MAPE-E Dialog_ID = 731
  MAPE-E Src = 1D
  MAPE-E Dst = 15
  MAPE-E Rsp_req = 0 Class = 0 Status = 0 Err_info = 0 *Nxt =
0000
        - - - - - - - - - - PARAMETER - AREA - - - - - - - - - - -
    PA_Len = 40
    MAP-OPEN-REQ
     MAPPN_dest_address(Q713)(1)
      L = 013
      Data: Routing on GT, Global Title included(4),
          Signaling Point Code (ITU) = 4-160-7 ( 9479)
          Subsystem Number = HLR(6),
          Global Title :
            Translation Type = 0,
            Numbering Plan = ISDN/Telephony(E164),
            Nature Address Indicator = International number,
            Address information = 85292395815
     MAPPN_orig_address(Q713)(3)
      L = 010
      Data: Routing on GT, Global Title included(4),
          No SPC in address
          Subsystem Number = MSC(8),
          Global Title :
            Translation Type = 0,
            Numbering Plan = ISDN/Telephony(E164),
            Nature Address Indicator = International number,
            Address information = 6596197979
     MAPPN_applic_context(11)
      L = 009
      Data: (Hex) 060704000001001402
          ShortMsgGatewayPackage_v1_or_v2 MAP V2
```

1.5   Mobility Procedures

The mobility procedures allow a cell phone to be used in another network (the visited network) if that network has a roaming agreement with its own network (the home network) and to have access to all services (originate and receive a voice call or a SMS, initiate a GPRS connection to the Internet services provided by his or her operator).

1.5.1 Update Location Procedure

Let us look at the sequence of exchanges that occurs when the cell phone is roaming in network B (Figure 1.6). When it is turned on, a MAP_UPDATE_LOCATION_ REQ procedure will be initiated, because the home network A (Hong Kong) needs to know that cell phone A is visiting network B (to forward his or her calls or messages) (see Table 1.14). Remember that visited network B *never knows the MSISDN*, that is, the ordinary number E164 = +8613601111234; it merely receives the IMSI (the E212 address = 460 00 1146012055) of the cell phone when it connects to his or her network over the radio channel (the IMSI is on the SIM card, not the MSISDN).

So the visited VLR computes an E214 address from the IMSI in order to get an address for the home HLR as explained here (for example, China Telecom):

E212   460 00 1146012055 MCC=460 MNC=00;

E214   86 139 1146012055, that is, the MCC is replaced by the *country code* (CC) = 86 and the MNC by the *mobile global title* (MGT) = 139.

Then the MAP_UPDATE_LOCATION_REQ is sent by the VLR (let's say, Hong Kong) to this E214 called party address through the international SS7 network, giving the IMSI and the GT of the VLR and MSC.

The GMSC A of the home network has two main functions, which make use of two tables: an E214 translation facility (for roaming) and an E164 translation facility (to receive SMS-MT) in order to address the request to the HLR concerned:

| (E214) +86 139 1146000000 to 861 139 1146999999 | ➤HLR GT = 861340273 |
| (E164) +86 136 0011 0000 to 861 136 99119999 | |

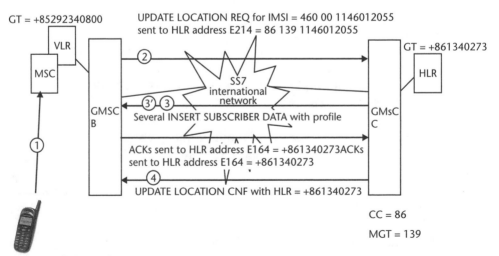

Figure 1.6 Mobility and roaming procedure.

Table 1.14 UPDATE_LOCATION_REQ and CNF Message Details

UPDATE_LOCATION_REQ

Parameter	Class	Context
Primitive type octet	M	V1
Invoke ID	M	V1
IMSI	M	V1
VLR number	M	V1
MSC number	M	V3

UPDATE_LOCATION_CNF

Parameter	Class	Context
Primitive type octet	M	V1
Invoke ID	M	V1
User error	M	V1
HLR number	M	V1

Then the HLR will respond with an INSERT_SUBSCRIBER_DATA message, which contains the subscriber profile (subscribed teleservices, forwarding conditions, and so on) *and* the E164 address (+861340273) of the HLR (see Table 1.15). In all subsequent exchanges, only the E164 address of the HLR will be used now by the VLR of network B!

If roaming visitor A has a CAMEL subscription, the INSERT SUBSCRIBER DATA sent by his HLR will contain the CAMEL subscription information parameter, the SCF GT and the service key. So whenever he makes a call from the visited network, the MSC can address the concerned SCP with the good GT and the good service key. The subscriber profile includes all of the parameters shown in Table 1.16.

Thus, in order that all roaming functions will work, the barring operator, operator A, must allow the E214 translation facility to be available and let the MAP messages addressed to the MGT range [that is the addresses of all its network equipment (HLR, MSC)] go through. So the VLR sends this service to the HLR, with an E214 address—33609123456—for an IMSI 20810123456 of SFR. The HLR sends the subscriber profile.

Table 1.15 INSERT_SUBSCRIBER_DATA_REQ and CNF Message Details

INSERT_SUBSCRIBER_DATA_REQ

Parameter	Class	Context
Primitive type octet	M	V1
Invoke ID	M	V1
IMSI	C	V1
MSISDN	C	V1
Full subscriber profile	C	V1

INSERT_SUBSCRIBER_DATA_CNF

Parameter	Class	Context
Primitive type octet	M	V1
Invoke ID	M	V1
User error	M	V1
Provider	C	V1

Table 1.16   Parameters for INSERT_SUBSCRIBER_DATA Profile

Typical Values	Example of Values
CATEGORY Q.763	Ordinary calling subscriber
SUBSCRIBER STATUS	Service granted
BEARER SERVICE LIST	No bearer services info
TELESERVICE LIST	telephony_+_short Message MT-PP_+_short Message MO-PP
LINE ID INFORMATION LIST	CW (call waiting), Quiescent-Provisioned-Registered-Active, Hold (call hold), Quiescent-Provisioned-Registered-Active, CLIR (calling line ID restriction), Quiescent-Provisioned-Registered-Active, CLIP (calling line ID presentation), Quiescent-Provisioned-Registered-Active
FORWARDING INFORMATION LIST	CFU (call forwarding unconditional) of all speech transmission services, Quiescent-Provisioned-Not Registered-Not active,_+_CFB (call forwarding on mobile subs busy) of all speech transmission services, Quiescent-Provisioned-Registered-Active,_to:_+85292339233,_+_CFNRy (call forwarding on mobile subs no reply) of all speech transmission services, Quiescent-Provisioned-Registered-Active,_to:_+85292339233,_ +_ cfnrc(call forwarding on mobile subs not reachable) all Speech Transmission Services,Quiescent-Provisioned-Registered-Active,_to:_+85292339233
CALL BARRING INFORMATION LIST	BAOC (barring of all outgoing calls) of all speech transmission services, Quiescent-Provisioned-Not Registered-Not active,_+_all Short Message Services, Quiescent-Provisioned-Not Registered-Not active,_+_BOIC (barring of outgoing international calls) of all speech transmission services, Quiescent-Provisioned-Not Registered-Not active,_+_BOIC (barring of outgoing international calls) all SMSs, Quiescent-Provisioned-Not Registered-Not active,_+_BOIC-exHC (barring of outgoing international calls
VLR CAMEL SUBSCRIPTION INFO	Trigger detection point (e.g., collected information)
	Service key to be invoked for call setup (see Chapter 11)
	GSM SCF (service control function) address [service control point (SCP) GT]
	Default call handling (e.g., RELEASE CALL)

When the cell phone registers on a new VLR, the previous one will receive a CANCEL_LOCATION message, which frees the profile recorded in the VLR (see Table 1.17).

1.5.2   Making a Telephone Call to a Mobile

Figure 1.7 is a diagram of the network entity level, which gives all the exchanges concerning the implementation of the roaming. When a call is received from a given telephone, the GMSC issues a SEND_ROUTING_INFO to the HLR (see Table 1.18). The main parameter answered by the HLR is the *mobile subscriber roaming number* (MSRN), so that the GMSC may call this number.

This primitive is not an SMS primitive; however, in a nonstandard SMS procedure, obtaining the roaming number could be useful. To do this, the HLR knows the VLR of the cell phone and sends it a PROVIDE_ROAMING_NUMBER message (see Table 1.19), which allocates a temporary number (it changes at each call).

Table 1.17   CANCEL_LOCATION_REQ and CNF
Message Details

CANCEL_LOCATION_REQ		
Parameter	*Class*	*Context*
Primitive type octet	M	V1
Invoke ID	M	V1
IMSI	M	V1
CANCEL_LOCATION_CNF		
Parameter	*Class*	*Context*
Primitive type octet	M	V1
Invoke ID	M	V1
User error	M	V1

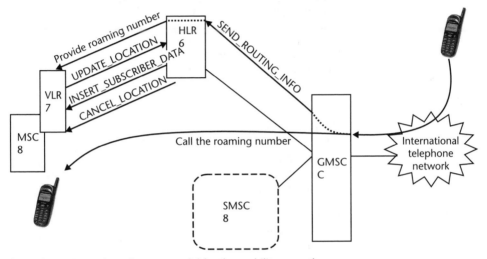

Figure 1.7   Network equipment model for the mobility procedure.

1.6   GPRS Procedures: The Gc Interface

The chapter on MMS includes a description of the GPRS network, which is used to exchange data with cell phones that are GPRS capable. The SGSNs are the equivalent of the MSCs, which are used for conducting communication over circuits. Note that in MAP V3, it is possible to send the SMS over the GPRS connection (through the SGSN) instead of over the circuit connection (through the MSC) (see Table 1.20). This allows the *GPRS gateway* (GGSN) to interrogate the HLR to obtain the SGSN address. It will be implemented later when it will be possible to initiate a WAP session from outside. Currently, a WAP session can only be established by a cell phone! This will be used to push data over a GPRS connection with the cell phone, but initiated from the network, *not* by the cell phone. A general implementation is still a ways off.

1.7   SMS Billing Records and Methods

Regardless of the marketing policy used, the billing source will be the CDRs.

Table 1.18 SEND_ROUTING_INFO_REQ and CNF
Message Details

SEND_ROUTING_INFO

Parameter	Class	Context
Primitive type octet	M	V1,V2
Timeout	O	V1,V2
Invoke ID	M	V1,V2
MSISDN	M	V1,V2
CUG interlock	O	V2
CUG outgoing access	O	V2
Number of forwarding	O	V1,V2
Network signal information	O	V1,V2

SEND_ROUTING_INFO_CNF

Parameter	Class	Context
Primitive type octet	M	V1,V2
Timeout	O	V1,V2
Invoke ID	M	V1,V2
Where User Error Included:		
User error	M	V1,V2
CUG reject cause	O	V2

Where User Error Not Included (Version 1):

Parameter	Class	Context
IMSI	M	V1
Roaming number	C	V1
Forwarded to number	C	V1
Forwarding options	O	V1

Where User Error Not Included (Version 2):

Parameter	Class	Context
IMSI	M	V2
Roaming number	C	V2
Forwarded to number	O	V2
Forwarded to subscriber address	O	V2
Forwarding options	O	V2
CUG interlock	O	V2
CUG outgoing access	O	V2

Table 1.19 PROVIDE_ROAMING_NUMBER_REQ
and CNF Message Details

PROVIDE_ROAMING_NUMBER_REQ

Parameter	Class	Context
Primitive type octet	M	V1
Invoke ID	M	V1
IMSI	M	V1

PROVIDE_ROAMING_NUMBER_CNF

Parameter	Class	Context
Primitive type octet	M	V1
Invoke ID	M	V1
MSRN	M	V1

Table   1.20   SEND_ROUTING_INFO_FOR_GPRS_REQ
and CNF Message Details

SEND_ROUTING_INFO_FOR_GPRS_REQ		
Parameter	*Class*	*Context*
Primitive type octet	M	V3
Timeout	O	V3
Invoke ID	M	V3
IMSI	M	V3
GGSN address	O	V3
GGSN number	M	V3
SEND_ROUTING_INFO_FOR GPRS_CNF		
Parameter	*Class*	*Context*
Primitive type octet	M	V3
Invoke ID	M	V3
Where User Error Not Included:		
SGSN address	M	V3
GGSN number	O	V3
Absent subscriber diagnostic	O	V3
Where User Error Included:		
User error	M	V3
Unknown subscriber diagnostic	O	V3
Absent subscriber reason	O	V3

1.7.1   SMS-MO CDRs

SMS-MO CDRs are generated in the MSC and sent to the billing system for processing. An example follows of a Siemens billing record. The main fields are the A number, which is the cell phone number, and the B number, which is the service center GT.

A CALLING NUMBER	REC TYPE	BEG DATE
B CALLED NUMBER	SEQ NO	
A 33608091234	SMS MO	02/12/18
B 33689004000	3112609	
A 3747210075	SMS MO	02/12/18
B 3746200000	3114773	

The first billing record corresponds to a French subscriber sending an SMS-MO to the SMSC. It will be used by the visited network to charge the home network.

The second record is for my own subscriber (Armenia), which sends an SMS-MO to its own SMSC (+3746200000). It can be used to bill one's own subscriber.

Note that, unless additional software is used, the MSC SMC-MO CDR does not include the real destination cell phone number, which makes it impossible to charge different prices for different categories of destinations.

A popular billing method is to use SMS-MO CDRs created in the SMSC. Then it is possible to invoice only the successful SMS. However, to charge the SMS-MO of the roaming visitors, it is necessary to use the MSC SMC-MO CDRs. This is why many operators use the two sources of SMS-MO CDR in their billing system.

1.7.2 SMS-MT CDRs

In the case of AA19 agreements, the network receiving an SMS for one of his own subscribers charges the sending HPLMN. However, if the SMS-MT is sent to a roaming subscriber of this HPLMN, it is not charged. How can the billing system make the difference?

A CALLING NUMBER	REC TYPE	BEG DATE
B CALLED NUMBER	SEQ NO SS	
B 2080112345678	SMS MT	02/12/18
A 33689004000	45234	
B 20301253025	SMS MO	02/12/18
A 33689004000	45237	

In these CDRs, the A number is the GT of the sending SMSC. With a country code of 33 and a *network destination code* (NDC) of 689 for Orange France, we can identify the sender. The B number in these CDRs is the IMSI of the receiving cell phone. The *mobile country code* (MCC) is 208, designating France, and the *mobile network code* (MNC) is 01, identifying Orange France.

So the first CDR is an SMS-MT sent by Orange France to one of its subscribers roaming in Armenia and it is not charged. The second is also sent by Orange France to an Armenia subscriber (MCC = 283, MNC = 04), but it will be charged by the visited network (if an agreement exists).

1.8 Load Test of an SMSC

It is very important to make accurate calculations and provide a direct measure of the throughput of an SMSC. As we will see, it is directly proportional to the number of SS7 signaling links. With modern servers, the processing is not a limitation. We must also state what we measure.

1.8.1 SMS-MT Test Configuration

You must create a file of 500 (it is enough) numbers (all different) (Figure 1.8). Because a SMSC creates a queue for each number, it will not send another SMS until it has a confirmation. Otherwise the buffer in the MSC, one for each number, would overflow.

The SMS sending speed of the PC, which is connected to the SMSC by means of the standard SMPP protocol (see Chapter 8), may be adjusted. We use the logs of the SMSC to obtain the time between the first and the last SMC confirmation. We redo the measure with different numbers of SS7 links between the SMS and the network (1 to 4). We also look in the logs to see if any warnings about congestion on the links are present.

1.8.2 Results and Performance Model

In Figure 1.9, the dotted curves indicate a congestion limit that is observed on the sender side; at this value no messages are lost, but higher values may lose messages!

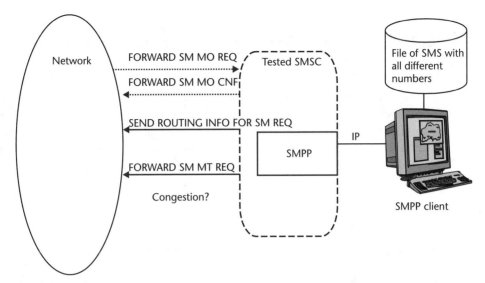

Figure 1.8   SMS-MT test configuration.

For very low sending rates (less than 25), the number of links does not matter: The three experimental curves follow the theoretical one, and the measured speed matches exactly the sending speed of the client PC.

When the number of SS7 links increases, the maximum reachable speed—without any congestion—also increases. From our results, one SS7 link gives a maximum speed of about 27 messages per second, two links give almost 50 messages per second without any congestion and 80 otherwise, and four links give around 100 messages per second without any congestion.

For an operator, even the smallest congestion has dramatic effects in that some messages may be lost! So sending at the appropriate rate, as determined by the number of SS7 links, is extremely important and is the reason for the great quality of service that is expected from an operator.

When sending messages at speeds higher than the congestion limit, we observe a saturation in the sending speed: The bandwidth of the SS7 links (64 Kbps) in the outgoing direction is limiting the traffic.

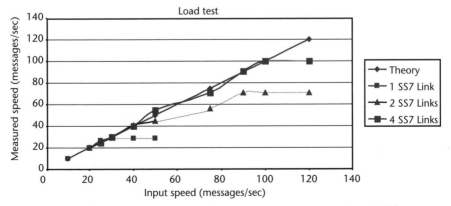

Figure 1.9   Measurement of the SMS per second as a function of the SS7 bandwidth.

In the preceding test, there was no SMS-MO, which would also occupy the bandwidth. Then, the maximum speed for four links is adjusted by estimating that the outgoing bandwidth of the CNF to a SMS-MO is the same as for a SEND_ROUTING_INFO_FOR_SM_REQ. This gives an adjusted value of $100 \times 2/3 = 67$ messages per second with four links [SMS-MO + SMS-MT (without retry)]. You should be wary if someone tells you that their equipment does 200 SMS per second with only six signaling links. The correct value is 17 messages per second per SS7 link. Note that with a SIGTRAN connection (see Chapter 2), there is no limit coming from the link capacity of the SMSC.

Exercises

1.1   Name all of the MAP services that allow a subscriber to obtain the IMSI from the MSISDN (see [1, 3]).

1.2   Name the MAP services that allow us to obtain the MSRN knowing the IMSI.

1.3   Name the MAP services that give the MSRN from the MSISDN.

References

[1]   *Mobile Application Part (MAP) Specification,* GSM 09.02 (ETS 100 974, V7.14.0, 2003-3).

[2]   *Technical Realization of Short Message Service (SMS),* GSM 03.40 (ETS 100 901, V7.5.0, 2001-12)

[3]   *Mobile Application Part (MAP), Release 5,* 3GPP TS 29.002 (V5.2.0, 2002-6).

SS7 Network and Protocol Layers

The Moon begins where the sherry ends with the lemon
—André Breton

2.1 History

SS7 was first specified in the 1979–1980 Yellow Book of the *Consultative Committee on International Telegraphy and Telephony* (CCITT), now known as the *International Telecommunication Union–Telecommunications Standardization Sector* (ITU-T). SS7 has been gradually developed to meet the increasing signaling requirements of new services developed in the 1980s and 1990s. This standard has been released in the ITU-T's red (1984), blue (1988) and white (1992) recommendations (Q700 series) [1–16]. The Q700 series defines the procedures and protocols by which network elements in the *public switched telephone network* (PSTN) exchange services over a digital channel, such as call forwarding, calling party number display, conference calls.

2.2 Efficient and Secure Worldwide Telecommunications

The standardization of the signaling network is primarily concerned with the bottom three layers of the OSI model. This is the purpose of SS7 protocol stack in which the *message transfer part* (MTP) and *signaling connection control part* (SCCP) cover the layers 1 through 3 of the OSI model. The SCCP protocol can be viewed as an extension of the MTP layers that deals mainly with the integration of intelligent nodes (as SCPs or HLRs, for example), which are usually located outside the traffic-handling machines. SCCP belongs to OSI layer 3 and was first specified in 1984 (Recommendations Q711–Q714 [1–4]). SCCP also supports the use of the SS7 network as an advanced packet switched network for direct communications between two network entities without physical ties to traffic circuits through connection-oriented or connectionless communications.

The PLMN benefits from the use of this protocol for operations such a location updating, which the HLR to be updated with information showing the location of *mobile stations* (MSs). In such a case, the HLR needs to exchange data with different *visitor location registers* (VLRs), and this is done through the MAP protocol, which lies on the SCCP network layer and then on the SS7 protocol stack. The PSTN also benefits from the SCCP protocol for the execution of *intelligent network* (IN) services [free phone services, *virtual private networks* (VPN), and so on]. The

29

service logic is implemented in a network entity called the *service control point* (SCP). To execute these services, the *service switching points* (SSPs) must exchange data with the SCP in order to control the communication switch. It is implemented through the *intelligent network application part* (INAP) protocol (CAMEL in PLMNs; see Chapter 11) over *transaction capability application part* (TCAP) and SCCP (i.e., over the SS7 protocol stack).

As we can see from Figure 2.1, the SS7 protocol stack has been continuously modified and improved to fulfill the need to integrate intelligence into the SS7 signaling network in order to provide more and more complex telephony services. SCCP was first implemented to improve MTP routing layer 3 and provide more efficient routing capabilities (OSI layer 3, network). Then TCAP was implemented to provide a transactional layer [based on *association control service element* (ACSE) and *remote operation service element* (ROSE)] to applications externalizing the service logic from the SS7 network and then to provide a reliable, efficient, and far more effective way of implementing new services in a telephony network. (It is far more flexible to update and install one SCP node implementing a telephony service rather than updating or changing all the SSPs of a telephony network.)

2.3   MTP Protocol (OSI Layers 1–3)

The SS7 signaling network (Figure 2.2) is made of several network entities linked by signaling data links and uniquely identified by a numeric *point code* (PC). These point codes are carried in signaling messages exchanged between signaling points to identify the source and destination of each message. Each signaling point then uses a routing table to select the appropriate signaling link for each message. The three kinds of signaling points in the SS7 network are SSP (service switching point), STP (signal transfer point), and SCP (service control point).

The SSP switches originate and terminate calls. An SSP sends signaling messages to other SSPs to set up, manage, and release voice circuits required to complete a call. An SSP may also initiate a transaction to an SCP to determine the routing of a call (e.g., toll-free services).

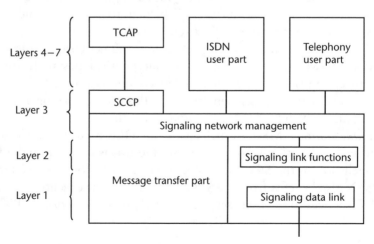

Figure 2.1   The main functions of each OSI layer.

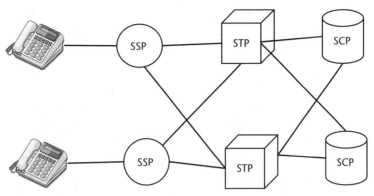

Figure 2.2   The SS7 signaling network.

Network traffic between SSPs may be routed via an STP packet switch. An STP routes each incoming message to an outgoing signaling link based on routing information contained in the SS7 message. Because it behaves as a network hub, there is no need for direct links between signaling points. An STP may perform a GT translation, a transformation by which the routing is based on digits present in the SS7 message (a dialed number or a mobile subscriber identification number). To achieve fault tolerance, STPs or SCPs are usually provided in pairs and signaling data links and are bound into linksets (a set of several links). In a linkset the traffic is shared across all of the links and can be rerouted to another link if one of them fails.

2.3.1   MTP Layer 1: Signaling Data Link Level

It is important to note that MTP layer 1 is not specified in SS7, but an SS7 network must, of course, include a physical layer that meets MTP's requirements (see Recommendation Q702 [7]).

This level defines the requirements that need to be met by the physical circuit (*pulse code modulation* [PCM] channel, for instance). SS7 messages are exchanged over 56- or 64-Kbps bidirectional channels called signaling links. Signaling occurs out of band on dedicated channels (signaling networks) rather than in band on voice channels. Out-of-band signaling provides faster call setup times, more efficient use of voice circuits, an improved control over network usage and support for IN services (SCPs do not manage voice trunks). Once the signaling terminals in the traffic switching or network intelligence nodes are interconnected by physical circuits, they represent a signaling link.

MTP layer 1 is equivalent to the OSI physical layer and defines physical, electrical, and functional characteristics of the digital signaling link. Physical interfaces defined include E-1 (2048 Kbps; 32 64-Kbps channels), DS-1 (1544 Kbps; 24 64-Kbps channels), V.35 (64 Kbps), DS-0 (64 Kbps), and DS-0A (56 Kbps).

2.3.2   MTP Layer 2: Signaling Link Functions

The signaling link represents the signaling points, the signaling data link, and the equipment that links the signaling points to the signaling link (Figure 2.3). MTP layer 2 ensures accurate end-to-end transmission of a message across a signaling

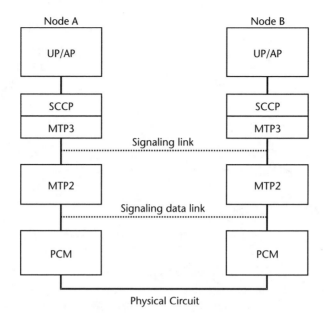

Figure 2.3   Signaling data link and signaling links and their relationship to the OSI model.

link. For this purpose, the signaling link functions implement separation of messages, error detection, error correction, and supervision. It is equivalent to OSI's
data link layer.

An SS7 message is called a *signal unit* (SU) (Figure 2.4). There are three kinds of
signal units: *fill-in signal units* (FISUs), *link status signal units* (LSSUs), and *message
signal units* (MSUs).

FISUs are transmitted continuously on a signaling link in both directions unless
other signal units (MSUs or LSSUs) are present. FISUs carry basic level 2 information only (e.g., acknowledgment of signal unit receipt by a remote signaling point).

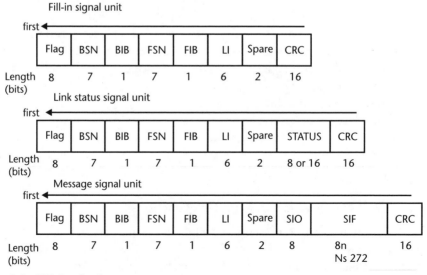

Figure 2.4   SS7 signal units.

LSSUs carry link status information between signaling points at either end of the link. Link status information is used to control link alignment and to indicate the status of a signaling point to the remote signaling point.

MSUs carry call control, query and response, network management, and network maintenance data in the *signaling information field* (SIF). MSUs have a signaling label contained into the SIF field that allows an originating signaling point to send information to a destination signaling point across the network.

The *length indicator* (LI) is a 6-bit field indicating the number of octets that follow the LI field up to the *cyclic redundancy check* (CRC) field, thus indicating which type of SU applies (the LI is a binary value between 0 and 63, where the 63 indicates 63 bytes or more):

LI = 0 means FISU;

LI = 1 or 2 means LSSU;

LI 2 means MSU.

If the number of octets that follows the LI up to the CRC is less than 63, the LI contains this number. Otherwise, the LI is set to 63, which indicates that the message length is equal to or greater than 63 (up to a maximum of 273 octets). This implies that the maximum length of a signal unit is 279 octets (the sum of all fields: flag, BSN, BIB, and so on plus 273 octets maximum for the SIF).

The flag indicates the beginning and end of a message. The opening flag serves as a closing flag for the previous SU (the bit pattern is 01111110). So as not to send false flags on the signaling link, MTP level 2 removes any 6-bit pattern of a message by adding a "0" bit after any sequence of five "1" bits. On receiving a signal unit and stripping the flag, MTP level 2 removes any "0" bit following a sequence of five "1" bits in order to restore the original content of the message.

Corr (the 16-bit error correction field) consists of four subfields:

- The *backward sequence number* (BSN) is used to acknowledge the receipt of SUs by the remote signaling point. The BSN contains the sequence number of the SU being acknowledged.
- The *backward indicator bit* (BIB) is used to indicate a negative acknowledgment by the remote signaling point when toggled.
- The *forward sequence number* (FSN) contains the sequence number of the signal unit.
- The *forward indicator bit* (FIB) is used in error recovery together with BIB.

When a signal unit is ready for transmission, the signaling point increments the FSN by 1. The CRC value is then calculated and appended to the forward message. On receiving the message, the remote signaling point checks the CRC and copies the value of the FSN into the BSN of the next available message scheduled for transmission back to the originating signaling point. If the CRC is correct, the backward message is transmitted (Figure 2.5). If the CRC is incorrect the remote signaling point toggles the BIB, indicating a negative acknowledgment, and sends the backward message. When the originating signaling point receives a negative acknowledgment, it retransmits all forward messages beginning with the corrupted one, with the FIB toggled.

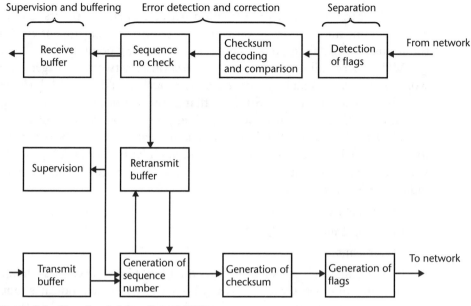

Figure 2.5   Signaling link functions (LSSU).

The 7-bit FSN can store values between 0 and 127. Then a signaling point can send up to 128 signal units before requiring acknowledgment from the remote signaling point. The BSN indicates the last in-sequence signal unit received correctly by the remote signaling point. The BSN acknowledges all previously received signal units as well. For example, if a signaling point receives a signal unit with BSN = 5 followed by another with BSN = 10 (and BIB is not toggled), the latter BSN implies successful receipt of signal units 6 through 9 as well as 10. The FIB field is followed by the LI field and 2 spare bits.

A *service information octet* (SIO) is found in MSUs only. It is divided into two 4-bit subfields: the *service indicator* (SI) and the *subservice field* (SSF). The SSF consists of 4 bits, the two most significant of which are called the *network indicator* (NI) (e.g. national, international), and the two least significant, which are called the message priority (0–3, with "3" being the highest priority). Message priority is only considered during congestion conditions. Low-priority messages may be discarded during periods of congestion.

The SI specifies the MTP user, thereby allowing the information contained in the SIF to be decoded. The SIF is found only in MSUs and contains the information carried by the MSU toward the MTP user part (Table 2.1). *Status field* (SF) is carried only by the LSSU message and indicates the state assumed by a signaling link side after a change of state. It consists of 8 or 16 bits. The CRC is used to detect correct data transmission errors.

2.3.3   MTP Layer 3: Signaling Network Functions

This layer provides message routing between signaling points in the SS7 network. MTP reroutes traffic from failed links and signaling points and controls traffic when congestion occurs. It is equivalent to the OSI network layer. Signaling network func-

Table 2.1   Service Indicator Field

Service Indicator	MTP User
0	Signaling network management message (SNM)
1	Maintenance regular message (MTN)
2	Maintenance special message (MTNS)
3	Signaling connection control part (SCCP)
4	Telephony user part (TUP)
5	ISDN user part (ISUP)
6	Data user part (call and circuit-related messages)
7	Data user part (facility registration/cancellation messages)

tions are divided into 2 categories: signaling message handling and signaling network management.

2.3.3.1   Signaling Message Handling

Signaling message handling means ensuring that the user data of the received MSU reaches the right user [*user point* (UP), e.g., ISUP; *application provider* (AP), e.g., TCAP, via SCCP] at a terminating signaling point, or that it is sent toward the next signaling point. If an MSU is addressed to another signaling point in the network, then it is routed to a suitable signaling link according to instructions from the signaling network management functions.

An MSU that is to be sent from a signaling point is routed to the signaling link—in a signaling linkset—selected for the transport of this MSU. The selection of the signaling link set is based on the MSU's *destination point code* (DPC). Incoming MSUs are separated. If an MSU addressed to another signaling point is received, it will be routed to the appropriate new linkset. A terminating MSU is sent onward to the correct user part, based on the content of the SIO.

2.3.3.2   Signaling Network Management

The signaling network management functions perform continuous supervision of the signaling network to detect errors and abnormal situations. Depending on the state of the network—such as performance affected by defective signaling links or signaling points—signaling traffic may have to be rerouted over alternative routes to reach its destination. In situations where rerouting is not possible, the traffic must be stopped or restricted at the source.

There are three types of MSUs, each having specific contents: (1) MSUs that have signaling information going to and from users, (2) MSUs containing signaling information for signaling network management (MSU-SNM), and (3) MSUs containing signaling information for signaling network testing and maintenance (MSU-SNT, SST). The main difference between these frames is the field *circuit identification code* (CIC), which is 12 bits long and includes *signaling link selection* (SLS), in MSUs for user parts. The CIC is replaced by the field *signaling link code* (SLC), which is 4 bits long in an MSU-SNM or MSU-SNT.

The SIF contains a routing label, a heading (UP or AP), and data. The SIF that is found in MSUs is transmitted to users in layers 4 through 7. The SIF in MSU-SNM and MSU-SNT is sent to the signal network management in layer 3.

2.3.3.3   Routing Label

The four types of routing labels are (1) type A for MTP management messages, (2) type B for TUP, (3) type C for ISUP (circuit-related) messages, and (4) type D for SCCP messages. They include the following fields (depending on the type of routing label) (Figure 2.6):

- *Destination point code (DPC):* Identifies the destination, that is, the signaling network node to which the message is addressed.
- *Originating point code (OPC):* Identifies the sender, that is, the signaling network node from which the message is sent.
- *Circuit identification code (CIC):* Includes the *signaling link selection* (SLS), directly binding the signaling network to the traffic circuit.
- *Signaling link code (SLC):* The SLC is the number of signaling links to which the MSU-SNM or MSU-SNT is related.
- *H0 (message group) and H1 (message type):* These are 24-bit headings indicating the type of information in the SIF for TUP messages.

Data contain up to 256 octets of information intended for users, network management, or testing.

MTP layer 3 routes messages based on the routing label in the SIF of MSUs. The routing label is composed of the destination and originating point codes and the SLS. Points codes are numeric addresses that identify each signaling point in the SS7 network. When the DPC in a message indicates the receiving signaling point, the message is forwarded to the appropriate user part (e.g., ISUP or SCCP) indicated by the service indicator in the SIO. Messages for which the DPC is different from the receiving one are routed to the correct DPC, provided that the receiving entity has message transfer capabilities (like an STP). The selection of the outgoing link is based on information in the DPC and the SLS.

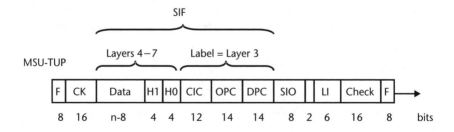

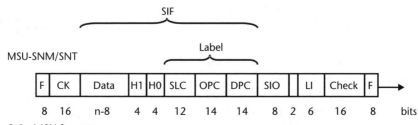

Figure 2.6   MSU formats.

ITU-T point codes and *American National Standards Institute* (ANSI) point codes are different. ANSI codes PCs with a 24-bit field, whereas ITU-T uses a 14-bit field. For this reason, signaling information exchanged between ANSI- and ITU-T-based networks must be routed through a gateway STP, protocol converter, or any other signaling point that has both ANSI and ITU-T PC definitions.

ITU-T point codes are binary numbers that may be stated in terms of zone, area, network, and signaling point identification numbers. For example, the point code 8321 is stated 4-016-1. ANSI point codes consist of network, cluster, and members octets (e.g., 001-044-230), which can contain any value between 0 and 255.

The SLS is used, together with the DPC, for the selection of the outgoing link. This ensures that message sequencing will occur (that is, any two messages sent with the same SLS will arrive at the DPC in the same order in which they were originally sent). Moreover, the SLS allows equal load sharing of traffic among all available links. If a user part sends messages at regular intervals and calculates the SLS values by means of a round-robin fashion, the traffic level should be equal among all links.

2.4 Signaling Connection Control Part

SCCP performs the following tasks: connection-oriented control, connectionless control, routing, and network management. When SCCP is added to MTP functionality, the first three functions provide powerful signaling message handling in the SS7 network. Signaling message handling can be made independent of the traffic part, and it is also possible to make connection-oriented transfers (in case many messages or long messages must be sent to the same destination).

SCCP improves the routing algorithm based on MTP layer 3 by providing additional routing information (Figure 2.7). *Subsystem numbers* (SSN) allow messages to be addressed to specific applications (called subsystems) on the DPC (e.g., 144 for an SCP, 6 for an HLR, 8 for an MSC). SCCP is used as the transport layer for the TCAP-based services.

SCCP also provides means by which an STP can perform *global title translation* (GTT), a procedure by which DPC and SSN may be obtained from digits (the GT) present in the signaling message. In such a case, the OPC does not need to know the

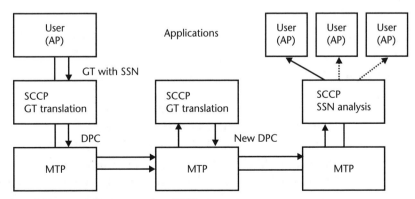

Figure 2.7 SS7 connection setup using SCCP.

DPC or SSN of an associated service. It uses only the destination GT of the network node performing the service, and the STP, with the help of maintained database linking the GT with the DPC and SSN, performs the translation and the routing in adequacy.

2.4.1 SCCP Message Format

The service indicator of the SIO is coded 3 for user SCCP. SCCP messages are encapsulated in the SIF field of the MSU. It begins with an octet field indicating the SCCP message type followed by data, the coding of which depends on the message type. In case of a connectionless SCCP transaction, all SCCP messages contain an SCCP message header (SCCP addressing information) including called and calling party addresses and local references.

The SCCP addressing scheme uses the four following separate elements (Recommendation Q713 [3]):

- *Address indicator (AI):* This is a flag indicating the type of translation that SCCP needs to perform.
- *Point code (PC):* The DPC in an SCCP address requires no translation and will determine if the message is destined for that SP or instead needs to be routed forward over the SS7 network via MTP. In the case of outgoing messages, the DPC is identical to the MTP DPC.
- *Global title (GT):* The GT may be made of dialed digits or another form of address that will not be recognized by the underlying SS7 network. Then if the associated message needs to be routed over the SS7 network, translation is required. This translation will result in a new DPC and possibly also a new SSN and GT.
- *Subsystem number (SSN):* SSN identifies a local subsystem accessed via SCCP within the node (e.g., VLR, MSC). When examination of the DPC has stated that the incoming message belongs to the node, the SSN will specify which one of the SCCP users is concerned with the message.

One, two, or all of these elements may be present in the calling or called party addresses depending on the AI field.

2.4.2 SCCP Layer Architecture

The architecture of the SCCP layer is made up of four entities: *SCCP routing control* (SCRC), *SCCP connection-oriented control* (SCOC), *SCCP connectionless control* (SCLC). and *SCCP management* (SCMG):

- SCRC is responsible for two major operations: routing and addressing. Routing is done by relaying the traffic to another user entity or to one of the three SCCP entities. Addressing is actually address translation.
- SCOC is responsible for setting up a connection between two users of SCCP, transferring traffic between them, and tearing down the connection. It supports several features such as segmentation, sequencing, and flow control.

- SCLC is responsible for transferring traffic between two SCCP users but it does not create a connection. It does not implement all SCOC additional features.
- SCMG is used for management and status operations.

Usually the DPC is not present in the SCCP called party address because it is redundant with MTP layer 3. So, if it is present it must be ignored.

The SCCP calling party address is used because it reveals the origin of the SCCP message. Address translation at an intermediate node results in the OPC of the MTP 3 routing label being changed to the point code of the intermediate node.

2.4.3 SCCP Routing

2.4.3.1 Receiving a Message from a User Layer Within the Node

If the SCRC receives a message from an upper layer within the node (through the SCOC or SCLC), it will receive one of the following combinations:

- DPC;
- DPC + (SSN and/or GT);
- GT;
- GT + SSN.

Then the translation is made by a mapping table created by the SS7 administrator.

If the DPC is present in the message and it does not identify this node, then SCCP must forward the message to MTP for relaying to another node. If the DPC is the node itself, it must send the message back to the accurate entity based on SSN through the SCOC or SCLC.

2.4.3.2 Receiving a Message from MTP Layer 3

When receiving a message from MTP layer 3, the DPC in the MTP 3 routing label identifies the signaling point code of this SCRC. Then the SCRC scans the AI field of the incoming signal unit. This field includes a bit called the *routing indicator field,* which indicates whether the routing must be performed based on the GT of the SCCP called party or on the MTP 3 routing label DPC and the SSN of the SCCP called party address. In the first case, SCRC must use the GT value to find a new DPC, possibly a new SSN or GT, and also possibly a new address routing indicator in a mapping table. It sends the message back to MTP 3, which puts the DPC in its DPC field of the routing label. In the second case, it means the message has reached the terminating SCCP node and further routing must be performed based on SSN.

2.4.3.3 Examples of Addressing and Routing

Successful Operation with One Translation. As illustrated in Figure 2.8, the translation takes place at signaling point B. Three nodes are involved: A, B, and C. Node A is sending traffic from its MAP user application, identified by SSN 8. To do so, A sets its OPC to A and the DPC to B in MTP layer 3. In the SCCP header, it

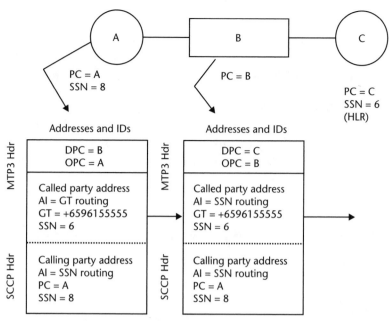

Figure 2.8   Example of address/identifier translations.

codes the called party address with an AI indicating that routing must be performed on the GT. The SSN is set to 6. The calling party address is not used in this case for the address translation in B.

This message is routed to signaling point B (for example, an STP). MTP 3 forwards the message to SCCP because DPC B identifies the B node and SI identifies the SCCP user. In this case, SCCP sublayer SCRC has a translation table indicating GT +6596155555, which must be translated to SSN routing, PC = C and SSN = 6 (HLR function). It is important to note that the calling party address is not modified by the successful translation. Then the message is forwarded to node C through MTP 3, which indicates OPC B in the MTP 3 routing header. In node C, the SCCP layer computes the SSN routing AI and forwards the message to the MAP layer.

Unsuccessful Operation.   This case is illustrated by Figure 2.9 and is similar to the successful one except that node B does not have any translation for GT +6596155555. So, on node B, the routing fails. Node B responds to node A with a message indicating an error and specifying as called party address, node A with routing on SSN and SSN of the calling party address. In an unsuccessful case, node B's SCCP layer indicates the former called party address as calling party address for the unsuccessful response, thus indicating to node A that the translation it has asked for is not possible.

Successful Operation with Two Translations.   The case of successful operation with two translations is similar to the first successful case (Figure 2.10) except that two translations occur at STPs B and D. At node B the GT is translated into DPC D and GT +6596161234; at this node no SSN is processed since we have not arrived at the destination node. At node D, the translation computes GT +6596161234 to DPC C

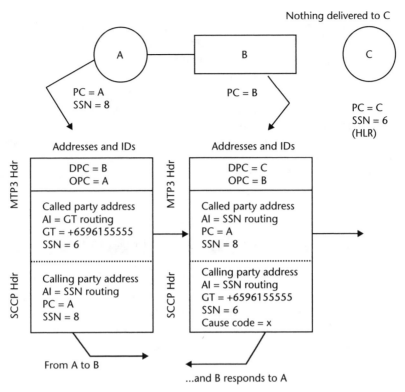

Figure 2.9 Unsuccessful operation.

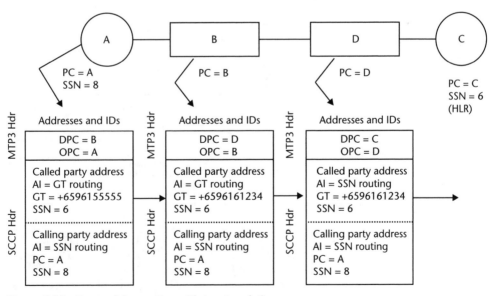

Figure 2.10 Successful operation with two translations.

and SSN 0110. This message is then sent to node C, which delivers it to the HLR application.

Successful Operation with Two Translations in Two Directions. In this case, illustrated by Figure 2.11, the successful operation for the forward routing is similar to the previously reverted one, except the content of the SCCP calling address indicates a route on GT with an SSN and the GT of SPA. This address allows the SPC to route back an answer to node A if needed by implementing the GT translation for GT +6596155556 toward node D. Nodes D and B present the same GT translation rule as for the forward case, allowing the SCCP message to be routed toward node A. If the original GT translation rule in node C is not present, the backward path cannot be followed if node A puts an OPC in the calling party address. Then as specified in ETSI recommendations, *one should not set the original PC in the SCCP calling party address.*

2.5 Transaction Capability Application Part (TCAP)

As we have seen before, SCCP routing always perform the final routing procedure based on the SSN. This number uniquely identifies a user part on a signaling point corresponding to a functional entity of the telecommunications network. To abstract the routing mechanisms that, as we have seen before, may involve several routing nodes to perform an operation on a remote system, the TCAP protocol has been implemented to provide a protocol-oriented transactional layer over SCCP,

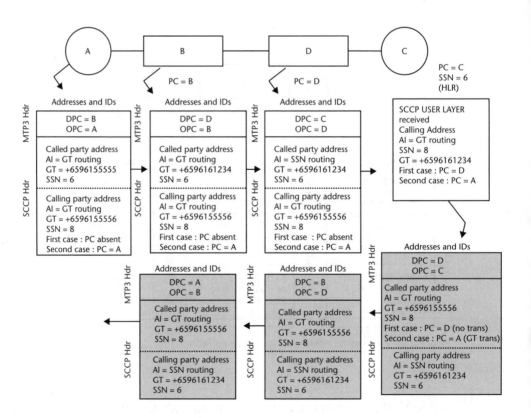

Figure 2.11 Successful operation with two translations in two directions.

which allows functional entities to perform data-based operations over the SS7 signaling network.

TCAP has been implemented over ACSE and ROSE protocols. ACSE provides a connection-oriented mechanism (associate, release, and so on) and ROSE provides Invoke-Result operations. They are part of the OSI presentation layer. This layer provides to upper protocol layers a session-oriented transactional service, which is needed to perform the functional operations that entities such as SCP, HLR, and MSC may need.

ACSE is a fundamental base for the TCAP layer and then for upper user layers (MAP, INAP) because it is used to associate a particular function with a remote identical function through the use of an application context.

The TCAP layer also provides SSN-based routing, which is used to specify which upper layer protocol must be used for a specific SSN. For example, SSN 144, which identifies an SCP functional entity, must be routed toward the INAP protocol, and SSN 6, which identifies an HLR. must be routed to the MAP protocol.

It is also possible within the TCAP layer to compute on the same node several functions based on different application parts (such as SCP and HLR or MSC). The TCAP layer provides an abstraction level for application parts that completely masks the SS7 routing and addressing problems.

2.5.1   Main Features of TCAP

The TCAP layer provides the following services on behalf of applications:

- Multiple transactions per application/subsystem number;
- Assembly of application components (operation invokes and replies) and parameters into TCAP messages;
- Association of results with invokes;
- Optional timing for results to invokes (ITU-T only);
- Handling of abnormal conditions: protocol errors, timeouts, aborted transactions;
- Use of the ITU-T standard TCAP with the ANSI-standard MTP/SCCP stack and vice versa.

The TCAP layer uses the SCCP connectionless transport service. In addition to the services described above, the TCAP layer also makes the following SCCP layer services available to the application: (1) addressing by point code/subsystem number and/or GT and (2) subsystem and point code status change indications.

2.5.2   TCAP Architecture

TCAP is designed in two separate layers, the component and the transaction layers, based, respectively, on ROSE and ACSE.

The transaction layer is responsible for establishing an association between a local application part and a remote one. It uses BEGIN, CONTINUE, END, and ABORT TCAP messages (ASSOCIATE and RELEASE ACSE messages). A transaction is identified by a Transaction ID. Between the BEGIN and END (or ABORT), the transaction layer is responsible for data exchange between the two application parts through CONTINUE. There may be several CONTINUE frames between BEGIN and END. Each of them is linked to a unique transaction ID through the use of an Invoke ID and Linked ID.

The component layer formats Invoke, Result, Error, Reject, Return-Result, and Cancel messages, which carry data and machine state invocations between the two application parts into transactional frames (BEGIN or CONTINUE). The Invoke ID identifies a specific Invoke operation that requires a remote application to perform processing and to send back the result with a Result, Error, and so on, depending on the success of the operation. During the processing of this operation, the remote node can also initiate a Invoke on the originator node as we will see in the following example.

2.5.3 TCAP Operation Invocation Example

Figure 2.12 shows a TCAP transaction corresponding to an invocation of a service on a remote switch which could be described as follows: A telephone subscriber asks for the invocation of a service. The switch invoked realizes it does not have the necessary information it needs to compute the service, so it asks the subscriber to enter certain information (such as an access code). The subscriber then enters the information requested and the transaction ends.

On the TCAP level, this service invocation can be described as follows: On the SSP, the TCAP layer instantiates a TCAP transaction toward the service node (e.g., an SCP) through a BEGIN message containing the (encapsulated) Invoke. The Invoke contains the information needed to access the specific service on the remote node. This transaction is identified by its Transaction ID and the remote operation invoked by an Invoke ID. (Be careful the Invoke ID does not identify a specific operation, but merely an instance of this operation during the TCAP transaction process.) The remote node then realizes it does not have all of the information it needs to perform the operation. It then initiates an invocation on the originating node with a CONTINUE containing a Invoke component specifying that the originating node must perform an operation to collect information. This Invoke is identi-

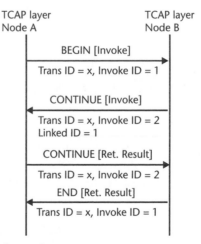

Figure 2.12 TCAP operation invocation.

fied by a different Invoke ID, and the Linked ID is set to say that this operation is linked to the previous operation it has sent. After completing the information collection, the originating node answers the invocation with a Return-Result in a CONTINUE message identified by the Invoke ID to which it corresponds. Then the remote node can end the TCAP transaction (it has collected all the information it needs) and return a Return-Result component in a END message with the result of the operation it has performed.

2.6   User-Level Application Parts: MAP, INAP, CAMEL

2.6.1   User Part Mapping onto TCAP: MAP Example

As described at the beginning of this chapter, one could consider MTP and SCCP to be equivalent to the lower three layers of the OSI model. Everything concerning the upper layers (i.e., TCAP and MAP or INAP or CAMEL, ISUP, or TUP) cannot be easily mapped to the OSI model because these parts do not follow its architecture and principles. We provide our view on that point in the following section by describing how GSM Recommendation 09.02 [17] specifies the mapping of MAP into the TCAP layer.

MAP is made of two different parts, the MAP User and MAP Provider levels. The MAP Provider level is completely masked for any user application based on MAP and deals with driving of the TCAP layer and with mapping the MAP protocol into the TCAP protocol. The TCAP driving state machine is outside the scope of this presentation, but we will give a short description of a problem with which the MAP User level has to deal. It concerns the driving of sublayers and shows that this protocol layer does not fill the requirements of an OSI model.

One can consider that any user part based on TCAP, for instance, MAP, is totally independent from the TCAP layer (as it should be in an OSI model). This is why the MAP protocol has been divided into the driving layer MAP Provider and the MAP User layers.

2.6.1.1   MAP Provider Layer

The MAP Provider layer will map the MAP protocol onto the TCAP protocol (Table 2.2). The first aspect of this is to define how to use the TCAP transactional

Table 2.2   Mapping MAP onto the TCAP Protocol

MAP Service Primitive	TC Service Primitive
MAP_OPEN request	
(+ any user-specific service primitives)	BEGIN request
+ MAP_DELIMITER request	(+ component handling primitives)
MAP_OPEN response	
(+ any user-specific service primitives)	CONTINUE request
+ MAP_DELIMITER request	(+ component handling primitives)
(+ any user-specific service primitives)	CONTINUE request
+ MAP_DELIMITER request	(+ component handling primitives)
(+any user-specific service primitives)	END request
+ MAP_CLOSE request	(+ component handling primitives)
+ MAP_U_ABORT request	U_ABORT request

primitives to implement the MAP services. Then GSM Recommendation 0902 [17] states how to map MAP_OPEN_REQ and MAP_SERVICE_REQ onto the TCAP protocol. Table 2.3 shows how this has been specified.

Moreover, some MAP primitives simply do not exist in the MAP protocol but are only a translation of TCAP primitives into MAP User or Provider ones. (For instance, MAP_U_ABORT or MAP_P_ABORT is merely the ABORT primitive.) Thus, many standard MAP parameter fields are not defined in the MAP protocol but are only translations of equivalent TCAP ones. (For instance, the MAP User error field, which is very common in MAP_SERVICE_RSPs, can only be a translation of a TC problem code in a U_REJECT; also a MAP Provider error is not a MAP-specific error code because it concerns only a TCAP REJECT.)

The MAP Protocol is defined through a complete ASN.1 specification in GSM Recommendation 0902 [17] and this specification cannot compile without the TCAP ASN.1 specification. Therefore, the MAP layer cannot be considered to be totally independent from TCAP but should be considered an extension of the TCAP layer (a specialization of the TCAP) to fulfill some specific services required by PLMNs.

2.6.1.2   MAP User Layer

The MAP User layer specifies all services that may be used in the GSM network for client/server-type operations (location updating, handover management, and so on). Its description is outside the scope of this presentation and may be found in the GSM Recommendation 0902 [17].

2.6.1.3   MAP, TCAP, SCCP, MTP, and the OSI Model

As described earlier, one cannot consider this layered architecture to follow the OSI model for several reasons. The first one has already been discussed: Every layer is quite encapsulated in the lower layer. To demonstrate this by means of a very simple example, we discuss the problem of segmented messages in the MAP protocol (Figure 2.13).

The MAP protocol includes a specific case to apply if, when mapping, a MAP_OPEN_REQ, a MAP_SERVICE_REQ, and a MAP_DELIMITER_REQ into a BEGIN need segmentation. (The rule for segmentation is explained next.) If segmentation is needed, the transaction must be listed as MAP_OPEN_REQ and MAP_DELIMITER_REQ mapped into a BEGIN followed by a MAP_SERVICE_REQ and a MAP_DELIMITER_REQ mapped into a CONTINUE, when the former request response has been accepted (MAP_OPEN_CNF ok). If no segmentation is needed (Figure 2.14), the MAP User layer must send the MAP_OPEN_REQ, the MAP_SERVICE_REQ, and the MAP_DELIMITER_REQ into a single BEGIN, which will be acknowledged by the remote point with, for instance, an END including the MAP_OPEN_CNF and MAP_SERVICE_CNF responses. Then the problem is to know which characteristic decides whether we need segmentation or not, because the characteristic could imply that two MSUs should be sent instead of one.

The response is quite surprising, but the discriminating factor of this problem is found in the SCCP layer. According to the MTP specification, the SCCP layer only allows a maximum length of 279 for MSUs, which allows 273 octets of data. Then,

Table 2.3   MAP + TCAP Operation Invocation (No Segmentation)

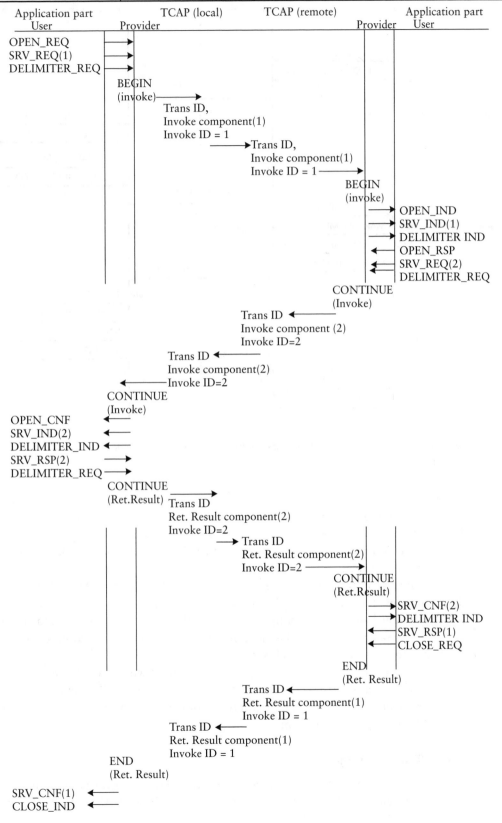

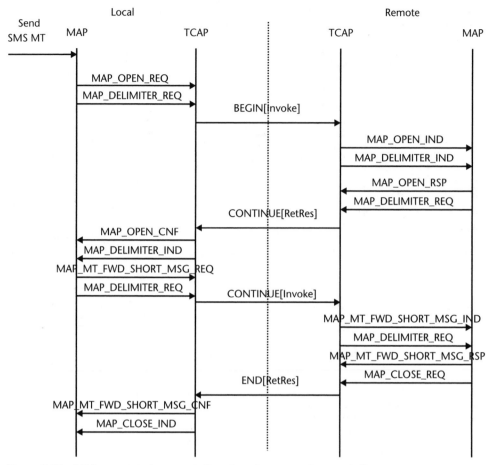

Figure 2.13   MAP segmented messages flowchart (segmentation needed).

because all TCAP-based applications use SCCP connectionless transactions, it is not possible for the SCCP layer to send more than 273 octets of data. Therefore, the TCAP layer is also limited by this characteristic and as a consequence so is the MAP layer. Then we see that even the mapping of the MAP protocol into the TCAP layer is run-time dependent on the length of data to be sent (segmentation needed or not). The calculus is quite simple: One must compute the length of the SCCP header (it is dependent on the presence of GT and on its length, for instance), then add the BEGIN header length, and finally add the length of the MAP request that is to be sent. This problem is perhaps the most obvious way to show how completely inadequate it is to make an OSI model comparison between SS7-based protocols and the OSI layer, since upper layers are dependent on the lowest ones. In that case, the absence of the transport layer is obvious; even if SCCP or TCAP seem to behave as transport layers, they are still quite different from an OSI transport layer that would deal with message segmentation.

2.6.2   Routing Design

As we saw earlier, SCCP provides a way to route a message toward a particular SSN, which represents an abstract functionality (e.g., MSC, HLR, VLR, SCP). Telecom-

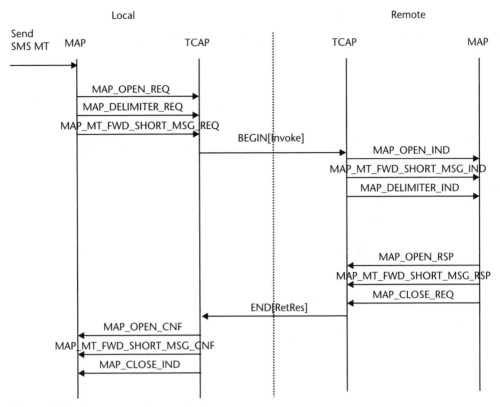

Figure 2.14   No segmentation needed.

munications standard recommend that this particular advantage not be linked with any physical implementation of the services it is supposed to provide. For instance, the recommendation specifies functions (SSF, SCF, and so on) as an abstract application part layer without specifying how to implement them. This principle is resolved in the way communications and transport layers have been specified (SCCP, TCAP) with the use of SSN for SCCP and application context in TCAP, MAP, or INAP layers. Then we see that this specification of SCCP and TCAP allows a particular signaling point to fulfill multiple services and to be considered—depending on how it is invoked—as a VLR, SCP, HLR, and so on.

The first step of the functional identification of a request is performed in the SCCP layer, which decides which upper application part must be invoked—the VLR function through TCAP and MAP, the SCP function through TCAP and INAP—and then in the TCAP and MAP layers which specific application service will be invoked through the use of the application context—roaming number inquiry between an HLR and a VLR. This allows the functional representation of a network entity to be based on a service-oriented design rather than a functionally oriented design. This feature is made possible through the ACSE base of TCAP layer, which, along with the ASSOCIATE command, creates a logical link between an invoking abstract entity and an invoked specific service on the remote signaling point.

Figure 2.15 shows a synoptic scheme for a protocol service-oriented design for telecommunications applications based on SS7 networking.

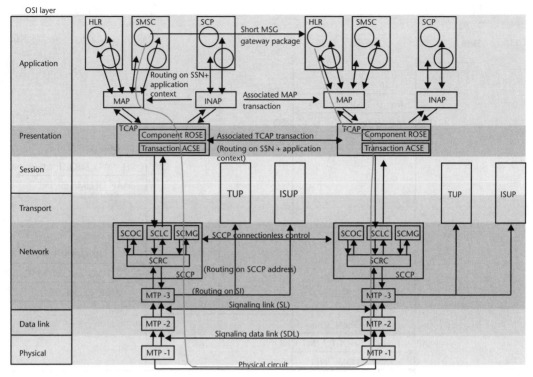

Figure 2.15   Service-oriented design.

2.6.3   Service-Oriented Design: Application to an SS7-Based Fault-Tolerant System

Suppose that we implement an HLR function to be installed in a GSM network. A major problem that will appear during the conception process is to provide continuous service to other entities of the network: The HLR must be fully operational 24 hours a day, every day. The most common way to fulfill this requirement is for each network entity to be provided with a pair, a master and a slave. When the master crashes, the slave relays it until the master comes up again. This method is used by most of the major telecommunications operators.

The precedent presentation shows another possibility for fulfilling this feature, which is to design for each node a redundant configuration. Instead of providing one HLR, we can provide two HLRs working together with the same PC and GT and sharing the SS7 data links with the SS7 network. Then the incoming traffic is equally loaded on the signaling links, invoking operations on the two network HLRs as long as the signaling data link is up (MTP layer 1 and 2 up). If one of the systems crashes, it is no longer invoked because the SS7 link is broken; in such a case, the SS7 signaling traffic is sent to the other one. When the crashed system starts up again, its SS7 links bind again and it receives traffic. This feature is made possible by the way upper abstract layers have been implemented. Regardless of what HLR function is performed through a distributed real-time system, we only invoke an operation from a GT or PC identified node. (In that architecture the SCCP layer must include a communication layer with its dual to exchange messages that it does not manage.)

2.7   SS7 and VoIP Interworking Overview SIGTRAN

With the development in the late 1990s of the Internet, solutions have appeared to provide a way to use the *Transmission Control Protocol/Internet Protocol* (TCP/IP) connection in order to make phone calls. *Voice-over-IP* (VoIP) technology has been developed to use IP connections as voice circuits. When this happened, it became crucial for this technology to have an interworking capability with the PSTN and PLMN. This interworking solution has been developed through an *SS7-based-over-IP* (SS7oIP) signaling protocol designed for the Internet, making then possible the interworking between an SS7 MTP-based PSTN or PLMN and SS7 IP-based networks. The key feature of this SS7oIP protocol is called the *Stream Control Transmission Protocol* (SCTP), which defines an association-based protocol to be used over IP that can replace MTP or MTP+SCCP or even MTP+SCCP+TCAP for the Internet network.

2.7.1   SCTP

The SCTP has been defined by the *Signal Transport* (SIGTRAN) working group of the *Internet Engineering Task Force* (IETF). It is a transport-level datagram transfer protocol that operates on top of a datagram service such as the IP. SCTP is based on association and streams, an association being made of multiple logical streams (Figure 2.16).

SCTP services include (1) acknowledged error-free nonduplicated transfer of user data, (2) segmentation of user datagram, (3) sequenced delivery of user datagrams within a stream, (4) optional bundling of multiple user datagrams into one SCTP datagram to improve bandwidth utilization, and (5) support for multiple transport addresses at either or both ends of an association.

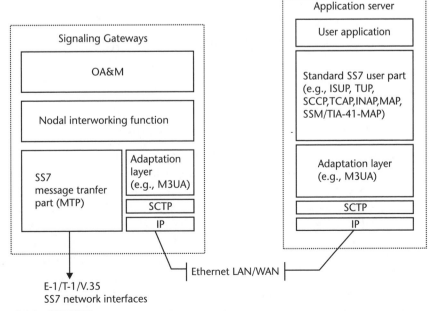

Figure 2.16   SIGTRAN.

2.7.2   Interworking with SS7

SCTP has been specified with *user adaptation* (UA) layers to provide interworking capabilities with the SS7 network. Then several UA layers were defined for an SS7-based service to interwork with the two signaling transport protocols (PSTN or IP):

- SUA (SCCP user adaptation layer),
- M3UA (MTP 3 user adaptation layer),
- M2UA (MTP 2 user adaptation layer), or
- TUA (TCAP user adaptation layer—not defined yet).

2.7.3   M3UA Layer

The MT3UA layer transports signaling messages from MTP 3 users (e.g., ISUP, SCCP, and TUP) over the IP network. M3UA supports two operational modes:

1. It bridges the boundaries of the SS7 network and IP network. In this mode M3UA transports the signaling messages from a signaling gateway located at the edge of the SS7 network to an IP-resident gateway.
2. It supports the SS7 call signaling protocols (SCCP and so on) to communicate within an IP network.

2.7.4   M2UA Layer

Same as M3UA but for the MTP 3 protocol layer.

2.7.5   SUA Layer

Same as M3UA but for SCCP user protocols such as TCAP. It introduces the routing on IP in addition to routing on PC and routing on GT.

2.7.6   TUA Layer

Same as M3UA but for TCAP user application parts, such as MAP or INAP.

2.8   Conclusions

2.8.1   Powerful, Efficient Network Architecture

This overview of the SS7 signaling network and application parts shows the efficiency of this solution for telecommunications services that necessarily deal with problems such as intelligent services deployment and GSM mobility management. The principal aspect of and interest in using the SS7 network is the use of a strongly specified protocol network that provides an efficient way to invoke operations on worldwide distributed systems through the use of MTP and SCCP routing (PC, GT, or SSN routing). The other key point of this specification is that it should be totally independent from the hardware architecture—in its conception but also through the use of a specification language such as *Specification and Description Language* (SDL) and *Abstract Syntax Notation 1* (ASN.1). This allows an operator to implement and install complex telecommunications services in a short time and to integrate them in any telecommunications network that respects the ITU-T recommendations.

2.8.2 Application to a Worldwide SMS Service Network

As a conclusion to this presentation, and as an adjunct to former presentations that have been made on global SMS services, we now consider a short problem that concentrates all the aspects presented before in a very synthetic way. The current global SMS service network is based on roaming agreements shared between several SMSCs. In the architecture described in Figure 2.17, the dynamic routing is performed by a virtual MSC, HLR, or SMSC function lying over MAP.

Figure 2.18 shows the next-generation network architecture that needs to be studied in order to provide a far more efficient way of routing and sharing roaming agreements between several network nodes. This architecture only requires a more complex SCCP GT translation function in order to route directly toward the destination network, the MAP primitives an origin network may invoke. This is quite similar to implementing an STP in the node. Then with this architecture, the upper services layers can be reduced to a single SMSC function that presents all roaming agreements of the whole network, making one node's behavior exactly like that of an SMSC + STP having roaming agreements with all operators in the network. This architecture is far more efficient in terms of computation (most of the transit traffic will not go beyond SCCP) and also in terms of service logic, because any node can concentrate on the basic SMSC functions (SMS MO processing and alerting).

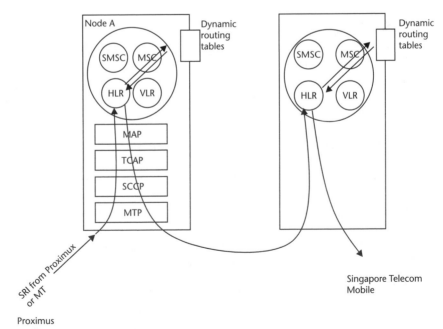

Figure 2.17 Current global SMS service network.

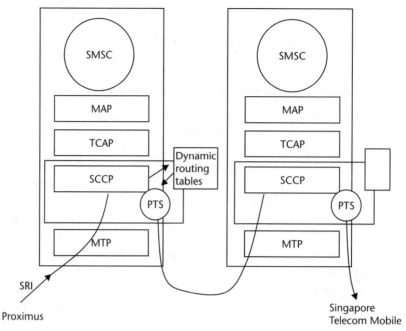

Figure 2.18   Next-generation SMSC for global SMS services.

References

[1]   *Q711: Specifications of Signaling System No. 7: Signaling Connection Control Part: Functional Description of the SCCP.*

[2]   *Q712: Specifications of Signaling System No. 7: Signaling Connection Control Part: Definition and Function of SCCP Messages.*

[3]   *Q713: Specifications of Signaling System No. 7: Signaling Connection Control Part: Signaling Connection Control Part Formats and Codes.*

[4]   *Q714: Specifications of Signaling System No. 7: Signaling Connection Control Part: Signaling Connection Control Part Procedures.*

[5]   *Q700: Specifications of Signaling System No. 7: Introduction to CCITT Signaling System No. 7.*

[6]   *Q701: Specifications of Signaling System No. 7: Functional Description of the Message Transfer Part (MTP).*

[7]   *Q702: Specifications of Signaling System No. 7: Signaling Data Link.*

[8]   *Q703: Specifications of Signaling System No. 7: Signaling Link.*

[9]   *Q704: Specifications of Signaling System No. 7: Signaling Network Functions and Messages.*

[10] *Q705: Specifications of Signaling System No. 7: Signaling Network Structure.*

[11] *Q706: Specifications of Signaling System No. 7: Message Transfer Part Signaling Performance.*

[12] *Q771: Specifications of Signaling System No. 7: Transaction Capabilities Application Part: Functional Description of Transaction Capabilities.*

[13] *Q772: Specifications of Signaling System No. 7: Transaction Capabilities Application Part: Transaction Capabilities Information Element Definitions.*

[14] *Q773: Specifications of Signaling System No. 7: Transaction Capabilities Application Part: Transaction Capabilities Formats and Encoding.*

[15]   *Q774: Specifications of Signaling System No. 7: Transaction Capabilities Application Part: Transaction Capabilities Procedures.*

[16]   *Q775: Specifications of Signaling System No. 7: Transaction Capabilities Application Part: Guideline for Using Transaction Capabilities.*

[17]   *GSM Recommendation 09.02: Mobile Application Part.*

## CHAPTER 3

Standard Procedures for SMS in IS-41 Networks

It is my ambition to say in ten sentences what everyone else says in a book—what everyone else does not say in a book.

—Friedrich Nietzsche, *Twilight of the Idols*

3.1 Introduction

3.1.1 IS-41 Networks

The earliest version of ANSI-41 was written to define a way for first-generation *Advanced Mobile Phone System* (AMPS) networks in the United States to communicate with each other in order to facilitate roaming. Roaming among AMPS networks before IS-41 was a nightmare of phone calls to help desks, arrangements with credit cards, and missed calls even after all the preparations before travel. The specification was eventually expanded and altered to include TDMA (IS-136) networks and then altered again to include functionalities for CDMA (IS-95) networks. In all of its versions, it specified mobility management and call establishment, which had to be enhanced in subsequent versions of ANSI-41. Until recently, it was seen as a specification limited to the E interface. It is, in fact, the equivalent of the MAP protocol for CDMA and TDMA networks. As the MAP protocol, it is an application part using TCAP for transaction capabilities and SCCP over the ANSI SS7 MTP protocol.

The functional entities involved in the IS-41 protocol are the same as for GSM networks, MSCs, VLRs, HLRs, and so on, although some of the functional architecture is quite different from that of GSMs, as discussed in the following sections.

3.1.2 Inefficient Handover Chain Procedure

3.1.2.1 Voice Call

In IS-41, when a cell phone receives or initiates a call, it visits an MSC, which is called the *anchor MSC* (Figure 3.1). At the beginning of the call, if the user moves while continuing this call, the MSC may change and the MSC that is currently being used is called the *serving MSC.*

In IS-41, the call is tromboned through each participating MSC, which establishes a voice call with the next MSC, thus creating a chain (of serving voice circuits) from the anchor MSC through tandem MSCs until it reaches the serving MSC. This process is quite inefficient, and if the call lasts a long time as the cell phone continues

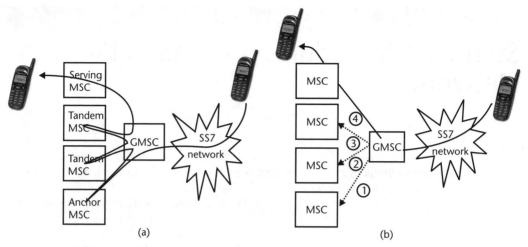

Figure 3.1 Mobility management in (a) an IS-41-initiated call versus (b) a GSM-initiated call.

to move, a loop may eventually be created! In comparison, for the same case using GSM, the route changes dynamically: All MSCs have circuits with a GMSC. Also, the visited MSC (serving MSC in the IS-41 case) is not updated in the HLR. It is still that of the anchor MSC.

3.1.2.2 SMS-MO Procedure

A consequence of the handoff chain in IS-41 networks is that the SMS-MO procedure between the visited MSC and the message center contains a chain of relay points among the several MSCs that may be involved in the handoff (Figure 3.2).

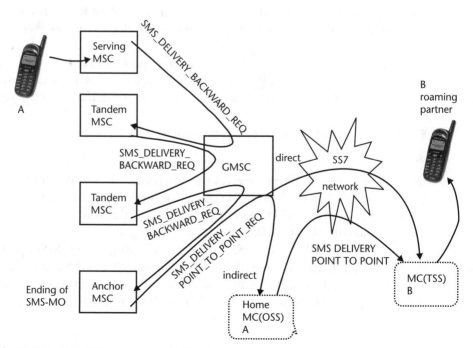

Figure 3.2 SMS-MO procedure in IS-41 networks.

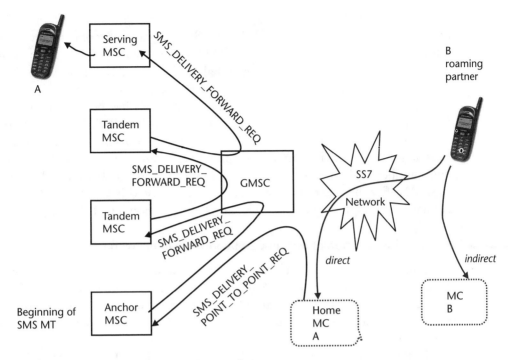

Figure 3.3   SMS-MT procedure in IS-41 networks.

3.1.2.3   SMS-MT Procedure

As explained earlier, because the HLR contains only the visited MSC address (SMS_Address) in IS-41, the anchor MSC, the SMS will have to be relayed through the various tandem MSCs until it reaches the serving MSC (Figure 3.3). However, because the CDMA customers are not supposed to move very much, this process may be considered acceptable!

3.1.3   MIN and IMSI for IS-41 Networks

3.1.3.1   Mobile Identity Number Is Not the Ordinary MSISDN

After explaining the comprehensive procedures required to implement SMS in a GSM network, here we list the major difference with the IS-41 network: In GSM, the keys used to address the HLR and obtain the visited MSC are a standard MSISDN and the MSISDN used to address the HLRs' GT (the routing is established by the home PLMN's GMSC). In IS-41 networks, the keys to address the HLR and obtain the visited MSC are a *mobile identity number* (MIN) assigned by an operator's proprietary table or function in the terminating servicing SMSC or the requirements that each HLR must be individually addressed by its point code (a full table of a range of numbers) and the point code of each HLR must be configured. For example, TDMA Argentina has a table of more than 3,000 lines to connect the MSISDN to the MIN, which allows it to interrogate the HLRs and send SMS-MT or perform the equivalent of the UPDATE_LOCATION_REQ in a GSM network, that is, REGISTRATION_NOTIFICATION_REQ.

The IS-41 SMS interworking architecture is a consequence of the SMS-MO being sent to the destination HPLMN's SMSC, as it can only compute the MIN, and a consequence of IS-41 networks (all TDMA and only a few CDMA, such as Globalstar satellite), and not implementing global titles and an E164 table MSISDN → GT of the HLR in a gateway MSC (as explained in Section 3.5).

The notion of *international roaming MIN* (IRM) has been introduced by the *International Forum on ANSI-41 Standards Technology* (IFAST) to allow roaming between IS-41 networks. For the networks of the *American Numbering Plan Association* (ANPA), which includes the United States and Canada, the IRM is the 10-digit number beginning with 1, so it is straightforward to obtain the IRM from the MSISDN. For the other areas that have IS-41, the IRM is a 10-digit number of the format 0-XXX + 6 digits or 1-XXX + 6 Digits, as shown in Table 3.1.

3.1.3.2   Introduction of IMSI in CDMA and TDMA Networks

IMSI has always been supported by GSM standards. It is currently supported by CDMA standards from IS-95 Revision A, TDMA standards from IS-136 Revision A, and by intersystem operations if the recommendations of IS-751 are incorporated in a TIA/EIA-41-D system. However, IMSI is not supported in any analog standards.

In the United States, a special IMSI format (310+00+MIN, where 310 is the MCC), the so-called MIN-based IMSI, has been defined in consideration of the backward compatibility. Although the concept of the MIN-based IMSI is useful in other countries, it is not possible to universally designate "00" as the IMSI_11_12 (the first 2 digits of MNC) for the MIN-based IMSI for all the MCCs, since the MNC numbering plan is a national matter. If this technique is not universally supported, when a mobile terminal with the MIN-based IMSI sends a registration request to the HLR, the HLR may not be able to recognize the ANSI-41 IMSI parameter and the registration could fail. A list of MIN-based IMSIs should be shared among roaming partners and programmed in their MSCs to solve this problem.

Currently, although IMSI is considered an ideal long-term solution, because of lack of support in the analog systems and incorrect implementation in some early digital mobiles, it is unlikely that IMSI will be implemented in the immediate future. However, as the demand for international roaming increases and availability of IRMs decreases, carriers may be forced to implement IMSI as the only viable alternative to the MIN problem. Further, since the MNC numbering plan is a national matter, the value of the IMSI_11_12 of the MIN-based IMSI is determined on a country-by-country basis.

Table 3.1   Example of International Roaming MIN

IRM Prefix	Mobile Operator	Country
0100	KT Freetel (CDMA)	Korea
0101-0104	SK Telecom (CDMA)	Korea
0119	Telesp	Brazil
0130-0138	Unicom (CDMA)	China
0226	TESAM (CDMA + Globalstar)	Peru

3.2   Implementation of SMS Services

3.2.1   SMS-MO Implementation

3.2.1.1   Basic Procedure

The SMS-MO is sent by the MS to the serving MSC, which is providing the current radio coverage. If this is not the anchor MSC, it will be relayed by an SMS_DELIVERY_BACKWARD message to the anchor MSC, eventually passing through several tandem MSCs (Table 3.2). Then, the final anchor MSC will forward the SMS-MO to the MC with an SMS_DELIVERY_POINT_TO_POINT (MO). Then, the final anchor MSC will send the SMS-MO to the MC using the SMS_DELIVERY_POINT_TO_POINT (Table 3.3). Note a huge difference with GSM: The SMS-MO does not contain the target message center address.

The anchor MSC must analyze the SMS_OriginalDestinationAddress, then it must have the numbering plan of all roaming partners so it can find the address of the destination message center. So all MSCs (which can be an anchor MSC in the procedure) must have access to this table, which makes maintenance very difficult.

3.2.1.2   SMS Interworking Issues with IS-41 Networks

The interworking procedure between IS-41-based networks has three different interworking possibilities. The first type, depicted in Figures 3.2 and 3.3, is called

Table 3.2   SMS_DELIVERY_BACKWARD_REQ and CNF Message Details

IS-41_SMS_DELIVERY_BACKWARD_REQ	
Parameter	*Class*
Primitive type octet	M
Invoke ID	M
InterMSCCircuitID	M
MobileIdentificationNumber	M
SMS_BearerData	M
SMS_TeleserviceIdentifier	M
ElectronicSerialNumber	O
SMS_ChargeIndicator	O
SMS_DestinationAddress	O
SMS_OriginalDestinationAddress	O
SMS_OriginalDestinationSubaddress	O
SMS_OriginalOriginationAddress	O
SMS_OriginalOriginationSubaddress	O
SMS_OriginationAddress	O
Timeout	O
IS-41_SMS_DELIVERY_BACKWARD_CNF	
Parameter	*Class*
Primitive type octet	M
Invoke ID	M
SMS_BearerData	O
SMS_CauseCode	O
SMS_FaultyParameter	O
User error	C
Provider error (CNF primitive only)	C

Table 3.3 SMS_DELIVERY_POINT_TO_POINT_REQ and CNF Message Details

IS-41_SMS_DELIVERY_POINT_TO_POINT_REQ (MO)	
Parameter	*Class*
Primitive type octet	M
Invoke ID	M
SMS_BearerData	M
SMS_TeleserviceIdentifier	M
ElectronicSerialNumber	O
MobileIdentificationNumber	O
SMS_ChargeIndicator	O
SMS_DestinationAddress	O
SMS_MessageCount	O
SMS_NotificationIndicator	O
SMS_OriginalDestinationAddress	O
SMS_OriginalDestinationSubaddress	O
SMS_OriginalOriginationAddress	O
SMS_OriginalOriginationSubaddress	O
SMS_OriginationAddress	O
Timeout	O
IS-41_SMS_DELIVERY_POINT_TO_POINT_CNF	
Parameter	*Class*
Primitive type octet	M
Invoke ID	M
SMS_BearerData	O
SMS_CauseCode	O
SMS_FaultyParameter	O
User error	C
Provider error (CNF primitive only)	C

the direct type. In the direct case, the anchor MSC of the originator has a MIN-to-MC translation table. In such a case, it may relay the SMS through an SMS_DELIVERY_POINT_TO_POINT to the destination's handset message center. The second one is an indirect type. In the indirect case, the anchor MSC of the originator must end the SMS at the originator MC, which will make the translation and send the SMS to the destination's handset message center. The third type is referred to as "GSM-like," in which case the originator MC directly interrogates the destination's handset HLR and sends the SMS to the destination anchor MSC.

For the SMS interworking procedure, there is only one way in which all SMSs must be relayed to a third party because the IS-41 network does have roaming privileges with the destination network. The users must add a prefix to the destination number such as 639193412345 if they want to send an SMS to a Smart Telecoms (Philippines) cell phone with a number such as 1415. This number is such that it cannot be confused with any other CC-NDC of another IS-41 roaming partner.

For example, to send an SMS abroad, SK Telecom (CDMA, Korea) subscribers will send to 1415639193412345. SK Telecom has programmed the tables of all of their MSCs so that the SMS_DELIVERY_POINT_TO_POINT will go to their MC. In the MC, they strip off the 1415 and relay the SMS to the third party (the SMS interworking network) with which they have an SMS interworking agreement. The third party will send the SMS to the final destination network subscriber 639193412345: Smart Telecoms (GSM, Philippines) with the appropriate convention.

We have been challenged by Smart because their IS-41 SMS interworking part-ners had to add this 1415. Due to the IS-41 network architecture, there is no other way. To improve the ergonomics, we have added 1415 automatically in front of the originating address.

3.2.2   SMS-MT Implementation

The equivalent of the SEND_ROUTING_INFO_FOR_SM in GSM is SMS_ REQUEST for IS-41 (Table 3.4). However, the key is the MIN, which is the derived with a proprietary table from the MSISDN. Most often, the national form of the MSISDN must also be used, which the other network does not know because it is not standardized. The response from the HLR contains the SMS_Address, which is the anchor MSC.

The equivalent of the FORWARD_SHORT_MESSAGE (MT) is SMS_ DELIVERY_POINT_TO_POINT (MT) (Table 3.5). But it is sent by the message center (MC), the equivalent of the GSM SMSC, to the anchor MSC, which may not be the switch providing the radio termination to the cell phone (MS).

In a case where the serving MSC is not the anchor MSC, the anchor MSC will relay the SMS-MT (eventually through several tandem MSCs) by means of the SMS_DELIVERY_FORWARD message (Table 3.6). The equivalent message to the GSM ALERT_SERVICE_CENTER message is SMS_NOTIFICATION in IS-41 (Table 3.7).

3.3   IS-41 Procedure for SMS

The IS-41 standard is a specification for networking operations between network entities of a mobile phone network. This standard defines the IS-41 MAP protocol used to communicate among machines in order to execute all mobile phone services (call establishment, authentication, SMS services, and so on).

Table 3.4   SMS_REQUEST_REQ and CNF Message Details

IS-41_SMS_ REQUEST_REQ	
Parameter	Class
Primitive type octet	M
Invoke ID	M
MobileIdentificationNumber	M
ElectronicSerialNumber	O
SMS_NotificationIndicator	O
SMS_TeleserviceIdentifier	O
Timeout	O
IS-41_SMS_ REQUEST_CNF	
Parameter	Class
Primitive type octet	M
Invoke ID	M
ElectronicSerialNumber	O
SMS_AccessDeniedReason	O
SMS_Address	O
User error	C
Provider error (CNF primitive only)	C

Table 3.5 IS-41_SMS_ DELIVERY_POINT_TO_POINT_REQ and CNF Message Details

IS-41_SMS_DELIVERY_POINT_TO_POINT_REQ (MT)

Parameter	Class
Primitive type octet	M
Invoke ID	M
SMS_BearerData	M
SMS_TeleserviceIdentifier	M
ElectronicSerialNumber	O
MobileIdentificationNumber	O
SMS_ChargeIndicator	O
SMS_DestinationAddress	O
SMS_MessageCount	O
SMS_NotificationIndicator	O
SMS_OriginalDestinationAddress	O
SMS_OriginalDestinationSubaddress	O
SMS_OriginalOriginationAddress	O
SMS_OriginalOriginationSubaddress	O
SMS_OriginationAddress	O
Timeout	O

IS-41_SMS_ DELIVERY_POINT_TO_POINT_CNF

Parameter	Class
Primitive type octet	M
Invoke ID	M
SMS_BearerData	O
SMS_CauseCode	O
SMS_FaultyParameter	O
User error	C
Provider error (CNF primitive only)	C

This specification focuses only on the IS-41 SMS protocol. As a consequence, the protocol interface between the MSC and the MS-based SME is discussed in this section because it is dependent on the radio technology used by the operator (CDMA, AMPS, TDMA) and is not within the scope of the IS-41 standard. As specified in [1], the radio technology used in mobile phone networks does not impact the SMS protocol.

In the following discussion, the SMS interworking network router is considered to be the system and the GSM- or IS-41-based network is considered to be the environment in which the system resides.

3.3.1 Functional Description of IS-41 SMS Services

As specified in [1], the two messages SMS_DELIVERY_BACKWARD and SMS_DELIVERY_FORWARD are only intended to route the SMS delivery request of an MS-based SME from the serving MSC toward the handoff chain to the anchor MSC of the MS either for originating or terminating delivery. Because the SMS interworking network routers are completely free of handoff matters, they may only be involved in the routing process after the SMS delivery request has reached the anchor MSC. Therefore, the SMS interworking network router will not have to implement the SMS_DELIVERY_BACKWARD and SMS_DELIVERY_FORWARD operations.

Table 3.6 IS-41_SMS_ DELIVERY_FORWARD_REQ and CNF Message Details

IS-41_SMS_DELIVERY_FORWARD_REQ

Parameter	Class
Primitive type octet	M
Invoke ID	M
InterMSCCircuitID	M
MobileIdentificationNumber	M
SMS_BearerData	M
SMS_TeleserviceIdentifier	M
ElectronicSerialNumber	O
SMS_ChargeIndicator	O
SMS_DestinationAddress	O
SMS_OriginalDestinationAddress	O
SMS_OriginalDestinationSubaddress	O
SMS_OriginalOriginationAddress	O
SMS_OriginalOriginationSubaddress	O
SMS_OriginationAddress	O
Timeout	O

IS-41_SMS_DELIVERY_FORWARD_CNF

Parameter	Class
Primitive type octet	M
Invoke ID	M
SMS_BearerData	O
SMS_CauseCode	O
SMS_FaultyParameter	O
User error	C
Provider error (CNF primitive only)	C

For purposes of comprehension, functional descriptions of SMD_REQUEST, SMS_DELIVERY_BACKWARD, and SMS_DELIVERY_FORWARD are given only as an informative specification (if needed).

Table 3.7 IS-41_SMS_ NOTIFICATION_REQ and CNF Message Details

IS-41_SMS_NOTIFICATION_REQ

Parameter	Class
Primitive type octet	M
Invoke ID	M
ElectronicSerialNumber	O
MobileIdentificationNumber	O
SMS_AccessDeniedReason	O
SMS_Address (serving MSC)	O
Timeout	O

IS-41_SMS_NOTIFICATION_CNF

Parameter	Class
Primitive type octet	M
Invoke ID	M
User error	C
Provider error (CNF primitive only)	C

3.3.1.1   SMS Services

SMS is a service that enables an SME (which can be MS based or an MC, an HLR, an MSC, or an Internet client) to send and receive user-defined data to and from another SME. In most cases the user data are text messages. This service is not connected; that is to say, it uses only the signaling network to carry the user data but no data connections.

3.3.1.2   Functional Entities

The functional entities involved in the SMS service are used either to route the message from its origination to its destination (MSC, HLR, VLR, SMS router) or to initiate or furnish specific delivery services (SME and MC).

- *SME.* The SME is the entity responsible for initiating or receiving an SMS delivery.
- *MSC.* The MSC, as for a voice call, is the switching entity that will route the message to its destination. Three types of functional MSCs are used in CDMA mobile phone networks due to the handoff procedure. The serving MSC is the one that is currently serving the MS (the first MSC on the call path). If a handoff is necessary, the second MSC on the call path may be a tandem MSC (the one that was the serving MSC just before the handoff procedure succeeded). Then the anchor MSC is the first serving MSC of the MS (before any handoff).
- *MC.* The message center is the entity responsible for SMS delivery execution. It also includes supplementary services (on the originating or terminating side).
- *SMS router.* The IS-41 specification is rather elusive on this particular entity description. It seems to be an interworking facility to route SMS messages to network destinations different from the originating one.
- *HLR.* The HLR stores all information specific to an MS. It is essential to the routing process because it knows the VLR in which the invoked MS is roaming.
- *VLR.* The VLR stores temporary information for an MS. It is used in the routing process because it knows the MSC address of the destination SME.

3.3.1.3   SMS Procedures

This section gives a description of procedures and of how messages flow among network entities for providing SMS services. Here the term *handoff* refers to the definition given in [1]; that is to say, it can only happen when a MS has a dedicated voice channel with the network. The handoff procedure refers to the ability of the network to ask the MS to change its dedicated channel for a better one and eventually to change its serving MSC.

SMS Delivery Without Supplementary Services

MS-based SME initiating—no handoff—no originating supplementary services (OSS).
In this case the SMS is originated by an MS-based SME and the serving MSC is also the anchor MSC of the MS. The MS establishes a signaling link with its serving MSC and sends an SMD_REQUEST to it through the BSS. During the establishment of

the signaling connection, the MSC retrieves the profile information for the originating MS and more precisely the SMS_OriginationRestrictions parameter through an interrogation of the MS's VLR with a QualificationRequest operation. Then according to the MS SMS_OriginationRestrictions, it initiates an SMS_DELIVERY_ POINT_TO_POINT message to the destination home MC or if SMS_OriginationRestrictions is set to "Force Message Center" for the origination home MC. The destination MC sends to the HLR corresponding to the destination MIN an SMS_REQUEST message to ask for the address of the destination MSC. The HLR relays the SMS_REQUEST message to the VLR in which the destination MS is roaming, and the VLR relays the SMS_REQUEST message to the anchor MSC of the destination MS. The result of this invoke operation is then routed back to the destination MC, which can initiate the SMS_DELIVERY_POINT_TO_POINT invoke toward the destination MSC. On receipt of this invoke, the anchor MSC of the destination MS either sends an SMD_REQUEST message to the destination MS or initiates an SMS_DELIVERY_FORWARD operation toward the serving MSC (relayed by the tandem MSC if needed) of the MS, which then sends the SMD_REQUEST message to the destination MS.

MS-based SME initiating—handoff—no originating supplementary services (OSS).   In this case the SMS is originated by an MS-based SME and the MS has handed off the message. The serving MSC is different from the anchor MSC and there may be a tandem MSC on the handoff chain. In this case when the serving MSC receives the SMD_REQUEST message sent from the MS-based SME, it initiates an SMS_DELIVERY_BACKWARD message toward the anchor MSC (or the tandem MSC if needed), which then will process the SMS_DELIVERY_POINT_TO_POINT process toward the originating home's MC or destination's one according to the profile information of the originating SME.

SME initiating (not MS)—no OSS.   In the case where the initiating SME is not an MS, the termination procedure is the same but the originating delivery starts with an SMS_DELIVERY_POINT_TO_POINT message sent by the SME to the destination MC, which then delivers the message to the destination MS. In this case there are no restrictions on the originating profile.

SMS Delivery with Supplementary Services   If supplementary services have been requested either at the origination side or destination side, the message is indirectly routed to the origination's home MC (OSS) or directly to the destination's MC *terminating supplementary services* (TSSs).

Originating MC supplementary services.   When OSSs have been requested, the MC performs first any service indicated by the originating message or the subscriber's SMS profile (delayed delivery, distribution list, and so on). Then the MC loops on the services requested until they all have been performed.

Terminating MC supplementary services.   This case is similar to that for OSSs, but the *terminating supplementary services* (TSSs) are different (delivery notification, for example).

SMS Delivery to an Unreachable SME   When the originating SME has requested an SMS_Delivery and set the parameter SMS_NotificationIndicator to "Notify When

Available," if message delivery cannot be processed (for example, if the destination SME is not available), either the HLR or the destination MSC sets a delivery pending flag for the destination MS. At this point, the destination's MC enters a loop on the reception of the SMS_Notification message from the HLR or the MSC. If the SMS_Notification message does not arrives at the MC before the validity timer expires, the message delivery is discarded; otherwise, the message delivery is processed.

Address Translation　　The principle used to route the message toward the destination is as follows. To be delivered to the destination MS, the message must first be sent to the destination MS's MC. To do so, the originating anchor MSC or the originating MS's MC conducts a MIN-to-MC translation. When the destination MS's MC receives the SMS_DELIVERY_POINT_TO_POINT message, it performs the HLR interrogation with a MIN-to-HLR address translation to send the SMS_Request message to the HLR to which the destination's MIN belongs. Then the HLR answers back with the destination MS's serving MSC address for the SMS_Delivery.

3.3.2　IS-41 SMS Protocol Description

The IS-41 standard defines the protocols for interworking facilities between mobile phone network entities (MSC, HLR, VLR, MC, and so on). As the scope of this chapter is to describe only the specifications for SMS service, we will focus on the IS-41 SMS protocol.

3.3.2.1　Functional Entities

The SMS protocol defines all messages exchanged by network entities in order to provide full SMS services. The functional entities involved in this protocol are MSC, MC, SMS router, HLR, VLR, and SME.

3.3.2.2　Messages

All messages involved in the IS-41 SMS protocol are processed through the TCAP layer, which is based on the ACSE/ROSE protocol. TCAP will then in particular provide to IS-41 the operations of ROSE: Invoke, Return Result, Return Error, and Reject. For any SMS protocol message, there will be an Invoke for invoking the service on the remote machine, which will reply with a Return Result, Return Error, or Reject to notify the client machine of the result of its invoking operation.

- *SMS_DELIVERY_BACKWARD*. This message is described in Section 6.4.2.41 of [1] and is used to convey an MS-originated short message toward the anchor MSC when the originating MS has handed it off. It then can only be used by MSCs in the handoff chain (among the serving, tandem, and anchor MSCs).

- *SMS_DELIVERY_FORWARD*. This message is described in Section 6.4.2.42 of [1] and is used to convey an MS-terminated short message toward the serving MSC when the terminating MS has handed it off. It then can only be used by MSCs in the handoff chain (among the serving, tandem, and anchor MSCs).

- *SMS_DELIVERY_POINT_TO_POINT.* This message is described in Section 6.4.2.43 of [1] and is used to convey a short message between any entities of a network that are not MS-based SMEs or serving or tandem MSCs.

- *SMS_NOTIFICATION.* This message is described in Section 6.4.2.44 of [1] and is used to report a change in an MS's ability to receive SMS messages between an MSC or HLR and an MC.

- *SMS_REQUEST.* This message is described in Section 6.4.2.45 of [1] and is used to request an MS's current SMS routing address from an MC to an HLR then VLR and then MSC. It is the equivalent of a GSM SEND_ROUTING_INFO_ FOR_SM_REQ.

- *SMD_REQUEST.* This message is described in Annex D of [1] and is used in the air interface between a serving MSC and an MS-based SME. Its format is merely informative since the air interface depends on the radio protocol used (CDMA, AMPS, or TDMA).

3.3.2.3   Information Elements

Information elements (IEs) define the various fields contained in an SMS protocol message. They may be mandatory or optional.

SMS-Specific IEs.   These particular IEs are only present in SMS protocol messages:

- *SMS_AccessDeniedReason.* This IE is described in Section 6.5.2.122 of [1] and indicates why short message delivery is not currently allowed to an MS-based SME.

- *SMS_Address.* This IE is described in Section 6.5.2.123 of [1] and is used to convey the current routing address of the serving MSC for the purpose of a short message termination to an MS-based SME. It can be encoded in BCD format, IP address format, or SS7 point code format.

- *SMS_BearerData.* This IE is described in Section 6.5.2.124 of [1] and is to be used and interpreted by an SMS teleservice. This IE will contain the user data (message text).

- *SMS_CauseCode.* This IE is described in Section 6.5.2.125 of [1] and indicates why an SMS message has not been delivered.

- *SMS_ChargeIndicator.* This IE is described in Section 6.5.2.126 of [1] and specifies the charging option for an SMS message.

- *SMS_DestinationAddress.* This IE is described in Section 6.5.2.127 of [1] and conveys the address of a destination SME. It may be encoded in BCD format or IP address format.

- *SMS_MessageCount.* This IE is described in Section 6.5.2.128 of [1] and is used to indicate the number of SMS messages pending delivery.

- *SMS_MessageWaitingIndicator.* This IE is described in Section 6.5.2.129 of [1] and prompts the serving MSC and the HLR to be prepared to launch an SMS_Notification when the MS-based SME becomes available.

- *SMS_NotificationIndicator.* This IE is described in Section 6.5.2.130 of [1] and is used to control sending of SMS_Notification messages.
- *SMS_OriginalDestinationAddress.* This IE is described in Section 6.5.2.131 of [1] and is the address of the original message destination. It may be encoded in BCD format or IP address format.
- *SMS_OriginalDestinationSubaddress.* This IE is described in Section 6.5.2.132 of [1] and is the subaddress of the original message destination (see Section 6.5.3.13 of [2]).
- *SMS_OriginalOriginatingAddress.* This IE is described in Section 6.5.2.133 of [1] and is the address of the original message sender. It may be encoded in BCD format, IA5 format, or IP address format.
- *SMS_OriginalOriginatingSubaddress.* This IE is described in Section 6.5.2.134 of [1] and is the subaddress of the original message sender (see Section 6.5.3.13 of [2]).
- *SMS_OriginatingAddress.* This IE is described in Section 6.5.2.135 of [1] and is used to convey the current routing address of an MS-based SME. It may be encoded in BCD format or IP address format.
- *SMS_OriginationRestrictions.* This IE is described in Section 6.5.2.136 of [1] and defines the type of message the MS is allowed to originate.
- *SMS_TeleserviceIdentifier.* This IE is described in Section 6.5.2.137 of [1] and indicates the teleservice to which the SMS message applies.
- *SMS_TerminationRestrictions.* This IE is described in Section 6.5.2.138 of [1] and defines the type of messages the MS is allowed to receive.

Nonspecific IEs. The following IEs are used by SMS protocol messages in the same way they are used other IS-41 protocol messages.

- *MobileIdentificationNumber.* This IE is described in Section 6.5.2.81 of [1] and is a 10-digit representation of the MS's MIN in BCD format.
- *ElectronicSerialNumber.* This IE is described in Section 6.5.2.63 of [1] and is used to indicate the unique 32-bit electronic serial number of an MS.
- *InterMSCCircuitID.* This IE is described in Section 6.5.2.72 of [1] and is used to identify a specific trunk in a dedicated trunk group between two MSCs.

3.3.3 Specification of the SMS Interworking Network IS-41 SMS Router

The goal of this section is to specify the global behavior of the SMS interworking network IS-41 SMS router in terms of its integration in both GSM- and IS-41-based networks. The functional needs of the SMS interworking network router are far simpler than that of the IS-41 SMS router: It must allow any IS-41-designed SME to send and receive SMS to and from any IS-41- or GSM-designed network. As described earlier, the IS-41 specification [1] includes a functional entity called the SMS router, in which the SMS interworking network IS-41 router will settle. Moreover, the IS-41 SMS interworking network router must provide all functionalities available in the GSM version. It must also provide MC facility to other SMS interworking network routers of the network (for interoperator SMS delivery). Concerning interworking functionality between GSM and IS-41 networks, the router will have the ability to translate an SMS delivery originated from an IS-41 network to GSM and vice versa.

3.3.3.1 IS-41 SMS Protocol in the SMS Interworking Network Routers

Because the SMS interworking network router will not have to act as a serving, tandem, or anchor MSC, it does not need to implement the SMS_DELIVERY_FORWARD and SMS_DELIVERY_BACKWARD messages. Moreover it cannot receive or send SMD_Request messages from or to an MS-based SME. As a consequence, it will not be dependent on the type of radio technology used (CDMA, AMPS, or TDMA) except for the translation of the GSM 03.40 specification to the TDMA or CDMA interface. It will then only take into account the SMS_DELIVERY_POINT_TO_POINT, SMS_REQUEST, and SMS_NOTIFICATION messages. As an MC function, the SMS interworking network router must implement the QualificationRequest message sent to the HLR to retrieve the subscriber's profile and perform any process enclosed in the profile's description.

3.3.3.2 Data Model

We now take a look at the data model:

- *Address translation data.* To perform the address translations needed by the SMS protocol, the SMS interworking network router will have to store a table for MIN-to-MC address translation and another for MIN-to-HLR address translation.
- *SMS delivery context.* When computing an SMS delivery, the SMS interworking network router will have to allocate a call context for the corresponding process, and if the delivery is postponed or delayed, it has to keep the context alive for later delivery.
- *Profile information.* If the SMS interworking network router is operating as an MC, it must retrieve the subscriber's profile information from the HLR with a QualificationRequest. (We assume that the SMS interworking network router will be authorized for this operation by the subscriber's network operator.) It will then need to store this information during the processing of the SMS delivery.
- *Temporary information.* During the process of an SMS delivery, the SMS interworking network router must allocate temporary data for storing all specific information received or to be sent according to the IS-41 specification [1]. These data will be released on receipt of a result message based on the description given in [1].

3.3.3.3 IS-41-Based Network Router Specification

The SMS interworking network IS-41-based router integrated in an IS-41 network can fulfill the requirements of the IS-41 MC functional entity. Here is a scheme that represents the interactions of a specific SMS interworking network router with other functional entities of the IS-41 network:

SMS Interworking Network Router as MC Specification.   The main functionality of the SMS interworking network router is as a message center. Reference [1] specifies that in an IS-41 network the SMS delivery service may be processed via either indirect or direct routing. If indirect routing is invoked, the originating SME has required OSSs or the originating subscriber's profile field SMS_ OriginationRestrictions is set to "Force Message Center." In this case, at the anchor MSC, the SMS delivery is routed to the originating home MC. Then after performing the OSS the message delivery is routed to the destination's home MC to process the TSSs according to the destination subscriber's profile field SMS_ TerminationRestrictions.

As a consequence, the SMS interworking network router has to provide the OSS and TSS for a network and then must implement the QualificationRequest to get back the subscriber's profile from the HLR. It is important to note that [1] lets any MC provide OSS and TSS. It will have, as described in the data model, two address translation tables for MIN-to-HLR and MIN-to-MC translations.

Moreover, if OSS is required and the MC providing OSS is not an SMS interworking network router, it must translate the destination MIN to the SMS interworking network router (as TSS). Then the SMS interworking network router must be registered in the target network.

SMS Interworking Network Router as SMS Router.   Section 4.46.10 of [1] includes a network facility called SMS router that allows an SMS message to be routed to a different IS-41 network. The SMS interworking network router will then provide this feature to the originating network.

3.3.3.4   Analogy of IS-41 with GSM Network

GSM Network MAP SMS Protocol in the SMS Interworking Network Router.   The SMS interworking network routers do not use the entire SMS protocol defined in [3]. It uses just the following MAP services:

- MAP_FORWARD_SHORT_MESSAGE for SMS delivery invokes.
- MAP_SEND_ROUTING_INFO_FOR_SM for routing information retrieval from HLR.
- MAP_ALERT_SERVICE_CENTER for notifying of a change in the delivery status of the destination MS.
- MAP_REPORT_SM_DELIVERY_STATUS to inform the HLR of the result of a delivery attempt.

For a complete description of the SMS interworking network's implementation of the SMS MAP protocol refer to [3].

Concerning SMS services, the MAP protocol represents the sublayer of the SMS protocol defined in [4]. This recommendation differentiates the MO and MT point of view and defines the TPDU set in the SM-RP-UI field of MAP messages. Indeed, the GSM SMS specification defines several layer levels for the SMS protocol, the MAP layer and above it three encapsulated layers: the *short message transfer layer* (SM-TL), the *short message application layer* (SM-AL), and the *short message relay layer* (SM-RL). The SM-TL is described in [4]. The SM-TL introduces the notion of

SUBMIT and DELIVER protocol messages to differentiate MO and MT short messages.

To allow interworking between MAP and IS-41 networks, we have to map and define a precise match between the MAP-defined requests and the IS-41 ones and vice versa. In the following list we cover requests from [1] for all SMS procedures and give the equivalent requests for the GSM protocol:

SMS_DELIVERY_POINT_TO_POINT Invoke.   This operation is to be mapped on the MAP_FORWARD_SHORT_MESSAGE: SM-TL (SUBMIT or DELIVER). The IS-41 parameters for this procedure are as follows:

- *SMS_BearerData.* This field represents the user data and must be interpreted according to the teleservice in use. It is to be linked to the SM-RP-UI field of the MAP_FORWARD_SHORT_MSG_REQ.

- *SMS_TeleserviceIdentifier.* This field indicates the teleservice for which the SMS message applies. It cannot be linked to any concept within the GSM MAP protocol since GSM is not a teleservice-oriented protocol.

- *ElectronicSerialNumber (ESN).* This field represents the hardware serial number of the MS (origination or destination MS depending on the functional entity processing the message). It must be linked to the *international mobile station equipment identity* (IMEI) in the GSM recommendation. But this field is not used in the GSM SMS protocol.

- *MobileIdentifierNumber.* This field is the numbering representation of the SME (origination or destination SME depending on the functional entity processing the message). It must be linked to the MSISDN or IMSI fields used in the GSM recommendation.

- *SMS_ChargeIndicator.* This field indicates to the network the charging policy to use for delivery of the message. It does not have an equivalent in the GSM architecture because in the GSM protocol the calling entity is always charged.

- *SMS_DestinationAddress.* This field contains the destination address for the message. It must be linked either to the SM-RP-DA field of the MAP_FORWARD_SHORT_MESSAGE or to the TP-Destination-Address of the TPDU SMS-SUBMIT depending on the processing of the message (MO or MT).

- *SMS_MessageCount.* This field indicates the number of messages pending delivery. It has no equivalent in the GSM recommendation.

- *SMS_NotificationIndicator.* This field indicates whether the message must be stored and forwarded. It has no equivalent in the GSM recommendation (Store&Forward operator dependent).

- *SMS_OriginalDestinationAddress.* This field conveys the original message destination in case the message is routed through a network entity for processing (the SMS_DestinationAddress could then be modified). It is to be linked with the TP-Destination-Address of the TPDU SMS-SUBMIT or to the SP-RP-DA field of the MAP_FORWARD_SHORT_MESSAGE depending on the processing involved.

- *SMS_OriginalDestinationSubaddress.* This field is to be inserted if applicable. It contains the subaddress of the original message destination. This field does not have any equivalent in the GSM SMS protocol.

- *SMS_OriginalOriginatingAddress.* Same as for SMS_OriginalDestinationAddress for the originating side.

- *SMS_OriginalOriginatingSubaddress.* Same as for SMS_OriginalDestinationSubaddress.

- *SMS_OriginatingAddress.* This field contains the originating address of the message sender. It must be linked to the SM-RP-OA field of the MAP_FORWARD_SHORT_MESSAGE or to the TP-Originating-Address of the TPDU SMS-DELIVER depending on the processing.

SMS_DELIVERY_POINT_TO_POINT Return Result. If the SMS_DELIVERY_POINT_TO_POINT Invoke is successful, the network returns an SMSDPTP Return Result, which is to be mapped to the GSM message MAP_FORWARD_SHORT_MESSAGE-CNF or ...-RSP. The two fields are optional.

- *SMS_BearerData.* This field has no equivalent in GSM.

- *SMS_CauseCode.* This field should be compared to the GSM fields User_Error and Provider_Error.

SMS_REQUEST Invoke. This operation is to be mapped to the GSM MAP_SEND_ROUTING_INFO_FOR_SM_REQ message. The IS-41 parameters for this operation are as follows:

- *MobileIdentificationNumber.* This field is the identification number of the MS we are requesting for the current routing address. It is to be linked to the MSISDN field of the MAP_SEND_ROUTING_INFO_FOR_SM message.

- *ElectronicSerialNumber.* This field is the hardware signature of the MS. It is equivalent to the IMEI in the GSM recommendation but is not used in MAP_SEND_ROUTING_INFO_FOR_SM.

- *SMS_NotificationIndicator.* This field is used to define the way HLR will process the SMS_Notification messages. It has no equivalent in MAP_SEND_ROUTING_INFO_FOR_SM.

- *SMS_TeleserviceIdentifier.* This field identifies the teleservice to be used. It does not have an equivalent in MAP_SEND_ROUTING_INFO_FOR_SM.

SMS_REQUEST Return Result. This operation is a response to the SMS_REQUEST Invoke message. It is mapped to the MAP_SEND_ROUTING_INFO_FOR_SM_CNF message in the GSM recommendation. The IS-41 parameters for this operation are as follows:

- *Electronic Serial Number.* This field has no equivalent in the MAP_SEND_ROUTING_INFO_FOR_SM_CNF message.

- *SMS_AccessDeniedReason.* This field returns the reason the operation is denied if such an occurrence happens. It is equivalent to the two fields User_Error and Provider_Error in GSM.

- *SMS_Address.* This field returns the address of the MSC serving the destination SME. It is equivalent to the field MSC Number of the message MAP_SEND_ROUTING_INFO_FOR_SM_CNF, but it is not a GT. It is the MSCID.

SMS_NOTIFICATION Invoke.   This operation is used by the HLR or MSC to notify the message center that the ability of the MS to receive SMS messages has changed. It can be compared to the MAP_ALERT_SERVICE_CENTER operating in GSM service. The IS-41 fields for this operation are as follows:

- *Electronic Serial Number and Mobile Identification Number.* These fields are used to identify the MS for which the delivery status has changed. It must be linked to the MSISDN-Alert field of the GSM MAP_ALERT_SERVICE_CENTER message.

- *SMS_AccessDeniedReason.* Same as for SMS_REQUEST.

- *SMS_Address.* Same as for SMS_REQUEST.

SMS_NOTIFICATION Return Result.   This operation is the answer to the SMS_NOTIFICATION Invoke operation and must be linked to the MAP_ALERT_SERVICE_CENTER_CNF service. There are no IS-41 fields for this operation.

3.4   Interworking Between IS-41 and GSM

To fulfill the requirements of sending an SMS from an IS-41 to a MAP network and from a MAP to an IS-41 network (Figure 3.4), the SMS interworking network routers will have to translate the content of the SM-RP-UI of the MAP_FORWARD_SHORT_MSG_REQ into the SMS Bearer Data and an appropriate teleservice ID of the IS-41 protocol (and vice versa). We need to be concerned with the adaptation of GSM Recommendation 03.40 into CDMA IS-637 and TDMA IS-136 specifications. Depending on the destination handset network (TDMA or CDMA), the interworking facility will use the following teleservices: CMT for TDMA-based networks (IS-136-710 specification) and WMT for CDMA-based networks (IS-637-B specification).

3.4.1   GSM Specifications of User Information

The SM-RP-UI field of a MAP_FORWARD_SHORT_MSG_REQ is specified in GSM Recommendation 03.40. This specification introduces the four types of messages supported by GSM handset, SMS_DELIVER, SMS_SUBMIT, SMS_STATUS_REPORT, and SMS_COMMAND. Refer to Chapter 1 for a complete description of these messages.

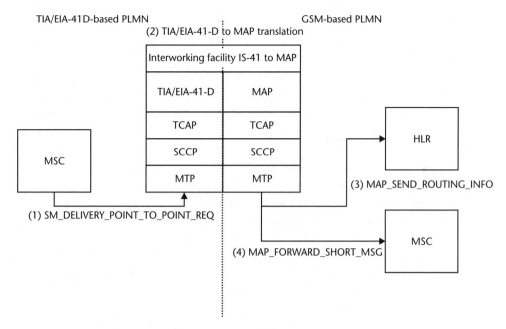

Figure 3.4   Interworking between IS-41 and GSM networks.

3.4.2   Mapping GSM to IS-637

The IS-637 specification describes the SM-RL, SM-TL, and SM-Teleservice layers used to send or receive SMSs to and from a CDMA-based mobile station. SM-RL is the relay layer between the message center and the base station subsystem, SM-TL is the transport layer relying on SM-RL, and the SM-Teleservice layer defines the appropriate encoding of the bearer service used between the MC and the MS. The problem of mapping GSM-based SMSs into appropriate IS-637-based SMSs is that of translation of the SM-TL layer defined by GSM Recommendation 03.40 into the SM-Teleservice layer of the IS-637 protocol. The IS-637 protocol defines several teleservices used in the CDMA architecture to address a mobile station. We will concentrate on the teleservice used for delivery of text messages, *wireless messaging teleservice* (WMT).

3.4.2.1   SM-Teleservice Layer WMT

This teleservice provides the same kind of messages as GSM Recommendation 03.40; that is to say, SMS_SUBMIT (Table 3.8), SMS_DELIVER (Table 3.9), and others that are specific to WMT such as SMS_CANCELLATION (Table 3.10), SMS_USER_ACKNOWLEDGMENT (Table 3.11), SMS_DELIVERY_ ACKNOWLEDGMENT (Table 3.12), and SMS_READ_ ACKNOWLEDGMENT (Table 3.13). For a complete description of fields encoding, refer to ANSI/TIA/ EIA-637-B [5].

3.4.2.2   Mapping GSM SM-TL and IS-637 WMT

The purpose of this section is not to give a field-to-field mapping of the two protocols, but rather to describe the philosophy of mapping. It is obvious that some fields

Table 3.8   SMS_SUBMIT WMT Details

Subparameter	Type
Message Identifier	Mandatory
User data	Optional
Validity period—absolute	Optional
Validity period—relative	Optional
Deferred delivery time—absolute	Optional
Deferred delivery time—relative	Optional
Priority indicator	Optional
Privacy indicator	Optional
Reply option	Optional
Alert on message delivery	Optional
Language indicator	Optional
Callback number	Optional
Multiple encoding user data	Optional
Message deposit index	Optional

Table 3.9   SMS_DELIVER WMT Details

Subparameter	Type
Message identifier	Mandatory
User data	Optional
Message center time stamp	Optional
Validity period—absolute	Optional
Validity period—relative	Optional
Priority indicator	Optional
Privacy indicator	Optional
Reply option	Optional
Number of messages	Optional
Alert on message delivery	Optional
Language indicator	Optional
Callback number	Optional
Message display mode	Optional
Multiple encoding user data	Optional
Message deposit index	Optional

Table 3.10   SMS_CANCELLATION WMT Details

Subparameter	Type
Message identifier	Mandatory

Table 3.11   SMS_USER_ACKNOWLEDGMENT WMT Details

Subparameter	Type
Message identifier	Mandatory
User data	Optional
User response code	Optional
Message center time stamp	Optional
Multiple encoding user data	Optional
Message deposit index	Optional

for WMT do not have an equivalent in the GSM recommendation, for instance, priority indicator and privacy indicator. For those fields when we translate a GSM-

Table 3.12   SMS_DELIVERY_ACKNOWLEDGMENT WMT Details

Subparameter	Type
Message identifier	Mandatory
User data	Optional
Message center time stamp	Optional
Multiple encoding user data	Optional
Message status	Optional

Table 3.13   SMS_READ_ACKNOWLEDGMENT WMT Details

Subparameter	Type
Message identifier	Mandatory
User data	Optional
Message center time stamp	Optional
Multiple encoding user data	Optional

based SMS into an IS-637 one, we must either omit the field if it is optional or set a default value that will disrupt the service.

For the other fields for which a direct correspondence can be made (for instance, the validity period), we need to use appropriate functions to perform a translation between the two different formats.

Then we can map SMS_DELIVER and SMS_SUBMIT directly from IS-637 to GSM 03.40. The SMS_CANCELLATION message can be mapped into an SMS_COMMAND of the GSM protocol. SMS_USER_ ACKNOWLEDGMENT, SMS_READ_ACKNOWLEDGMENT, and SMS_ DELIVERY_ACKNOWLEDGMENT can be mapped with SMS_STATUS_ REPORT operations of the GSM protocol.

3.4.3   Mapping GSM to IS-136-710

The IS-136-710 recommendation specifies the *cellular messaging teleservice* (CMT) used to provide SMS service in TDMA-based networks. As in the previously described WMT teleservice, it provides SMS_SUBMIT (Table 3.14) and SMS_DELIVER (Table 3.15) services and SMS_DELIVERY_ACK (Table 3.16), SMS_MANUAL_ACK (Table 3.17), and other services that will not be described because they do not fit into the GSM-IS-41 interworking specification.

The SMS_SUBMIT and SMS_DELIVER messages can be directly mapped from and to GSM Recommendation 03.40. Of course, as was true for WMT fields, fields that are not supported by one of the technologies used must be omitted or set to default values. SMS_STATUS_REPORT can be mapped into the SMS_DELIVERY_ ACK or SMS_MANUAL_ACK fields of IS-136.

3.4.4   SMS Delivery from IS-41 SME to MAP SME

3.4.4.1   Overview

We now consider the case in which an MS submits an SMS message to an IS-41 network MC and the message destination is an SME of an GSM network. When the IS-41 SMS interworking network MC has analyzed the received PDU SMS_ DELIVERY_POINT_TO_POINT Invoke, it must map its fields into the MAP_

Table 3.14   SMS_SUBMIT CMT Details

Information Element	Type	Length
Message type indicator	M	3
Message reference	M	13
Privacy indicator	M	3
Urgency indicator	M	2
Delivery acknowledgment request	M	1
Manual acknowledgment request	M	1
Message updating	M	1
User data unit	M	16
Validity period	O	13 or 49
Deferred delivery time	O	13 or 49
Callback number	O	20
Callback number presentation indicator	O	8
Callback number alpha tag	O	20
Multilingual callback number	O	28
Multilingual callback number alpha tag	O	32
Multilingual destination address	O	32

FORWARD_SHORT_MSG Indication protocol message. It must also translate the SMS_SUBMIT message into GSM Recommendation 03.40.

Then the SMS interworking network router will choose to which other SMS interworking network router it must send this request in order to deliver the SMS, depending on roaming agreements. When done, the message is treated as a GSM FORWARD_SHORT_MSG_MO received from any GSM MSC and appropriate interrogation of the HLR occurs. MAP_FORWARD_SHORT_MSG_MT is sent toward the destination MS's MSC. At this point, the IS-41 SMS interworking network router is expecting an answer from the visited MSC. When it receives the

Table 3.15   SMS_DELIVER CMT Details

Information Element	Type	Length
Message type indicator	M	3
Message reference	M	13
Privacy indicator	M	3
Urgency indicator	M	2
Delivery acknowledgment request	M	1
Manual acknowledgment request	M	1
Message updating	M	1
Validity	M	3
Display time	M	2
Reserved	M	3
User data unit	M	16
Message center time stamp	O	48
SMS signal	O	16
Callback number	O	20
Callback number presentation indicator	O	8
Callback number alpha tag	O	20
Multilingual callback number	O	28
Multilingual callback number alpha tag	O	32
Language identifier	O	16
Acknowledgment time	O	16

Table 3.16 SMS_DELIVERY_ACK CMT Details

Information Element	Type	Length
Message type indicator	M	3
Message reference	M	13
User data unit	M	16

Table 3.17 SMS_MANUAL_ACK CMT Details

Information Element	Type	Length
Message type indicator	M	3
Message reference	M	13
User data unit	M	16
Response code	O	8

MAP_FORWARD_SHORT_MSG_CNF from the visited MSC, it finishes its treatment according to the procedures defined in [1].

3.4.4.2   Procedure

The scenario of SMS delivery from the SME is as follows for the IS-41 side (refer to [1]):

Procedure 4.46.1 is executed: "SME Initiating SMS_Delivery." This procedure sets the following parameters:

- *If OSS required:* destination address := originating SME's address
- *If no OSS required:* destination address := destination SME's address
- SMS_NotificationIndicator (if needed)
- SMS_MessageCount (if needed)
- SMS_BearerData
- SMS_TeleserviceIdentifier

Procedure D.4 is then executed: "MS-Based SME Initiating SMD_REQUEST Toward an MSC." This procedure sets the following parameters (for simplicity, we assume no handoff):

- SMS_OriginalDestinationAddress := original destination address (destination SME)
- Transport layer originating address := MIN (originating MS)
- ElectronicSerialNumber
- MobileIdentificationNumber
- Send SMD_REQUEST to the serving MSC

Procedure D.5 is then executed: "Serving MSC Receiving an SMD_REQUEST." This procedure sets the following parameters:

- Destination address := anchor MSC
- Original destination address := SMS_OriginalDestinationAddress

- Originating address := MIN (originating MS)
- Original originating address := originating address := MIN (originating MS)
- We assume the serving MSC is also the anchor MSC.

Procedure 4.46.5 is then executed: "Anchor MSC Initiating SMS_DELIVERY_ POINT_TO_POINT." This procedure sets the following parameters:

- If SMS_OriginationRestrictions requires indirect routing, SMS_Destina- tionAddress := SMS_OriginalOriginatingAddress (originating MIN)

Procedure 4.46.2 is then executed: "Initiating SMS_DELIVERY_POINT_TO_ POINT." This procedure sets the following parameters:

- SMS_ChargeIndicator (if needed)
- Transport destination address := message destination (MC of SMS_Destina- tionAddress)
- SMS_DestinationAddress
- SMS_OriginalDestinationAddress := original destination address
- Transport originating address := Anchor MSC's address
- SMS_OriginatingAddress := originating address
- Send an SMS_DELIVERY_POINT_TO_POINT message to the Originating MS's MC

Procedure 4.46.6 is then executed: "MC Receiving an SMS_DELIVERY_ POINT_ TO_POINT Invoke." This is the point at which the SMS interworking network router, acting as the functional originating MS's MC, receives the SMS_DELIVERY _POINT_TO_POINT Invoke message and has to route it to a GSM SMS interworking network router with an IP transaction. Here are the delivery parameters received:

- Destination address := SMS_DestinationAddress (Originating MS's MIN)
- Original destination address := SMS_OriginalDestinationAddress
- Originating address := SMS_OriginatingAddress
- Original originating address := originating address
- If OSS have been required perform the supplementary services

Then we perform the mapping of IS-41 data to MAP_FORWARD_ SHORT_MSG Indication SMS-MO:

- Analyze original destination address (message destination) according to roaming agreements of the SMS interworking network routers to find the one to which the request must be sent.
- Set MFSMWR Dialogue ID.
- Set MFSMWR IP address of destination SMS interworking network router.

- Set MFSMWR SPC/ICP to −1.

- Set MFSMWR DA to original destination address.

- Set MFSMWR OA to originating address.

- Set MFSMWR Encoding.

- Translate Bearer Data according to teleservice into SMS_SUBMIT for GSM.

- Compute an SMS MAP_FORWARD_SHORT_MSG Indication SMS-MO.

- Interrogate HLR according to destination address.

- Send the SMS toward the destination MSC.

- Then when MAP_FORWARD_SHORT_MSG_CNF arrives, acknowledge invoking entities with Return Result, Return Error, or Reject.

3.4.5 SMS Delivery from MAP SME to IS-41 SME

3.4.5.1 Overview

In this case a GSM SMS interworking network router has received either by SS7 or by IP a request to deliver a short message toward an IS-41 destination. Then it sends an *MTU_FORWARD_SHORT_MSG_WWW_REQ* (MFSMWR) to the IS-41 SMS interworking network router according to its translation tables and an analysis of the IS-41 router's roaming agreements. The IS-41 SMS interworking network router receives a MAP_FORWARD_SHORT_MSG_MO_IND and must map the received parameters to the parameters of an SMS_DELIVERY_POINT_TO_POINT Invoke message, find the appropriate teleservice to use WMT for CDMA subscribers and CMT for TDMA subscribers, and execute the procedures as defined in [1].

3.4.5.2 Procedure

Let's look at the procedure for SMS delivery from a MAP SME to an IS-41 SME.

The SMS interworking network router on the MAP side receives a MAP_FORWARD_SHORT_MSG Indication SMS-MO message. It must extract the SM-TL SMS_SUBMIT message from the SM-RP-UI field and translate is according to the teleservice in use. Then it must build an SMS_DELIVERY_POINT_TO_POINT_REQ message as if it had received the SMS_SUBMIT from an IS-41-based network.

Then it must map these parameters to format an SMS_DELIVERY_POINT_TO_POINT (SMSDPTP) message to send to the destination's MC and execute Section 4.46.1 of [1], "SME Initiating SMS_DELIVERY":

- If the destination is an MS-based SME destination address := MS's MIN

- Original destination address := destination MS's MIN

- Else destination address := SME's destination address (We assume that the destination is MS based.)

- Optionally include SMS_NotificationIndicator and SMS_MessageCount

- Copy UI to SMS_BearerData

- Set SMS_TeleserviceIdentifier

Then it must map these parameters to format an SMSDPTP message to send to the destination. Execute Section 4.46.2 of [1], "Initiating SMS_DELIVERY_ POINT_TO_POINT":

- Transport destination address := destination address
- Set MIN to the destination MS's MIN
- SMS_OriginalDestinationAddress := original destination address
- Transport originating address := address of the IS-41 SMS interworking network router processing the message
- Send SMS_DELIVERY_POINT_TO_POINT to the destination MS's MC

Execute Section 4.46.6 of [1], "MC Receiving SMS_DELIVERY_POINT_ TO_POINT Invoke":

- Destination address := MIN
- Original destination address := SMS_OriginalDestinationAddress
- Originating address := SMS_OriginatingAddress
- Original originating address := SMS_OriginalOriginatingAddress
- Send Return Result to the sender of the Invoke

Execute Section 4.46.8 of [1], "Terminating MC Supplementary Services":

- Perform any TSS.

Execute Section 4.46.9 of [1], "MC Initiating SMS_DELIVERY_POINT_TO_ POINT to an MS-based SME."

The next steps in the process conform to the IS-41 specification [1].

3.4.6 IS-41 Numbering for SMS Delivery

The IS-41 recommendation uses the MIN format to retrieve information from the HLR for a specified MS. Because this number is not the same as the dialed number and, as a consequence, as the destination address field for the message, the way to retrieve the MIN from a dialed number has to be clarified. Instead of using the MIN, some operators allow the *mobile directory numbers* (MDNs) that fit the requirements of the MSISDN to be used.

3.5 Addressing HLRs in TDMA and CDMA Networks for SMS Interworking: Updating Point Code–Based Addressing Information

With GSM networks, which use GTs and gateway MSCs, once a roaming agreement has been set up, the partners do not need to change anything if one operator adds HLRs or MSCs, or if that operator changes the assignment of subscribers to HLRs. Everything is handled by the operator's GMSC so that the partners are not concerned as long as there is no change in the overall numbering plan, such as adding a number or an NDC (although it is very simple even in this case).

TDMA and most CDMA networks do not have GTs for their equipments and do not necessarily offer GMSCs for their roaming partners (with the exception of the Globalstar satellite/CDMA network which uses GTs like a GSM network). The addressing is "on PC," as illustrated by a part of the HLR levels of a TDMA operator (see Table 3.18). Thus, each of their roaming partners must install the table in each of their MSC/VLRs and address the correct point code of the HLR that corresponds to each interval.

When sending an IS-41 SMS_REQUEST (equivalent of the SEND_ROUTING_INFO_FOR_SM), the SMS address that is returned (that of the Visited MSC) is an *MSC identity* (MSCID). There must be a table that gives the point code of each MSC because routing on a PC must also be used. In a GSM network, one needs only to have the GT of the MSC, so that the work to keep the addressing information is much simpler than with an IS-41 network explaining why roaming is much more developed.

Table 3.18   SMS_READ_ACKNOWLEDGMENT_WMT

Location HLR	MIN Begin	MIN End	Point Code (ANSI)	CLLI Code
Anguilla	2642350000	2684649999	002-026-021	THVYAIAACM0
Anguilla	2684640000	2684649999	001-015-020	STJHANAFCM0
Barbados	2462300000	2462309999	001-141-011	GRZTBDAACM0
Barbados	2462340000	2462349999	001-141-011	GRZTBDAACM0
Trinidad	8686200000	8686209999	001-158-014	PTSPTRTHCM0

References

[1]   *ANSI/TIA/EIA-41-D: Cellular Radiocommunications Intersystem Operations,* http://www.tiaonline.org.

[2]   *ANSI/TIA/EIA-136-710-B: TDMA Third Generation—Short Message Service—Cellular Messaging Teleservice,* http://www.tiaonline.org.

[3]   *GSM 09.02: Digital Cellular Telecommunication System (Phase 2+); Mobile Application Part (MAP) Specification,* http://www.etsi.org.

[4]   *GSM 03.40: Digital Cellular Telecommunication System (Phase 2+); Technical Realization of the Short Message Service (SMS); Point to Point (PP),* http://www.etsi.org.

[5]   *ANSI/TIA/EIA-637-B: Short Message Services for Wideband Spread Spectrum Systems,* http://www.tiaonline.org.

## CHAPTER 4

Implementation of Mobile Number Portability and GSM-to-IS-41 Conversion

For always roaming with a hungry heart
Much have I seen and known; cities of men
And manners, climates, councils, governments,
Myself not least, but honour'd of them all;
And drunk delight of battle with my peers,
Far on the ringing plains of windy Troy.
           —Alfred Tennyson, *Ulysses*, 1883

4.1 Business Model

Mobile number portability offers to international gateway carriers a source of additional value if they choose to implement it for their subscriber operators. Handling the GSM-to-IS-41 protocol conversion is also a way to increase the value of their services beyond the transport of signaling packets. Various architectures offer several ways of deciding where the value is for international carriers or SMS/MMS interworking networks.

4.2 Basics of Roaming Agreement Implementation

In Chapter 2, we studied the SS7 network, which allows SCCP messages to be exchanged between two networks that are connected through *international gateway providers,* for instance, IGP 1 and IGP 2 in different countries in the example shown in Figure 4.1. In this figure, if B and C want to implement a roaming agreement, each carrier opens the NDC of its partner in its own gateway MSC (also called the *signaling transfer point,* STP) so that the SCCP messages with E164 format destination addresses can be sent out of its network to the roaming partner.

The carriers will also ask their own IGP to open the route (in most cases, this is done for contractual reasons), because if any other operator in the same country as B already has a roaming agreement with A, *then the route was already opened by the IGP.*

By acting on its own, because most of the international routes on the SS7 network are already opened, it is technically possible for an operator to open one side

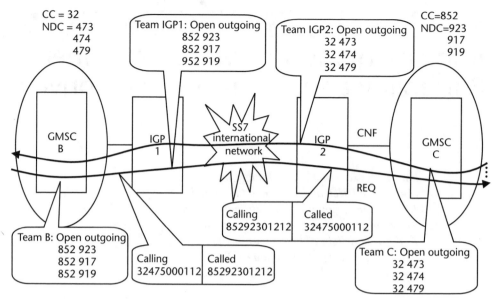

Figure 4.1 Opening the two outgoing SCCP routes.

of the roaming agreement from its network to another one. The other party can receive messages, but no response will come back. This will be used in the following.

In a GMSC, E214 routing tables are also used for the UPDATE_LOCATION message, which must be set up for full roaming (mobility). However, they are not used for SMS interworking.

4.3   Implementations of Number Portability

With *mobile number portability* (MNP), also called *wireless local number portability* (WLNP), or more generally for *local number portability* (LNP), which includes fixed lines, a customer can change its network (ported out) and keep the same number in another network (ported in) in the same country. How can a voice call or an SMS reach the customer and how can it be fairly accounted for (in the case of a payment of the HPLMN by the sending network of the SMS_MT)? One must be able not only to terminate the SMS, but to identify the network for accounting purposes. Also, what if the different mobile operators use different technologies: GSM, CDMA, TDMA, and so on? The solution depends on the country's regulations imposed by the regulatory authority:

- The end carrier (level N), which has the ported number in its allocated range, must take care of successfully terminating the voice call or the SMS (European case mainly).

- The gateway carrier (level $N - 1$), such as the international SCCP carrier, must handle it (so the level N has nothing to do in the United States and Canada).

- No obligation to terminate the traffic to a ported number is imposed (in unregulated countries such as Macau).

For SMS service, the same principle is used as is used for voice call service, but it is simpler because the question of the GSM-to-IS-41 protocol translation difficulty disappears, because the ISUP protocol, which is used for call setup, is identical.

4.3.1   MNP Handled by Each Individual Operator (Level *N*)

Each operator has a copy of a MNP database which tells for each ported number the regular network (old) and the new one, as illustrated by Figure 4.2. This database is queried by the gateway MSC (which has an STP function) of these operators and is then used to reroute the signal received (a SEND_ROUTING_INFO_FOR_SM in case of an SMS) or an incoming call signal for a voice call to the other network from which the number has been ported. This solution is rather expensive because each operator must equip its GMSCs with software that can interrogate the database and perform the routing based on the result. Otherwise, without MNP, the operators need merely to have routing based on the different range of numbers to address the concerned HLR level.

4.3.1.1   Direct Voice and Signaling Links Between Each Operator

The GMSC must have sufficient capacities with all other operators in the country through direct links (Figure 4.3). For example, Taiwan with six operators (in 2000) had to have five sets for each operator.

4.3.1.2   Use of a Common Gateway Carrier

To simplify the solution, it is convenient to use a common gateway carrier, which avoids the need for each operator to have direct links with each other operator.

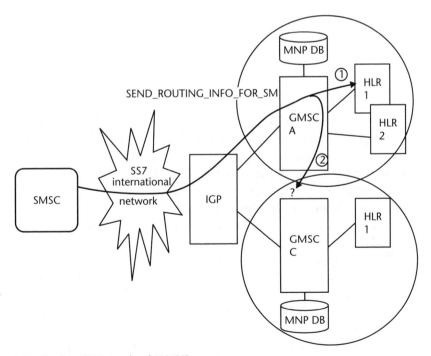

Figure 4.2   Sending SMS to a level *N* MNP country.

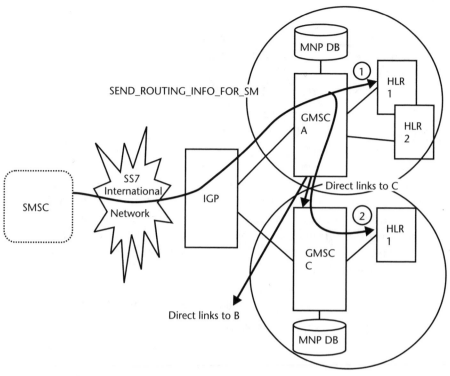

Figure 4.3   Direct links with competitors.

There is a simple trick for this. The regulator allocates a special additional code (which is like an NDC, but only used internally in the country) for each of the operators, such as

- 642 for Operator A;
- 643 for Operator B;
- 644 for Operator C.

Thus, if A receives a called party address SCCP [(1) in Figure 4.4] that corresponds to an MSISDN of C such as +33680123456 (which is in the range allocated to A, so the IGP routes it to A), it will replace it with +33643680123456 and send it back to the IGP (2). The IGP, without needing to have a database of the individual ported numbers, routes the SCCP message to C! Obviously, all of the MNP databases must be identical, otherwise damaging loops would occur.

One can see in Figure 4.4 that if the sending SMSC has a roaming agreement with C (but not with A), it will work for original numbers of C and *also for numbers of A ported in to C, even if the SMSC does not have a roaming agreement with A.* This is because A never has to answer to the sending SMSC; it only routes its SCCP requests to the other network C (direct links or relay through the IGP), which has roaming with the SMSC. So the SMSC will comply with the requirement that its subscribers be able to send SMS to all current subscribers of C. However, the SEND_ROUTING_INFO message of the SMS will fail (no response) when the tar-

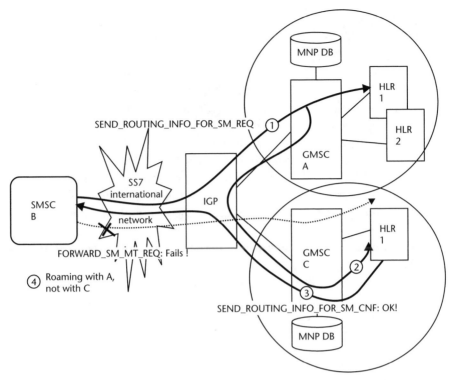

Figure 4.4   Receiving SMS from a ported-in subscriber's previous partners.

get is any subscriber of C ported out to another network. For example, France (IGP = France Telecom) and Hong Kong (IGP = Reach).

4.3.1.3   When Reception of SMS for Ported-In Numbers Cannot Be Maintained

If Operator C hopes that its new ported-in subscribers from, for example, Operator A continue to receive SMS from their friends' numbers (e.g., from B which had a roaming agreement with A), *C may think of opening all the outgoing SCCP routes to all of its competitors' roaming partners,* so the confirmation [(3) in Figure 4.4] to the SEND_ROUTING_INFO will return to the sending SMSC. It will also take out any barring policy (see Chapter 5) in its HLRs. However, this does not allow the SMS of B to C to work if B does not have a roaming agreement with C. If B does not have roaming with C, the FORWARD_SM_MT_REQ that is sent and addressed to a MSC with a specific GT of C *will not reach when the destination number is visiting C.* So the best that C can do is *to allow the SRI to work from any network,* but the FORWARD_SM_MT will fail [(4) in Figure 4.4] in general. *So the operator cannot create an SMS service from B to C (itself) without a specific agreement with B.* Now we see that B's rational behavior will be to open the route to C (without even telling C) even if it does not have a roaming agreement.

4.3.1.4   Maintaining Reception of SMS for Ported-In Numbers

Reception can be maintained work if the operators in the other countries have a rational commercial behavior. What is the attitude of an operator B that has agreements with at least one operator in a MNP country, but no access to that country's database. The operator claims in its publicity that it has roaming with A (so it must open the ranges of A in its gateway MSC). However, there could be subscribers of B and C who are ported to A, and Operator B must provide SMS-MT service to them. To ensure that the SEND_ROUTING_INFO (which could have a calling party equal to a number of B ported to A) reaches the destination country, the operator must then also open the ranges of B and C (even if it does not have roaming agreements with them!). So that, at the end, the rational behavior of B (sender) is to *open all the ranges to the destination MNP country*. This explains why the IR21 document, which gives all ranges of numbers of a given operator to its roaming partners, of an operator with MNP *also includes the ranges of all of its competitors in the same country*. See, for example, the IR21s of the Netherlands, Australia, Belgium, and so on. For simplicity, in almost all the cases, the operator will also open the exact range corresponding to the network equipment (because they belong to the MSISDN range).

Also, many IGPs (such as that of B) do not control the SCCP address, so all the routes to all SCP origin addresses of the MNP country will be opened in this case. So if the IGP or the receiving side does not control the origin address, all routes in both directions will, in fact, be opened. If you consider France (opened IGP) and Hong Kong (selectively opened routes depending on the origin address by the IGP), the communication will not work.

This section explains why operators seem to help their competitors in MNP countries.

4.3.1.5 Consequence: Toward an Open SMS World?

With the steady increase in the number of countries that have access to LNP or MNP, this has the following consequences. From Section 4.3.1.4, we know that operators such as A who are gaining ported-in subscribers in MNP countries will open all outgoing ranges (E164 addresses) as a simplification (because it is much simpler than to open only the ranges of their competitors' roaming partners). They will also open their HLRs.

Also all operators such as C who wish to send SMSs to MNP countries will open all routes to all the E164 ranges of these countries. The first factor primarily will allow any operator on the SS7 network to send SMSs to all operators in the MNP countries (a vast majority of the mobile numbers in the future, by just opening the outgoing ranges in their GMSC). It will work, as it is often the case, because the IGP of A and the IGP of B do not control the SCCP origin address for the routes they have already opened.

4.3.2 MNP Handled by the Entry International SCCP Gateway (Level *N* – 1)

Figure 4.5 shows how MNP is handled by the IGP at level *N* – 1. Consider, for example, Canada and the United States (IGP = TeleGlobe). With this method, if an SMSC tries to send an SMS to a nonexisting number, the SRI_FOR_SM will be timed out because no MSISDN translation exists in the database. Precisely as in the previous two cases, the sending SMSC can only send SMSs to numbers belonging to networks with which it has roaming agreements and all the previous remarks hold.

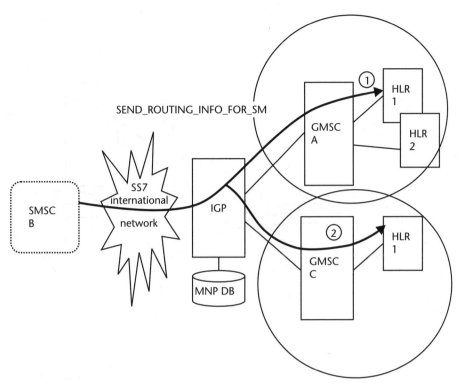

Figure 4.5 MNP handled by the IGP at level $N - 1$.

4.3.3 Unregulated Countries' MNP Process Must Be Handled by the SMS Interworking Network

In this case, nothing is implemented in the networks (GMSCs) or the IGP. The SMS interworking network database uses the search + cache method [1], which basically creates dynamically a MNP database. If no entry exists in the SMS interworking network MNP database for a certain country, we search the HLRs until the subscriber is found (Figure 4.6). (We must know the GT of each HLR and address them directly, not with the called party address = MSISDN.)

In this case only, the *SMSC of the interworking network must have roaming agreements with all networks in the unregulated country* if reliable service is to be provided; otherwise, if it has a roaming agreement only with A and it wants send an SMS to a number that belongs to C ported to A, it would not get the answer to the HLR interrogation of C, so it would not provide the SMS-MT servicefor subscribers of A ported in from C. An example is Macau (Hutchison, Ctm, Smartone).

4.4 SMS Routing Strategies for an SMS Interworking Operator to a Regulated MNP Country

Consider the preceding cases of a regulated MNP country (whether MNP is handled at level N or $N - 1$ does not make any difference) with an interworking operator

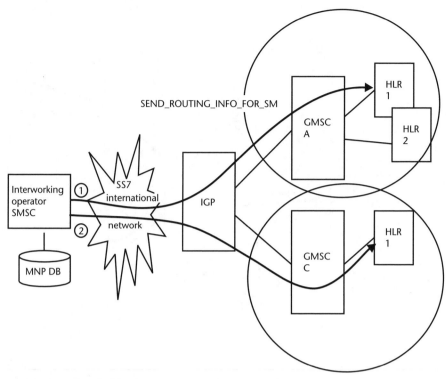

Figure 4.6   MNP handled by the SMS interworking network.

that has three SMSCs hosted in different networks, but does not have its own MNP database. Its SMSC 1 has roaming with A and C; SMSC 2 with A and B. Only SMSC 3 has roaming with all three (Figure 4.7). The requirement is *success for any number* whether it is in its original network or ported out to another network; this is a stronger requirement than for an ordinary operator.

When it must send an SMS to A, if the number is still in A or ported to C, it will succeed, but it will fail if it is ported to B. It could then try to use the SMSC 2, but this would complicate the procedure and makes it less reliable.

So the only simple strategy is to send all SMS traffic to SMSC 3, which has roaming with all operators in this country. If there is none, the *quality of service* (QoS) cannot be perfect. Also from the IMSI, it will be possible to make accurate billing records that indicate who the destination operator was.

4.5   MNP for SMS in Countries That Have Both GSM and IS-41 Operators

4.5.1   SMS-MT GSM to an IS-41 Destination

This is a very important case in the United States and Canada. There is also an intermediate, simpler case in countries that do not yet have MNP, but with IS-41 and GSM, as is true of several South American countries.

We will examine the case in which a GSM SM-MT wants to send an SMS to a Verizon customer in the United States (Verizon has both GSM and CDMA). But the

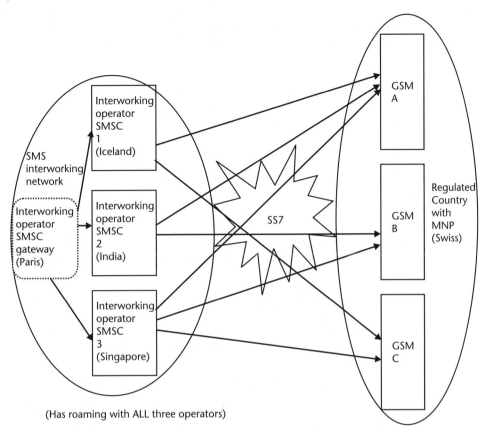

Figure 4.7   SMS routing strategies.

MNP available database specifies only Verizon (not type GSM or IS-41). The correct transparent operation includes the following:

- An SCCP gateway, which includes flexible routing software (such as MAP primitive type dependent);
- A GSM-to-IS-41 protocol converter;
- Additional codes for each mobile network (Verizon, which has GSM and CDMA with different GMSCs for each network, must have two, for example, 999 for GSM and 998 for IS-41).

The solution is such that the GSM-to-IS-41 converter does not have links with the customers of the IGPs (it would require a big system), only with the SCCP IGP.

A SEND_ROUTING_INFO_FOR_SM_REQ is received by the IGP. Possibly, it could look in its database. If the destination is GSM, then route directly. If the destination is IS-41 or unknown, then route to the protocol converter. (Verizon routes to a certain linkset for GSM traffic and to another for IS-41 traffic.)

In the case where routing tables in the protocol converter indicate IS-41 or a unique route, the process is straightforward: An IS-41 SMS_REQUEST is created and sent to the destination through the IGP. The protocol converter must have the MIN table for a given MSISDN so it can interrogate the IS-41 HLR.

In the case where the type is unknown, illustrated in Figure 4.8, the protocol converter starts with the most likely type, for example, GSM: no conversion, but the called party SCCP address is changed:

$$12364351234 \rightarrow + 19992364351234$$

so the IGP knows it has to take the route to the GSM gateway (*the IGP will strip 999*).

If the answer is an unknown subscriber from the GSM interrogation, it will then do a conversion to an IS-41 SMS_REQUEST and change

$$+12364351234 \rightarrow + 19982364351234$$

so the IGP will now route to the IS-41 gateway of Verizon. The IS-41 answer to the protocol converter will be converted to a SEND_ROUTING_INFO_CNF so that the sending GSM SMSC is happy. The protocol converter will return the GT of the visited MSC and some IMSI from a proprietary table that identifies each IS-41 network without confusion with the GSM MCC-MNC allocation. In the answer, one could also include the ESN, which identifies the type of cell phone, in an unused field such as the *local mobile station identity* (LMSI) (see Chapter 1).

4.5.1.1   Transparent Text SMS-MT Service for GSM-to-IS-41 Protocol Conversion

The GSM SMS-MT now sends a FORWARD_SHORT_MESSAGE_REQ. For simplicity, only text messages are handled (the protocol converter returns a GSM failure

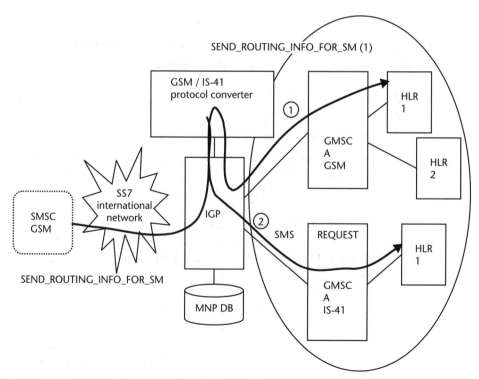

Figure 4.8   Interrogating the GSM and the IS-41 HLRs.

such as protocol error if it is binary), and text longer than the TDMA lengths is truncated. It is then converted to an IS-41 SMS_DELIVERY_POINT_TO_POINT message and sent as previously to the proper network. So this solution, although not perfect, allows us to use ordinary SMSCs with a reasonable transparent service.

4.5.1.2   Handling GSM-to-IS-41 Protocol Conversion in the Sending SMSC

Some vendors propose *multimode SMSCs* that can handle the GSM procedure as well as the IS-41 SMS procedure. They could use only the first part of the above procedure so as to obtain an identification and type of destination network and handle the protocol conversion, including whether the GSM SMS is longer than what the IS-41 network accepts, breaking it into two parts. Also it could convert the binary SMS (logos, ring tones) into their corresponding coding in the destination cell phone (the ESN identifies the type).

4.5.2   SMS-MT from an IS-41 Network to a GSM Destination

The procedure for sending an SMS-MT from an IS-41 network to a GSM destination is entirely different and simpler than the one just discussed, because the sending of SMS to another network does not work at all like it does in GSM.

The sending MC analyzes the destination MSISDN and, using a IS-41 SMS_DELIVERY_PONT_TO_POINT message, finds that the sending is handled by the IGP and will forward the SMS (like an SMS-MO) to the GT of the protocol converter. This is point (1), the SMS_DELIVERY_POINT_TO_POINT, in Figure 4.9, which will then work as an ordinary GSM SMSC (2) and (3) if the destination

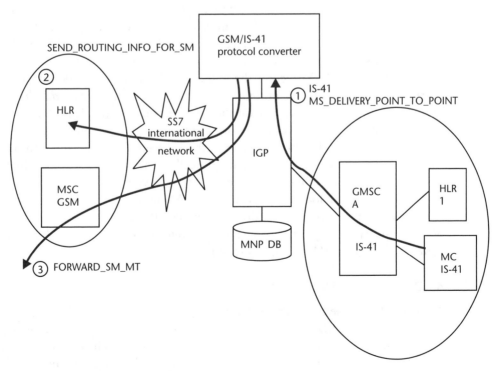

Figure 4.9   SMS from IS-41 to GSM.

network is known to be GSM. Otherwise, as in the previous paragraph, it can interrogate the destination network to determine the type and then proceed.

4.6   Identification of the Destination Network

The mobile SMS interconnection MMS and fixed-line SMS interconnection are completely different. In fixed-line SMS, transmission is from the originating node (MMSC or fixed-line MMSC) to the servicing node (MMSC or fixed-line SMSC). The key issue is which is the address of this servicing node (SS7 GT or IP address)?

With mobile SMS, one can expect to send blindly from one's own SMSC to the roaming partners and let the IGP and the GMSC of the destination do the proper addressing in the case of portability.

The purpose of this service is to provide any operator who wants to send an MMS with a fixed-line SMS (or a mobile SMS for a total coverage), with the identification and the address of the equipment in the destination network.

The service does not provide for sending itself, which remains the responsibility of the originating equipment. For a worldwide implementation, it may use a number of domain resolution servers with SS7 connections that are hosted by a small number of operators that together have a large number of roaming agreements in the GSM and IS-41 worlds.

4.6.1   MMS Interconnection

For MMS interconnection (see Chapter 9), it is very important to be able to have the service of a virtual HLR that, when interrogated for any worldwide MSISDN, will return some IMSI that identifies the network even if it was ported. The interrogating SMSC or MMSC (through the MM5 interface) gets back the identity of the network by extracting the MCC and MNC from the IMSI given by the real HLR that is hit. Note that a small number of operators offer (or charge!) on an IP access to their SMSC a partial DNS: You enter the number and it returns the identification of the destination network. *So for mobiles, we can expect not to need a database of individual numbers;* instead we rely on the interrogation of the HLRs (GSM or IS-41) and on the databases at level $N - 1$ or N in the destination countries.

The preceding solution answers exactly the need. And with the information on the destination network, and eventually the cell phone type, the sending MMSC, with a table, gets the domain name of the destination MMSC and can forward the MMS. In Chapter 9, we will look in detail at how the MMS service works and interconnects.

For IS-41 networks (which do not have IMSI), the code returned could be the MCC (country) or any other official code given to the network [such as the central *office code number* (OCN)] given by an official administration.

4.6.2   Fixed-Line SMS Interconnection

Now that fixed-line SMSs are becoming a business, the proper fixed-line telephone set is allowed to send or receive SMS to/from mobile cell phone or fixed-line phones. In Figure 4.10 we assume that the fixed-line SMSCs *a* and *b* have a connection to the international SS7 network (this is vendor specific); otherwise, a connection between

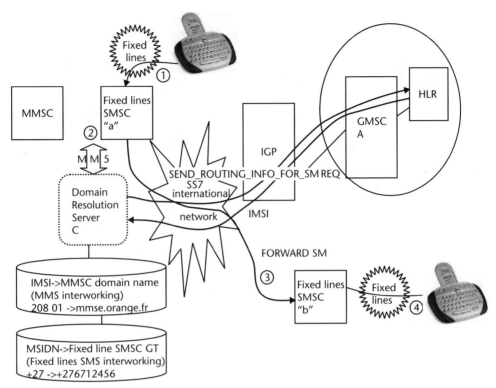

Figure 4.10   MMS and fixed-line SMS domain resolution server for interconnection.

them through the IP can also work. It uses some interface (MM5) to interrogate the domain resolution server and obtain with the (fixed-line) MSISDN the GT of the fixed-line SMSC. The transfer of the SMS from one fixed line to another can be done with an SMS-MO (some fixed-line SMSC vendors support it).

In Figure 4.10, the phone (1) sends an SMS to its fixed-line SMSC, which interrogates the *domain resolution server* (DRS) (2) with a protocol such as MM5 (see Chapter 9 on MMS). In this case the DRS would return as the visited MSC the GT of the fixed-line SMSC from a table or a database if the destination country has LNP. Then fixed-line SMSC *a* sends the SMS over the SS7 network to the destination SMSC *b* (3), which will send the SMS to the fixed-line phone (4). The compatibility between mobile SMS and fixed-line SMS is quite easy because the application level [2] for the fixed-line SMS also uses the GSM 3.40 standard for the formatting of SMS (160 characters).

4.6.3   MMS and Fixed-Line SMS Interconnection Business

To operate a DRS that can return some identifier and addressing information to a request from a MMSC or fixed-line SMSC can be a business for a third-party interworking operator. This operator just provides the information without having any role in the delivery. It is clearly an indispensable function so that a general interconnection exists among the various modes of messaging.

One can design an architecture using an agreement among several DRSs, each hosted by a GSM or IS-41 operator with many roaming agreements (Figure 4.11).

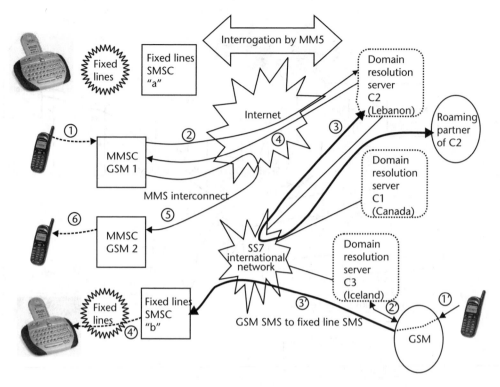

Figure 4.11   DRS architecture for telephone number mapping.

They would be interrogated using a standard protocol [3] and return identification and addressing information for the destination network. As explained in this chapter, *they do not need to have the number databases of their roaming partners;* instead they make use of the databases in the destination countries (case N or $N-1$ for number portability resolution). Revenues from each *telephone number mapping* (ENUM) request would be shared.

To explain, consider a cell phone sending a MMS to his MMSC [(1) in Figure 4.11], which interrogates (2) the DRS it has subscribed. This one (Lebanon GSM) interrogates the roaming partner (without knowing who it is because of MNP) with a SEND_ROUTING_INFO_FOR_SM. It gets back the IMSI and returns it (4) with additional routing information such as the domain name of the destination MMSC in GSM 2. MMSC in GSM 1 can then send the MMS to the MMSC in GSM 2 (5) using the MM4 interface (see Chapter 9).

When a GSM cell phone wants to send an SMS to a fixed-line phone [(1') in Figure 4.11], its SMS will also interrogate one of the DRS, C3 (2'), and get the GT of the destination fixed-line SMSC (or IP address if it does not have an SS7 connection). The SMSC will then send to the fixed-line SMS phone via (3') and then (4').

The partners of the DRS do not directly send messages (SMS or MMS); they merely provide the addressing information.

References

[1]   Henry-Labordére, A., *Procédé d'envoi de messages courts à des mobiles avec portabilité du numéro ou plan de numérotation incomplet,* Hong Kong Patent No. 1025476.

[2]   *ETSI 201 912 V1.1.1: Short Message Service for PSTN/ISDN.*

[3]   *IETF RFC 2916: Telephone Number Mapping (ENUM).*

## CHAPTER 5

Barring Inbound SMS-MT

One does not easily deceive Comrade Stalin.
—Josif Vissarionovitch Dzugashvili

5.1 Barring Inbound SMS-MT: An Important Business Issue

Some operators want to bar the SMS-MT from other networks, but they still want *their subscribers* to be able to roam in these networks, that is, to travel abroad and use their phones. This is the basic service that they feel is more important than letting them receive SMS from other networks. The reasons for barring include the following:

- *Economical:* Operators want to allow only the sending operators, which have signed a termination fee agreement for SMS (the AA19 of the GSM Association) to reach their network for SMS services.

- *Economical:* Operators want to reduce the cost of handling complaints from spammed customers, for which the handling costs can run to several U.S. dollars per complaint. Operators will bar other operators that they determine have supported spamming services. Or operators may use a better solution: They may install an SMS content filtering system. (We explain later how this works.)

- *Political:* To avoid the reception of news from abroad via SMS technology that could contain politically embarrassing information.

Offering a sophisticated way of barring inbound SMS is a service that international SCCP gateway providers can offer and, hence, it's part of their business. The SMS content filtering equipment described in Section 5.3.2 can be integrated with the gateway.

All of the methods discussed next make it impossible for the SMSC to obtain the IMSI and the visited MSC of the destination MSISDN, so that it is impossible to send an SMS. The first two methods only use processing at the SCCP level.

5.1.1 Filtering Service Offered by IGPs at the SCCP Level

This service may be provided by certain IGPs when requested by their customers who give them a *list of all their own network equipment* (HLR, MSC, VLR, SMSC) and a range of addresses for the SMSCs of the roaming partners with whom they have agreed to receive SMS-MT from for their own subscribers. Whenever the SCCP gateway receives an unauthorized E164 called party address, it will discard the SCCP message, which can be:

- For one of their customers who has requested the filtering service.
- A called party *who is not included in the list of their network equipment*. (It is then a MSISDN called party address, so it corresponds to a SEND_ROUTING_INFO_FOR_SM or a REPORT_STATUS_DELIVERY). This is of concern to the SMS service. This remark is the key to understanding how some of these filtering methods work: It is possible to recognize a SEND_ROUTING_INFO_FOR_SM at the SCCP level without any MAP processing.
- One for which the E164 origin calling party address is not a piece of equipment (SMSC GT) from an authorized roaming partner.

It is easy to see that the roaming function is not affected, because it uses an E214 format address (see Chapter 1). The SMS-MT service of unauthorized roaming partners cannot access the routing information.

5.1.2 Selective E164 Translation Facility Barring of the SMS-MT at the GMSCs SCCP Level

Let's say that the standard SMSC of B wants to send an SMS to cell phone A. It must address the HLR with the E164 MSISDN address of the cell phone, which is +8613601111234, and it expects the GMSC A to route the request to the PC of the corresponding HLR:

$$+8613601111234 \rightarrow \text{Point Code of } +86136011400$$

This is because the SMSC cannot guess which HLR corresponds to this particular MSISDN. (The correspondence is not published, and it can change quite often.) Consider that an operator such as Vodafone Germany has about 100 HLRs. So the maintenance of a remote table is completely impractical. Also the E164 translation table is often random with very small ranges pointing to the HLR number.

So, the method consists of setting in the GMSC a white list of *calling party addresses* so that the E164 translation facility is activated only by selected operators (for example, those having a country code starting with +86). The roaming facility will still work because it uses the E214 translation facility only!

A similar method consists of having a black list, where the country codes of the foreign partners that have the most incoming SMS traffic are entered, while letting the small countries be opened. For example, some operators just do not install any E164 translation table in their GMSC (this used to be the case of Morocco)! So they cannot receive SMS-MT from any other network.

Note that if the sending network knows the HLR GT for a given MSISDN, this barring method will not be sufficient.

5.1.3   HLR Barring

5.1.3.1   No Response from HLR When Addressing a Nonexistent Subscriber

When the HLR receives a SEND_ROUTING_INFO_FOR_SM for an MSISDN that does not belong to that HLR, it does not answer. Instead, as in the GSM MAP standard, it responds with a user error = 1 (wrong number). This is to avoid HLR searches. In this case, of course, no data are returned.

5.1.3.2   White List Control on the Service Center Address Field of the SEND_ROUTING_INFO_FOR_SM (SRI) Message

This MAP message contains, as main fields, MSISDN (E164) and the originating service center address (E164). If the service center has not been explicitly added to the white list, the MAP message is rejected with various user error codes. For example, Nortel HLR returns a user error = 36 (meaning "Bad Data").

5.1.3.3   Barring All SRI from Foreign Networks

This option may be used when an operator does not want to receive any SMS from foreign SMSCs. For example, Alcatel returns a provider error = 2 (meaning "Service Not Provided")

5.1.3.4   MAP Policy Software (SCCP-Level Barring)

This is a refinement of the white list control point from above. The control is performed on the service center address and on the GT in the SCCP calling party address. For example, Ericsson (this is an option) returns a user error = 21 (meaning "Incompatible Version"), Alcatel returns a provider error = 2 (meaning "Teleservice Not Provisioned").

In conclusion, these HLR barring functions are very efficient.

5.1.4   Origin Address Type Barring at the MSC Level

Some operators want to receive SMS-MT only from foreign subscribers' cell phones, and not from Internet service providers. They want to bar SMS-MTs that have an *origin address* (OA) of type unknown or alphanumeric. For example, Turkcell with a Siemens MSC, which returns a user error = 34 (meaning "System Failure").

5.1.5   MAP Barring by the GMSC

Doing the SMS-MT barring in the GMSC is simpler and more economical than installing barring software in each HLR, especially if there are several vendors. As discussed, the first method (at the E164 translation facility level) leaves potential holes.

A GMSC has a full SS7 stack and it may implement a MAP policy whereby it recognizes the HLR interrogation services (SEND_ROUTINE_INFO_FOR_SM) and will drop (no answer) any such addressing of one's own HLRs that comes from an SCCP calling party address in another country.

We can recognize this method by realizing that there is no response (with MAP policy at the HLR level, there is an answer). For example, consider the method used by China Mobile to bar SMS-MT coming from another country.

5.2   Barring or Restricting the SMS-MO of One's Own Subscribers

The first reason an operator might want to bar the SMS-MO of one of its own subscribers could be (quite rare) a political reason that is made to bar all SMS-MOs sending to other networks, which has been the case with China Mobile Telecom for a long time.

In the second case (quite rare), an operator will allow *their subscribers to use only its own SMSCs* to send or reply to SMS-MO. This is a billing issue. Many operators today use the CDRs of their SMSCs to bill their subscribers; others use the CDRs of the MSCs or both. Using the CDRs of the MSCs (in particular while roaming) is necessary so that the SMS-MO roaming charge may be added when the customer is sending an SMS while roaming in another country.

This could be justified when there were still some SMSCs at the beginning of 1999 that accepted SMS-MO from nonsubscribers. Now with operators restricting usage to their own subscribers, the restriction to one's own SMSC is not as justifiable any more, because the SMS-MO will not work with any other SMSC. So, the only reason is to bar the reply path and the Reply SMS-MO, which is not charged to the sender if the MSC SMS-MO CDRs are not used, only the SMSC CDRs.

To implement this, a MAP policy software feature in the MSCs will control the SMS-MO sent by *their* subscribers (based on the IMSI), so that the service center is that of the network. For example, consider SFR and Vodafone.

5.3   Intelligent Barring of SMS-MT

5.3.1   Origin Address-Based Barring

The French operators, among others, bar the reception of the SMS-MT from their roaming partners, unless an AA19 is signed (average price per SMS-MT: 0.06 euros). This is frustrating because they freely send SMS-MT to their small roaming partners, who do not bar them. This is easy, but the cost of the software feature in the HLRs or the SMS filtering function offered by some SCCP IGPs is not justified.

Thus, they would like every SMS to come from a third party (which pays them for the termination and has very flexible filtering, flow limitation, and antispamming software). Some large operators are also interested because it simplifies their SMS billing system. How can this be implemented without the intervention of the SMS sending operator with only an agreement with the SMS receiving operator (and the configuration instructions given to its IGP)?

In Figure 5.1, the GSM A is sending an SMS to the subscriber +261323451234 of GSM B. GSM B wants to receive the SMS, but with the sending SMSC address of the SMS interworking network, which will be the proxy for the reception of all SMS-MT sent *by GSM D's roaming partners as well as nonpartners.*

GT = +33899990000 AND GT = +261321000033 (virtual MSC
of B for the same equipment !)

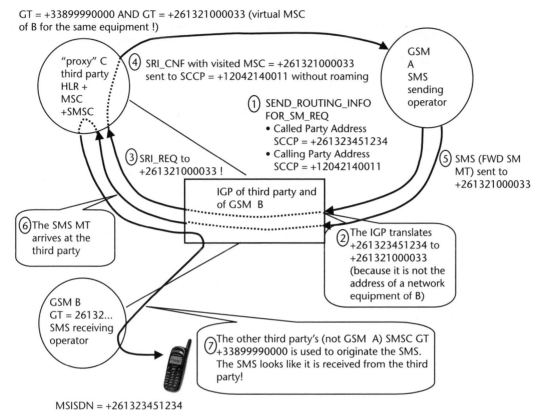

MSISDN = +261323451234

Figure 5.1   Origin address-based SMS-MT barring.

- D and the third-party patent have an agreement (and an SCCP connection).
- B and C have a roaming agreement.
- C and the SMS interworking network do not necessarily have an agreement.
- D and the SMS interworking network (for simplification of the explanation) have the same supplier for international SCCP service. This supplier provides the basic service of translating to GSM D's HLRs addresses the called party addresses = MSISDN used in the MAP_SEND ROUTING_INFO_FOR_SM. In that case (GSM agreeing), the translation table will point to a virtual HLR/MSC address that belongs to GSM D , but *physically corresponds to SMS interworking network equipment.*

Because the voice roaming only deals with SCCP addresses that are pieces of individual network equipment (HLR, VLR, MSC, SMSC), it is not impacted at all.

5.3.2   Filtering Based on Content of Incoming SMS-MT

SMS represents for operators an important source of revenue. This fact alone makes it vitally important to control inbound SMS traffic in order to reap the associated income revenue for this traffic (Interco fee, AA19, and so on). In addition it is also important to filter the content of certain messages that are not desirable (spamming, inappropriate adult content) for mobile subscribers.

We next explain the principle of an *intelligent SMS filtering system* (ISFS) (Figure 5.2), which provides the ability to filter the origin of the inbound messages and their content according to the different agreements in place and the operator service policy.

5.3.2.1 Principle Behind These Filters

A mobile network B wants to filter all incoming SMS MT based on:

- Barring all foreign numbers (the China Mobile Telecom case);
- Originating network (barring completely some origins or a certain percentage);
- The content (based on the presence of certain keywords or more sophisticated text analysis).

The recognition of the meaning of the content is a research topic, and today only crude solutions exist.

The recognition of the language and of the SMS is not trivial. For the 7-bit default alphabet, one must use language recognition software. When it uses 16 bits (UCS2 UNICODE), one can recognize the language from the range of UNICODE characters (Table 5.1). For example, Japanese characters are included from the 3041 position to the 32FE position (hexadecimal positions). Cyrillic characters are included from the position 0401 to the position 04CC.

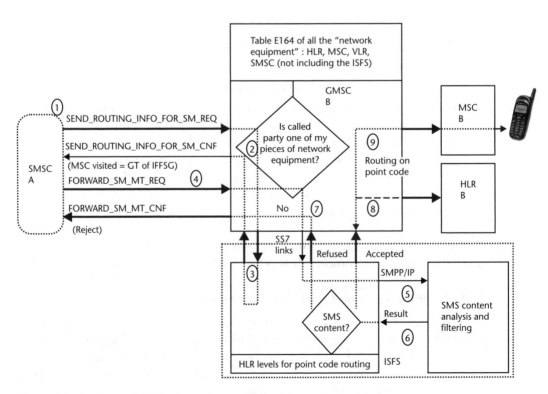

Figure 5.2 Intelligent SMS filtering system in relay mode.

Table 5.1   Range of Positions Corresponding to a Sample of Non-Latin Alphabets

Range of Positions in Hexadecimal	Alphabets
0401 to 04CC	Cyrillic
05B0 to 05F4	Hebrew
0970 to 09FA	Bengali
3041 to 309E	Hiragana Japanese
30A1 to 30FE and 32D0 to 32FE	Katakana Japanese

The purpose of the ISFS is to allow all incoming SMS-MT from the SMS roaming partners of B, while providing selective filtering (and reject) before they reach their destination. With this filtering, Operator B keeps the revenue (because most of the SMS-MT have legal content) that may have from SMS-MT terminating charges.

5.3.2.2   How Does It Work?

The ISFS is a piece of MAP-capable equipment to which the GMSC reroutes the SEND_ROUTING_INFO_FOR_SM message that it receives [(1) in Figure 5.2]. This is done, using only the SCCP level of the SS7 protocol and *only for the E164 addresses*.

So the voice roaming agreements for the outbound roamers, which use another table (E214) to route the UPDATE_LOCATION_REQ to the HLRs, are not disrupted by this setup. Chapter 7 explains in details the relay mode that is used to connect a mobile operator to a third-party interworking network. The SEND_ROUTING_INFO_FOR_SM are recognized by GMSC A because the SCCP called party address is not one of the GTs of the network equipment, which routes all of them to the ISFS (2). This is done by properly setting up of the E164 incoming routing tables in GMSC A.

The ISFS then returns its own GT as a visited MSC [(3) in Figure 5.2] to the SRI, without interrogating the HLR (this is the principle of the relay mode explained in detail in Chapter 7). It also returns a dummy IMSI (because there is no control) that hides the real IMSI. So the SMS-MT sent by the SMSC A (4) is received by the ISFS. It passes it to a filtering system (5), which validates the origin, the contents, and so on, and returns the response (6). The ISFS then returns either a reject response "Bad Data" to the SMSC A (7) following the normal GSM protocol or it forwards the SMS-MT to the hand phone, with a SEND_ROUTING_INFO_FOR_SM (8) and a FORWARD_SM_MT (9) returning the result (received, not reachable, and so on) to the sending SMSC.

To forward the SMS-MT, the ISFS includes an SMSC function. It must address the HLRs directly with routing on the point code (not with the MSISDN and routing on the GT), so it must have a GMSC function with a table of all the HLR levels and their point codes; otherwise, the SEND_ROUTING_INFO_FOR_SM sent by the ISFS would be routed back by the GMSC to itself. The E164 incoming routing table in the GMSC B applies to all SCCP messages, whether they come from a foreign SMSC or their own SMSCs (such as the ISFS).

For a foreign roaming partner, the system is almost transparent, enabling it to operate the normal roaming agreements to send SMS to B. It is *not* completely transparent because of the relay mode, which does not return the real IMSI of B's subscribers. The scheme can also work with the transparent mode (see Chapter 7) and then return the real IMSI; then the service is fully transparent.

On average, the incoming SCCP traffic is 50% of the total traffic, so that the ISFS should have an appropriate number of SS7 links (half of the number of international links) and enough processing capacity and redundancy.

So that all of the inbound SMS-MT traffic isn't flooding the ISFS at the beginning, it is better to introduce the service only for a limited range of customers. This is easy by setting the MSISDN rerouting of the SEND_ROUTING_INFO_FOR_SM to a limited range.

5.3.2.3   Filtering the Outgoing SMS from One's Own Subscribers

The preceding setup could also be used by GSM operators to enable an ISFS to reroute its SMS to other networks. Use of the content analysis server also provides the filtering of illegal outgoing SMS. It is the outgoing E164 routing table in the GSMSC that needs to be changed.

## CHAPTER 6

Virtual SMSC Implementation and Transit Agreements

Ich weiss nicht was soll es bedeuten   *I don't know what's the meaning of the fact*
Dass ich so traurig bin;   *That I'm in such a sad mood;*
Ein Märchen aus alten Zeiten   *A legend from olden days*
Das kommt mir nicht aus dem Sinn   *I can't get it out of my mind*

—Heinrich Heine,1823, *Die Lorelei*

6.1   Business Model

Small or new telecommunications operators may want to minimize their initial capital expenditures. The virtual SMSC offers these operators a way to provide SMS service to their customers without buying an SMSC. They use a third-party SMSC that charges them on the basis of the amount of SMS traffic, so revenues can be generated without any investment.

Having transit agreements is a way for large or medium operators to get additional revenues by renting their roaming agreements to a third party (the SMS interworking network), but they retain full, legal responsibility for their use. The SMS interworking network will pay the operator for any SMS sent within its roaming agreements.

The subjects of virtual SMSC and transit agreements are treated within the same chapter because any third party implementing a virtual SMSC for an operator (main SMSC or backup SMSC) has access to all roaming agreements for the sending of SMS-MT as explained in this chapter. The first part of the chapter covers SMS-MO and virtual SMSCs; the second covers the various implementation of transit agreements. The only setup is that of STPs or GMSCs to provide specific routing based on a GT. The third part concerns the implementation of multiple transit agreements in the same equipment and explains how several international gateways can be addressed.

6.2   Principle of the Virtual SMSC: Architecture and Billing of SMS-MO

6.2.1   Architecture

The provision of SMS service can be entirely subcontracted to an SMS interworking network. The real SMSC may be in another country, as illustrated by Figure 6.1. In

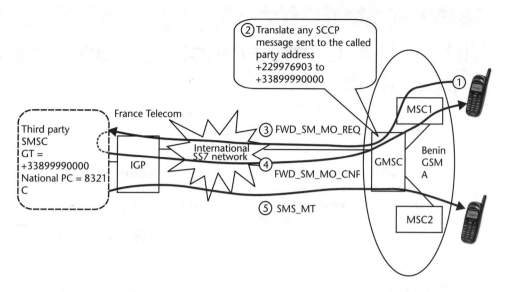

Figure 6.1 Principle of the virtual SMSC architecture.

this example, the SMSC, which is used for all of the traffic (internal and from/to other networks), is a third-party SMSC [1]. The message center (in the SIM cards) used is virtual +229 976 903: The GT is one chosen by the operator, for example, in Africa, which has installed in its gateway switch a GT translation table so that the local address (it does not correspond to any physical element) is translated to an another belonging to a third party. Very often, the service cost may be better than an investment, particularly when operation and maintenance costs are included.

This also means that if a roaming partner of A, that is, B, is replying to an SMS sent by virtual SMSC +229976903, it bounces back to C because the translation table of A works both for incoming messages from its own network and from other networks, such as B.

As illustrated by Figure 6.1, a subscriber of A sends an SMS-MO to his SMSC +229976903. It reaches the visited MSC, MSC1, then the GMSC (the MSC that interfaces with the international SS7 network). The GT is now translated and the SMS-MO reaches the third party's real SMSC, which acknowledges it. So the subscriber has sent a message on his cell phone. The only setup required is the translation from +229976903 to +33899990000, which is done entirely in the GMSC of A and under its control. There is no need for any setup to the international SCCP gateway. If Operator A is not happy, it can just take out the translation from its gateway switch.

6.2.2 Payment Issues

If the virtual SMSC can support CAMEL for prepaid SMS, which is very important because some operators have a majority, even exclusive, share of prepaid customers, a fully transparent service can be offered. A condition is that the home operator must have an IN that supports CAMEL. Also with this method, a prepaid or post-paid subscriber can send an SMS-MO from any visited network [e.g., Libancell

(Lebanon)] that has a roaming agreement with the home operator (even if this network has no roaming agreement with the third party), so the SMS-MO service is fully transparent from the operator's viewpoint.

6.2.3   Billing Coherence: Dynamic Originating SMSC GT

The business model is that whenever the virtual SMSC third party sends an SMS-MT to a subscriber of operator A that was originated by one of its own subscribers, the third party will not pay the SMS-MT. On the other hand, if the SMS-MT was originated by another network, the third party will pay the receiving operators.

To do this so that the operator's billing system can handle the process easily for Party A, the third-party SMSC sets the virtual SMSC GT (+229976903) in the SC_ADDR field of the FORWARD_SM_MT (example: +229976903), that is, at the MAP protocol level, when the SMS is from Operator A (i.e., one of the subscribers of A) to operator A (i.e., another subscriber of A):

| Party A (Sender) | 229976903 | SMS MT(type) | 02/12/18 (date) |
| Party B (Receiver) | 3747210005 | 3112995(ticket number) | 16:33:12 (hour) |

It sets its own SMSC GT (for example, +338999990000) when the SMS is originated by another network:

| Party A (Sender) | 33899990000 | SMS MT | 02/12/18 |
| Party B (Receiver) | 3747210005 | 3112995 | 16:33:12 |

Therefore, the *standard* CDR in the MSC will tell whether the third party (which has the GT +33899990000) is charged or not by Operator A. In the CDR, Party B is the receiving number. However, in this virtual SMSC setup, the third party should never use GT +229976903 as an SSCP calling party address.

6.2.4   Use of a Local Virtual SMSC GT in the SIM Card

A local virtual SMSC GT must be used in the SIM card (Figure 6.2). Take the case of a Benin subscriber. If she is in Benin, she could use the real SMSC GT (which is in France for example), that is, +33899990000, in her SIM card because her third-party supplier offers roaming with her network, so the SMS-MO will reach the third party. (Though in terms of marketing, this is not a good solution because the subscriber will not understand that the service center address in her SIM card is in France.)

However, if the subscriber is visiting Lebanon (which has roaming with his or her network; see Figure 6.2), the SMS-MO would not reach the third party (assuming it does have roaming with Lebanon!); but with the SMS-MO addressed to a virtual SMSC GT in Benin, it will reach Benin, bounce back to France, and reach the third-party SMSC. The SMSC will send the SMS-MT to the subscriber's friend in Benin or to any other country where she could be roaming. Thus, complete transparency is achieved: The subscriber can send an SMS-MO from any country with which her operator (Benin) has roaming agreements.

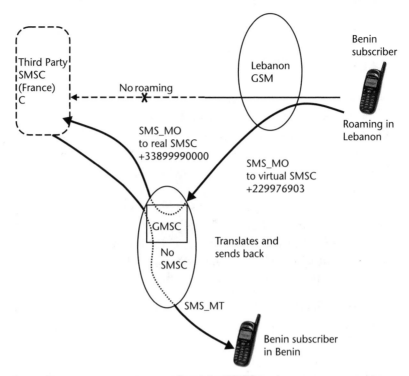

Figure 6.2 A virtual SMSC GT must be used in the SIM cards.

6.3 Detailed Implementation of the Virtual SMSC

We assume here that the real SMSC of the third party has an IGP, which provides the SCCP gateway service, that is, the routing on GT, based on the called party address in the SCCP envelope (see Chapter 2). The MTP transit service that requires the third party to have its own international PC is not used. For all MAP services used in this chapter, refer to their definition in Chapter 1.

6.3.1 Half-SCCP Roaming for SMS-MO

We now assume that the IGP of the third party, which provides an SMS interworking network (an example of such an IGP is France Telecom), provides its customers with a large number of opened outgoing SCCP routes. This means that any outgoing SCCP message sent by the third party will reach the destination network in most cases. However, if the recipient does not have roaming with the third party, it will not have opened the route to the third party and the response to a MAP request sent by the third party will not be sent back. Hence, this so-called "half-SCCP roaming" (Figure 6.3) cannot be used by the third party to interrogate a network with which it does not have roaming agreements. However, an answer sent by the third party to a request generated by such a network will reach it.

In Figure 6.3, a subscriber of A, which has a virtual SMSC with C, is sending (1) an SMS to his virtual SMSC +989347691001 (+98 is the country code for Iran). It reaches network A and bounces back to the third party (3 and 4). The IGP sends it to the national PC. The virtual SMSC answers (5) by inverting the called party address

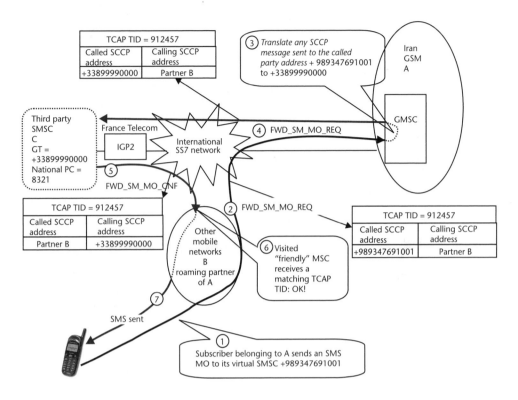

Figure 6.3   Half-SCCP roaming.

and the calling party address. Because of half-SCCP roaming, the answer will reach network B. The initiating cell phone will see the confirmation message "SENT."

6.3.2   Failure of Half-SCCP Roaming for SMS-MO

When the sending network B received the confirmation, it checked only the *TCAP transaction identity* (TID), which does match. This depends on the vendor's software in the MSC. Some of them will also check that the called party address of their request is the same as the received calling party address. Otherwise they discard the answer. The SMS will have been sent, but the sender (after a 30-second time out) will see "FAILED" on his or her cell phone, as shown in Figure 6.4.

6.3.3   Solving This Failure Case

It is now necessary to solve this failure case. *The third-party SMSC must have now a short global title* (for example, a number starting with +338 or +8885) so that when it is added in front of any existing GT (the maximum current length is 14 digits), it will not exceed 18 digits, which is the maximum length in the E164 standard for international numbers. *So the third-party GT should not exceed 18 – 14 = 4 digits.*

At (5) in Figure 6.5, the third-party SMSC takes out 338 from the called party address, yielding +989347591001. As shown, the SCCP calling party address returned by C to the sender (network B) is now the called party address (in network

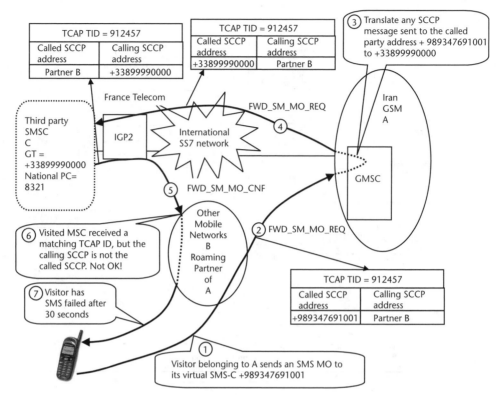

Figure 6.4 Control of the SCCP calling party address in the case of a failure.

A), so it will match! And the sender's cell phone will now get back a "SENT" message after his or her SMS-MO has reached the virtual SMSC of the third-party C network.

6.4 Implementation of Transit Agreements (SMS-MT)

6.4.1 Cases When a Virtual SMSC Has All Roaming Agreements of the Operator

Let's assume that C is the third party that provides a virtual SMSC to A and has been allocated one of A's GTs for the service.

6.4.1.1 Case in Which C Has Half-SCCP Roaming with B

When C has half-SCCP roaming with B, the SCCP path between C and B is such that, using any SCCP calling party address, the message sent by C to B will reach network B [(1) in Figure 6.6]. This occurs frequently because many SCCP gateways (1) do not control the originating SCCP address and (2) have a lot of forward roaming routes opened (e.g., France Telecom).

Thus, if C is using the GT allocated by A to address a request to B, the confirmation of B [(2) in Figure 6.6] will be sent to +65985440018. It will reach A, which will translate it to +33899990000 (3) and send to C (4). *So the confirmation will reach C*

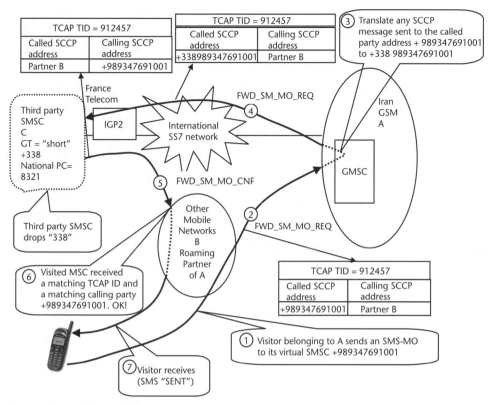

Figure 6.5   Solving the failure case.

even if B has not opened an outbound SCCP route to C! This means that (with only an agreement between C and A), C can use the roaming agreements of A with B (provided it has half-SCCP roaming). This is called a *transit agreement*: C pays A for every SMS that it sent using A's roaming agreements.

In the international SS7 network, many of the outbound routes are opened. If C' (another mobile operator in the same country as C) has roaming with B, the route is opened. But if B does not have roaming with C, it will not respond. In our scheme, B answers to A, with which it has roaming agreements.

- *Remark 1: How can A check the quantity of transit SMSs sent by C, using A's roaming agreements?* In the STP front end of network A, it is easy to count the confirmation messages: They are addressed to +65985440018. The easiest is then a payment formula, in which the operator pays X for each SCCP messages sent by C using an originating address of A. To relate to the number of SMS-MTs, one can use five SCCP messages per successful SMS. However, the current STPs (Ericsson, Siemens, Alcatel, Tekelec, and so on) do not provide SCCP counting based on the originating address. Most often, if counting is provided, other outside equipment (with probes) is used, which is expensive.

- *Remark 2: How can C protect himself from A's subscribers trying to send free SMS?* This could happen if they use +65985440018 as a message center number. These SMS-MOs will reach C, but the SCCP calling party address will not be an SMSC of A, it will be an MSC, and C will bar them.

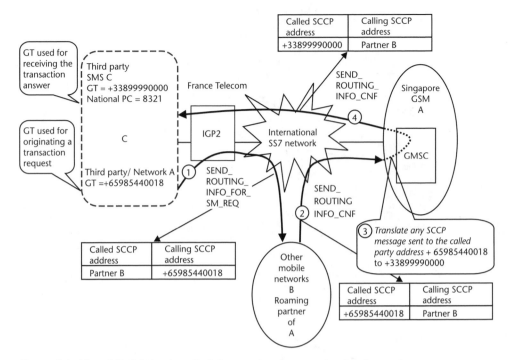

Figure 6.6 Virtual SMSC that has all of the roaming agreements of customer A.

- *Remark 3: What if A does not have a GT translation facility (or does not know how to use it)?* If A has an STP function, and is its own SCCP gateway (which is the case for several medium size operators), it can set it to reroute to 2-020-6 any message sent to +65985440018 (Figure 6.7). But it must instruct the SCCP provider of C to route to the third party the incoming SCCP messages to this GT. So the setup involves instructions to the SCCP provider of C. It is a little more complicated on the administration side. When A is capable of a GT translation, the setup was only between A and C and did not technically involve the SCCP carriers. For example, consider A = Iceland Telecom and C = Nilcom.

6.4.1.2   Case in Which C Does Not Have Half-SCCP Roaming with B

C might not have half-SCCP roaming privileges with B because no other operator, such as C' in the same country as C, has yet opened any route to B. Then C = B will fail and the scheme will not work.

6.4.1.3   Implementing Relay SCCP Roaming

If C does not have half-SCCP roaming with B, in Iran, for example, with the help of A, in Iran also, this can be turned around using only the SCCP layer. Operator A has allocated to C a full GT for the virtual SMSC, +989347691001 in the example. Assume C wants to reach B (GT = +9891) in Iran. To do so, it takes out +9891 (like in an E214 translation) and inserts +989345 (an unused GT of A), which A has allocated to C (Figure 6.8):

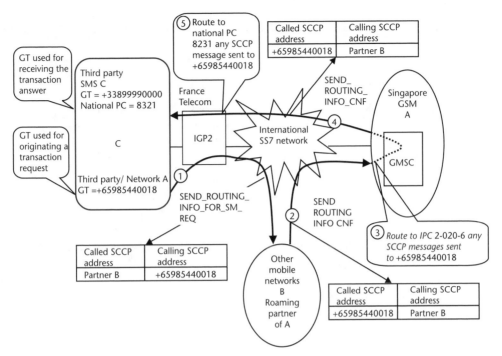

Figure 6.7 Use of one's own SCCP gateway to perform rerouting.

$$+9891\ 12345 \rightarrow 12345$$
$$12345 \rightarrow +989345\ 12345 \qquad \text{done by third-party C}$$

$$+989345\ 12345 \rightarrow 12345$$
$$12345 \rightarrow +9891\ 12345 \qquad \text{done by partner A of C}$$

The originating SCCP calling party set by C remains +989347691001 after the rerouting done by A. For example, Telecom Kish (Iran) = A, TCI (Iran) = B, and Nilcom = C.

The request sent by C will reach A then B [(2) and (3) in Figure 6.8]. As with half-SCCP roaming, the confirmation from B (4) will reach A (sent to +989347691001). A will translate it to +33899990000 (5) and it will go to A (6). This requires cooperation only between C and A; A has a roaming agreement only with B.

A simpler alternative (it depends only of C) can be implemented for the setup to B if A has a MTP transit service through an SS7 provider. Note that this is one of the most beautiful setups.

6.4.1.4 Restricting A's Roaming Agreements to Its Own Usage

If A has only a virtual SMSC agreement with an SMS interworking network A, but not a transit agreement, it will not want its roaming agreements to be used by C for usage other than its own. This is easily implemented by C, because the OA of an SMS-MO may be identified as a subscriber of A, and in that case only the roaming agreements of C are used. For non-A subscribers, another route (and other agree-

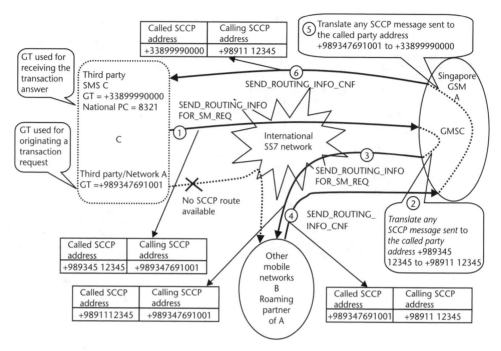

Figure 6.8 Relay SCCP roaming.

ment) should be used to reach B. The optimal routing algorithms given in Chapter 10 may make this possible.

6.4.2 Optimization of the Implementation of a Transit Agreement

If C and A have the same SCCP IGP, it is possible to optimize the traffic so that none of the confirmations to the requests issued by C to the roaming partners B of A will reach A's network! This is true even if the SCCP provider cannot do any address translation. In this case (Figure 6.9), A, instead of receiving the confirmation from B, has instructed its SCCP provider to reroute to C any incoming message to GT: +2309976903, which it has allocated to the third party.

Thus, there will not be any incoming, then outgoing traffic (charged to A) coming from B. Figure 6.9 is the same diagram as before, but the SS7 links IGP → A are never used when C uses the roaming agreements of A. For example, Emtel (Mauritius) = A. With C = Nilcom, they have France Telecom as their common international SS7 SSCP provider.

6.4.3 Use of an International Point Code: The Solution in Difficult Setup Cases

6.4.3.1 Interconnecting an SMSC to the Third-Party Network

Chapter 7 explains how the SMSC of an operator will connect via the SS7 network with an access point of the SMS interworking network. In some difficult setup cases, we may need to position a node in the international semaphore network (e.g., for operators acting as their own SCCP gateway). In such cases, the node will not use an SCCP IGP, but will directly implement the SCCP routing. It uses SCCP addressing

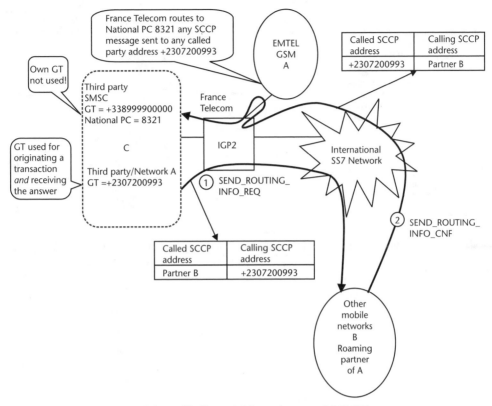

Figure 6.9 Optimization of the traffic if A and C have the same SCCP IGP.

that indicates a routing on GT for the called party address and then sets the international semaphore point of the destination's network IGP. Concerning the calling party A address, the node uses SCCP addressing to indicate a routing on PC with the international point code of the originating node.

In the third-party SMSCs, this configuration is set up with a virtual SMSC, the IGP point code of which is set to be dynamic. Then each time an SS7 path uses this virtual SMSC as an exit toward the destination network, the MTP DPC is set to the point code of the destination's network IGP. For this case, we also need to change the SCCP calling party address to fulfill a routing on PC with the international point code of the SMSC itself. Note that this implies that all destination networks' MTP routes have been opened on the MTP level. (This is very heavy work to maintain the SS7 MTP configuration.)

6.4.3.2 International PC Addressing Instead of Relay SCCP Roaming

Assume C has a MTP transit service with an SS7 provider (which may not necessarily be the same as the SCCP provider); that is, it can address directly any international PC, in particular, the gateway of the MSC belonging to A (which itself has a route with B). Chapter 2 describes the differences between an international PC, allocated by the national regulatory body, and a national PC, allocated by an SCCP gateway provider.

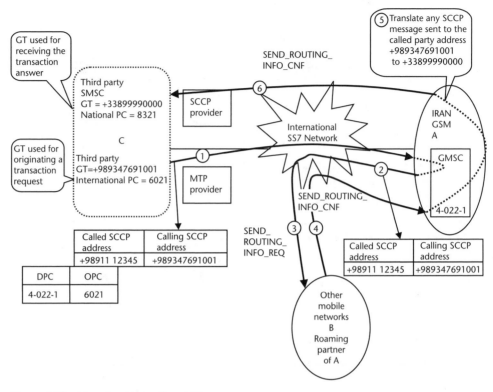

Figure 6.10 Use of an international PC.

If, using the international PC of C (Figure 6.10), an MTP route has been created from the MTP provider of C to the gateway of network A, A will use an originating calling party address = B (with routing on GT) at the SCCP level. It will use as the called party address that of B (without any modification to the GT). But the destination PC is the international PC of the gateway of B, not its (national) SS7 provider.

It will work even if C did not have an international PC with the MTP provider (just a national PC), because it never receives any MTP message addressed to it (half-MTP roaming). Note, however, that the MTP provider most often only connects international PCs (the service indicator of the linkset connecting C is international). For this reason, to use the above system, C must also have an international PC. This is a constraint from the MTP2 layer, not the MTP3 layer.

Note that in all of the previous cases, it is always routing on GT, which is used in the SCCP address even with half-MTP roaming.

6.5 Super-Routing Gateway and Multiple Virtual SMSCs in the Same Equipment

Up to now, we have seen the various possibilities for routing that could be available with one's own SCCP or MTP gateway. If the third-party SMS interworking network has several SCCP or MTP providers, it can act as a super-routing gateway.

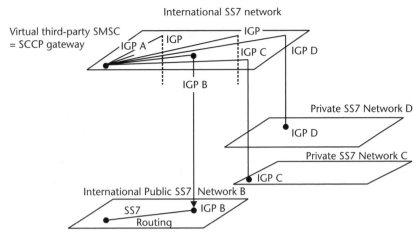

Figure 6.11 Multiple virtual SMSCs in the same equipment.

This means that, depending on the destination, it can select the SCCP or MTP provider it wants to use. (We use Chapter 10's concepts to optimize the margin.)

Then with these different kinds of virtual SMSCs, it becomes possible to implement on a single piece of equipment several virtual nodes as either part of a national SS7 signaling network or as part of an international semaphore network (Figure 6.11). It is assumed that the third-party SMSC has different routes (physical links) that allow it to reach different entry points on the SS7 network.

Depending on the destination to be reached, the third-party SMSC may select one of the following:

- An originating SCCP calling party address;
- The target destination point code of an IGP;
- The proper route to reach it.

At the routing application level, it is sufficient to set the two first parameters as the MTP configuration of the SMSC defines a route for each target point code.

For a third-party SMS interworking network, it is compelling to use the virtual SMSC method because of the investment saving: Only one system is necessary to have many transit agreements, and it can be located in its own premises.

For a client that is a small operator, using a virtual SMSC service increases the Opex (it has to pay for SS7 traffic) even for SMS in its own network, and also requires payments to the third party, but it avoids the initial Capex. So it is likely that after a few months, when the operator knows the service is successful, the operator would be better off to buy its own equipment.

Reference

[1]   *Système de messages courts, notamment de messages prépayés*, French patent FR 0107052.

## CHAPTER 7

Connecting Mobile Operators for SMS-MO

What was that snaky-headed Gorgon shield
That wise Minerva wore unconquered virgin,
Where with she freezed her foes to congealed stone?
But rigid looks of chaste austerity,
And noble grace, that dashed brute violence
With sudden adoration and blank awe.

                                  —Milton

7.1 Business Need for an SMS Interworking Operator to Connect Multiple Mobile Operators

We saw in Chapter 6 how a third-party SMS interworking network may send SMS to other networks. The third party needs to get the revenues from as many senders as providers, especially if it pays for the SMS-MT termination. So it needs to connect quickly, using the standard procedures of their SMSCs, as many mobile operators as possible and charge them for the SMS-MO sent by their subscribers.

It is very important to have the mobile operators as customers. They have a stable business structure (unlike application service providers), regular revenues, and stable administrative structures. This chapter explains how to connect the SMSCs of mobile operators so that they can use a third party to transmit their SMS to other networks with which they are not able to interwork (Figure 7.1). The chapter is divided into three parts: (1) the virtual HLR/MSC approach, (2) a discussion of how to route SMS traffic to a third party (the setup of the SMSC or the GMSC), and (3) how to create SCCP routing for SMS between the SMSC of the mobile operator and the third party when a GT translation is not available in the SMSC or GMSC.

7.2 Principle of the Virtual HLR/MSC Approach

7.2.1 Relay Mode

Mobile operators may connect their SMSCs to an SMS interworking network using the relay mode. They consider the SMS interworking network to be a worldwide HLR and MSC, and through a proper GT translation in their SMSCs or GMSC, the MAP signals addressed to networks with which they do not have roaming agreements are redirected to the SMS interworking network (Figure 7.2). We call this the

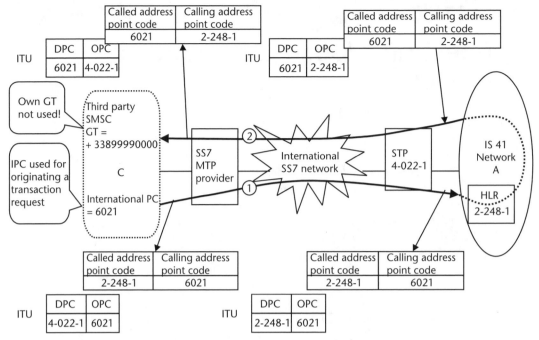

Figure 7.1 Transmission of signaling between a GSM and an IS-41 network.

relay mode because if the SMS interworking network finds that a retry is necessary (for example, because a network is not reachable or memory is full) it responds successfully, and the SMS interworking network will handle the retry. The reason is economy in terms of SS7 traffic. The retry is performed in the farthest servicing node.

All of the retries are handled by the servicing node (which has roaming with the destination), so no SS7 traffic is generated between nodes. This allows the SMS to be more cheaply priced, but has the following limitations:

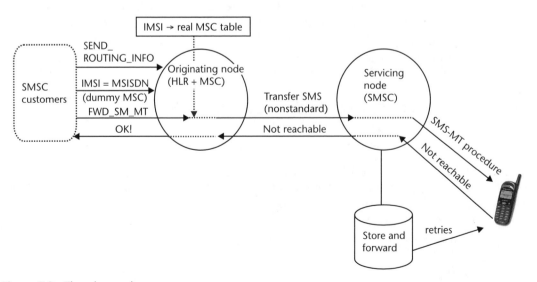

Figure 7.2 The relay mode.

- The SMSC sees "OK" when it may not be okay.

- The request for a status report is not transmitted to the SMS interworking network (it is not a parameter in the MAP SMS-MT service), so the sender will not receive a real status report.

The answer to the SRI is based only on the reachability, as detailed in the SMS interworking network routing table of the destination network. The real HLR interrogation is only performed when the SMS-MT is received from the SMSC with an IMSI = MSISDN, which is returned by the third party in this relay mode, so the SMSC never receives the real IMSI (which is not really a problem).

7.2.2 Transparent Mode

Operators may select the transparent mode (the price could be higher because of the additional SS7 traffic) when connecting to a third-party SMS interworking network (Figure 7.3). It is called *transparent mode* because the result is that the service behaves exactly as if the operators had direct roaming agreements with the operators that they reach through the third party. They get the real IMSI when they do a SEND_ROUTING_INFO_FOR_SM message (see Chapter 1) so they can create accurate reports that include the name of the destination network. They get the exact error back for each SMS-MT attempt; their own SMSC handles the retries and receives the ALERT from the HLRs with which they do not have roaming agreements. So it is much better for them, for the third party, because there is more SS7 traffic.

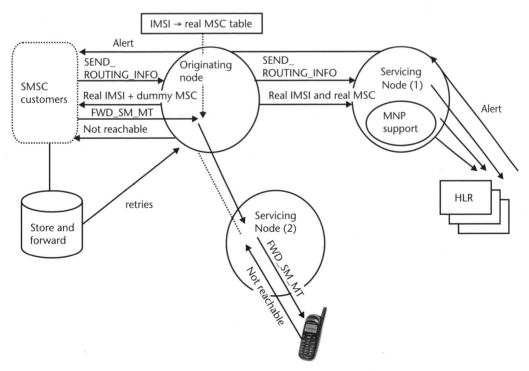

Figure 7.3 The transparent mode.

The GT of the sending SMSC [which is included in SRI, *FORWARD_ SHORT_MESSAGE_MT* (FWD_SM_MT), REPORT_STATUS_DELIVERY, and ALERT_ SERVICE_CENTER] is used by the originating node and the servicing node to determine whether this request is for the relay or the new transparent mode. In the latter case, *the SRI is relayed to a node that has roaming with the destination network*. If the number has MNP, it is routed to a node that has roaming agreements with all of the operators in the country. (For every country that has MNP, there must be at least one SMS interworking network node that has roaming with all of them.)

The distant servicing node will find the proper network and return the IMSI and the visited MSC. Then the SMSC will send the FWD_SM_MT, which may be sent by another servicing node. The *real status is returned*. So the retries are handled by the SMSC, as is the sending to the HLR of the STATUS_DELIVERY_REPORT, which will trigger an ALERT_SERVICE_CENTER from the HLR.

The ALERT_SERVICE_CENTER will be sent by the HLR to the servicing node. A mapping to the real SMSC is performed and the servicing node will send the ALERT_SERVICE_CENTER through the originating node, using the dynamically created MSISDN-to-SMSC table.

This transparent service is designed for major operators who want to fully handle by themselves the delivery of SMS.

Figure 7.2 shows an "IMSI → real MSC" table. The originating node has returned a dummy MSC (its own GT) to the SRI, so that the FWD_SM_MT is sent to it (not directly by the SMSC; otherwise the third party would lose this traffic!). So the FWD_SM_MT does not address the real MSC. The originating node may again find the real MSC with the table.

The system allows customers to connect either in relay mode or transparent mode. They distinguish this by means of the "SMSC GT → Mode" table in the originating node. How do the nodes know that they must provide the transparent mode or not for a received SRI? If a node receives an SRI from another node (it has a table of their GTs), it knows that it must be the transparent service.

7.2.3 Direct Interrogation of the HLR by the Client Operator

An operator α that wants to optimize its costs will perform directly the SEND_ROUTING_INFO_FOR_SM to interrogate the HLRs of its roaming partners, so as to avoid the charges of the SMS interworking network. However, if its SMSC finds that it does not have a roaming agreement with the visited operator , it *must send the FORWARD_SHORT_MESSAGE_FOR_MT to the SMS interworking network and pass the IMSI and the visited MSC GT*. With standard SMSC software, the only way to do this is to insert in front of the SCCP called party address (the visited MSC GT) a *short suffix* belonging to the SMS interworking network, as explained for the 338 insertion method in Section 7.3.1. Then the SMS interworking network will drop the suffix 338 and find the original visited MSC GT. It can send the SMS-MT *without interrogating the HLR* (it does not have necessarily roaming with it).

The SMS interworking network must return the FWD_SM_MT_CNF to the SMSC, because, since the MSISDN is not known (it just has the IMSI), the retry procedure must be handled by the sending SMSC. Thus, when the SMS interworking network receives a FWD_SM_MT with a called party address that is not the SMS interworking network (after dropping the suffix), *it must always use the transparent mode.*

7.2.4   SMS Interworking Network and the Status Report

Normally connected operators use their own SMSCs to send SMS to their own subscribers. There are a few exceptions such as Hutchison 3G, which uses an SMS interworking network for their outbound roamers. When H3G wants to send a STATUS_REPORT (with the FWD_SM_MT procedure) to one of its outbound roamers, its SMSC interrogates their HLR, finds that the visited MSC is not H3G, and sends the FWD_SM_MT (STATUS_REPORT) to the SMS interworking network.

The STATUS_REPORT type SMS is recognized from a COMMAND type SMS (up to MAP V2) by the value of the calling party GT.

7.3   Configuration of the SMSC or GMSC to Route to the Third Party

SMSC or GMSC equipment can be configured to route SMS traffic to a third party in four different ways:

1. Do an address translation in the GMSC.
2. Do an address translation in the SMSC.
3. Use a private conversion unit, which is a private STP that is able to do a GT translation.
4. Do nothing; use the services of a smart intelligent gateway provider.

7.3.1   GT Address Translation in the GMSC

When using the SMS interworking network, the operator may want to optimize its costs, which it can do by sending its SMS-MT directly if it has a roaming (and SMS interworking) agreement with the network of the destination handset *and* with the visited network. Most of them will also send to the SMS interworking network when it is too complicated to update all of the changes. For example, the United States has a total numbering plan of more than 65,000 lines.

A setup has been developed, called the *338 method*, that is optimum and that also resolves the paradigm (Figure 7.4): "α has roaming with network β, but the subscriber of β is roaming in network ε and α does not have roaming with ε. Note that many big operators do not handle this and, of course, still charge their subscribers for the SMS that could not be delivered!

7.3.1.1   How Does It Work?

The third party must have a short global title such as +338 or +3204 that includes all of the numbers that follow. So the SMSC or the GMSC (whichever is configured to do the GT translation; it is simpler in the SMSC) will simply translate the SCCP called party address *with any GT*, regardless of any roaming agreements, whether it is the MAP service SEND_ROUTING_INFO_FOR_SM or the FORWARD_SHORT_MESSAGE_MT used in the sending of SMS or anything else:

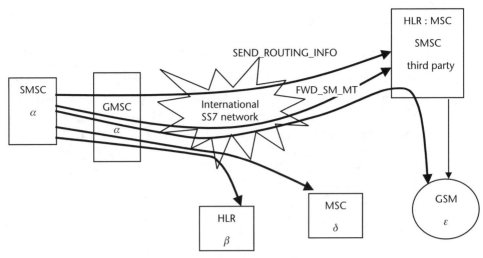

Figure 7.4 The 338 method.

$$+5372541212 \text{ (Cuba)} \rightarrow +3385372541212$$

Thus, any such SCCP signal will be sent to the SMS interworking network!

If this GT (case of an FWD_SM_MT from α to ε) is the called party address of a visited MSC, by stripping 338, the SMS interworking network routers get back the real MSC GT and can send it (using a path of the network). This process works because by adding 3 digits to the longest current GT (14 digits), the resulting GT address is no longer than 18 digits, which is the current maximum length in many international SCCP gateways (and the maximum limit in the current GSM specification for E164 numbers). So it works even with Chinese numbers (14 digits), and +3204 would also work.

7.3.1.2 Reason to Do the Address Translation in the GMSC

Address translation can be done in the GMSC for those operators that:

1. Have an SMSC that does not have a numbering plan (also called *HLR tables*). It can only send the SEND_ROUTING_INFO_FOR_SM to a GMSC (Gateway MSC), which then "translates" the MSISDN send in the called party address of SCCP to the corresponding HLR number.
2. Have an SMSC that is unable to perform any address translation at the MAP level (destination address) or at the SCCP level (called party addresses) in the SRI.

For example, Logica SMSC (before Release 2700) and CMG could send a SEND_ROUTING_INFO_FOR_SM in one of two formats only (what we call the standard SMSC addressing mode).

1. *GT SCCP addresses.* Routing indicator is set to route on GT. The SCCP called address is the GT of the MSISDN and SSN, and the SCCP calling address is the GT and SSN of the SMSC.

2. *National PC addresses.* Routing indicator is set to route on DPC. The SCCP called address is the DPC of the local HLR and SSN (as defined in MAPs table), and the SCCP calling address is the DPC and SSN of the SMSC.

The GT translations (MSISDN-to-SMS interworking network HLR GT) *must be performed in one of the GMSCs in this case.*

7.3.1.3   Method to Do the Address Translation in the GMSC

To do the address translation in the GMSC, the GT translation is performed by the GMSC to which the SMSC is connected. The setup of the global SMS service will not impact the existing voice roaming agreement. Table 7.1 shows the old table, without translation, prior to setting up the address translation in the GMSC.

Note that there is a common habit in the network department of creating *all NDCs of their roaming partners* in the GMSC's routing table for outgoing SCCP traffic. In fact, only the MGT (one of the NDCs) is used for voice roaming (update-location) and FWD_SM_MO (when their outbound roaming subscribers are sending an SMS-MO to their SMSC). All of the other NDCs such as 661, 662, and 663 *are used only to allow the transmission of SEND_ROUTING_INFO_FOR_SM in order to send SMS-MT to their subscribers.* So, if you do not have an SMS interworking agreement with them, *it is completely useless.* You should remove these ranges, which also allows you to avoid the work of maintaining these useless ranges.

Table 7.2 shows the new table, which takes translation into account: Send SMS to the SMS interworking network for this network and all others. You find that the MSCs and HLRs GT in your partner's IR-21 (the standard document that gives the numbering scheme of their equipment) is:

33   6600   Bouygues (France)

33   68900   Orange France

The setup of the GMSC is very simple because the pattern matching is such that it looks for the longest matching GT chain. After it has failed to match the number +33612123456 (SFR France) with one of the GTs in the table, it will translate to the third-party HLR. You need to take these steps:

Table 7.1   Old Table: Set Up of Your GMSC for a Voice Roaming Agreement

	CC	NDC	Translation to	Destination Point Code
	33	660 (MGT)	NONE	IGP
No translation	33	661	NONE	IGP
	33	662	NONE	IGP
	33	670	NONE	IGP
	33	689 (MGT)	NONE	IGP
	33	607	NONE	IGP
	33	608	NONE	IGP
	33	680	NONE	IGP
	33	681	NONE	IGP
	46	707(MGT)	NONE	IGP
	46	704	NONE	IGP
	46	739	NONE	IGP
	46	736	NONE	IGP
	46	7016	NONE	IGP

Table 7.2   New GMSC E164 Translation Table

	CC	NDC	Translation to	Destination Point Code
1: No translation	33	6600 (no SMS interworking)	NONE (All MSCs and HLRs will remain untranslated so voice roaming is not a concern.)	IGP
	33	68900 (no SMS interworking)	NONE	IGP
	46	707 (MGT) (SMS interworking OK)	NONE	IGP
	46	704	NONE	IGP
	46	739	NONE	IGP
	46	736	NONE	IGP
	46	7016	NONE	IGP
2: Translate to the SMS working network	1	Any other ranges	HLR of the third party (All SMSs sent to a number starting with 1 will go to the third party.)	IGP
	2	Any other ranges	HLR of the third party (All SMSs sent to +225, +20, and so on, will go to the third party.)	IGP
	3		And so on...	
	4			
	5			
	6			
	7			
	8			
	9			

1. Add at the end 1, 2, …, 9.
2. Take out the NDCs of the networks with which you do not have SMS interworking, while retaining only the MGT extended to include only the network equipment GT.

Note that this has no effect on the UPDATE_LOCATION, which uses separate E214 tables. Operators using this method include Cell C (South Africa) and Celltel (Sri Lanka).

7.3.2   Doing the Address Translation in the SMSC

7.3.2.1   GT Address Translation

These SMSCs are able to address directly a HLR in another (foreign) network using GTs. The third-party HLR will be addressed by the GT that the SMS interworking network gives you.

Principle.   Separate tables exist for the HLR and MSC/SGSN addressing in the SMSC. Based on the destination party MSISDN, a translation to either SPC or *another GT* (using the SMS interworking network's GT) is available. If no translation has been found, the MSISDN itself is used as a GT; simply set the translation to the SMS interworking network GT with routing on GT.

In conclusion, you can route routing information requests to specific entities for different number ranges, for example, to the SMS interworking network routers for Irish subscribers and to Orange for Orange subscribers.

Typical Layout of GT Translation in an SMSC.   Here is a typical layout for GT translation in an SMSC:

	GT	PC
Your numbers➜	HLR GT	HLR PC
Default (sent to the third party)➜	HLR third party assigned to you	IGP PC

Operators using this method include Mobitai (Comverse), FT Dominica (Logica), and Hutchison (Logica).

7.3.2.2   Destination Address Modification at the MAP Level by the SMSC

The SMSC is configured so that for any international (all the destination MSISDNs) number starting with a plus sign (+), it inserts 338 (the third-party MGT):

$$\text{MSISDN } +33608123456 \rightarrow \quad +338\ 33608123456$$

The SMSC will then address the HLR with this SCCP called party address +33833608123456 and routing on GT. The GT is 18 digits, which is the maximum (current ITU standard) for an E164 GT. So, because the SMS interworking network has a very short MGT, we are able with this method to address U.K. MSISDNs, which have 12 digits.

The GMSC/STP will route to France Telecom International (the SMS interworking network IGP supplier in this example), which will then route it to the SMS interworking network (France) because the GT is +338xxxxxx.

The SMS interworking network HLR (also MSC and SMSC) will take out the 338, giving back +33608123456, and then send the SMS-MT to the final network throughout our SMSC. Also, all of the international SMS traffic will be routed to the SMS interworking network but this method is very simple.

Operators using this method include New World PCS (Nokia) and Malitel (Alcatel).

7.3.3   Use of a Private Conversion Unit

A conversion unit is a "SS7 box" capable of performing a GT translation and of behaving as a HLR, MSC, and SMSC. It is installed in an operator's own networks. It has the same function as an originating node in the relay mode (see Figure 7.1) and the same procedure. It is a connection solution when there is no possibility to do any address translation in the GMSC (which is rare) or in the SMSC (either at the SCCP level or at the MAP level; see Chapter 3).

7.3.3.1   Consider the Conversion Unit to Be One of Your HLRs

One of the only possibilities in the SMSC is to declare another (internal) HLR and address it with routing on PC. So the idea is to create a new HLR, that is, the conversion unit.

Configuration Principle of the SMSC. The SMSC has a numbering plan (also called a HLR levels table), like that shown in Table 7.3, which is used for the address translation for the MAP_SEND_ROUTING_INFO_FOR_SM function. Table 7.3 shows that subscriber +436640312345 is assigned to the Salzburg HLR.

Assume that one wants to send an SMS to other subscribers (foreign) that are not in this table, but are accessible by the SMS interworking network. Usually the default SCCP called party address, when a number is not found in the table of HLR levels for the SRI, is

Routing indicator = route on GT
GT = MSISDN
SSN = 6

and is in the routed (MTP3 level) to the international gateway.

One could add the SMS interworking network router like one of your new HLRs just by adding a new line in this table, after assigning one of your GTs and PCs to the conversion unit in your network either (depending on your SMSC) like this:

default   +436649000050   The SPC that you adjust to the conversion unit

or like this:

+000000000000   +99999999999   +436649000050   7-004-8

(After scanning with your subscriber numbers, for any other number, it will find the conversion unit and the routing will go *through your network*, so you must give its SPC.) This HLR will return *its own address* as the localization MSC address (+436649000050 in the example), so that it will also receive the SMS-MT sent by your SMSC. (Remember that SMS-MT is a two-phase process: First is the interrogation of the HLR to obtain the MSC localization and second is the sending of an SMS-MT to this MSC.) With this method, the SMS interworking network SS7/IP router behaves like a HLR and a MSC on your network. Note that the operator can select the traffic that is routed through the conversion unit then to the third party by simply configuring the HLR levels of the new HLR.

Operators using this method include Telkomsel and SINGTEL.

7.3.3.2   Consider the Conversion Unit as the GMSC of the SMSC

One of the only other things you can do when you configure any SMSC, including CMG, is to set the point code of the GMSC. If you set it to be the conversion unit, any   SEND_ROUTING_INFO_FOR_SM   for   a   foreign   subscriber   or   any

Table 7.3   Numbering Plan for Address Translation

Min	Max	HLR GT(GT)	HLR (SPC)	Comment
+436640000000	+436640149999	+436649000010	7-002-1	Vienna 1
+436640150000	+436640299999	+436649000020	7-003-1	Vienna 2
+436640300000	+436640449999	+436649000030	7-004-2	Salzburg
+436640450000	+436640599999	+436649000040	7-004-3	Innsbruck

FORWARD_SHORT_MESSAGE_MT to one of your subscribers roaming in another network will address the conversion unit.

The configuration is very simple, but the third party is used even for one's own subscribers when they are roaming abroad. So, the method of Section 7.3.3.1 would be preferred in terms of cost reduction.

One operator using this method is Hutchison 3G (Hong Kong) (CMG).

7.3.4   Intelligent SCCP Routing by Your IGP

Using a third party for the sending of SMS to other networks requires a the SMSC or GMSC to be set up to do so. Here is an architecture in which the IGP provides the service with the set up *only on its side* (Figure 7.5). Remember that the method was to reroute all signaling messages going to nonroaming partners to a specialized third-party router. This rerouting could be handled transparently by the IGP with its own roaming agreements and an SCCP gateway that holds a database of the roaming agreements of all their customers. It involves only SCCP levels—it does not need to analyze higher MAP levels—so that the equipment remains a simple SCCP gateway with all the international SCCP traffic transiting through.

Because GSM A does not have roaming with GSM B, no mobility MAP messages will be attempted to be sent by GSM A to GSM F through the IGP. So, no undue load is created on the IGP. Whenever the IGP intelligent SCCP gateway, which processes only SCCP, sees a calling party address GT it checks (from the call-

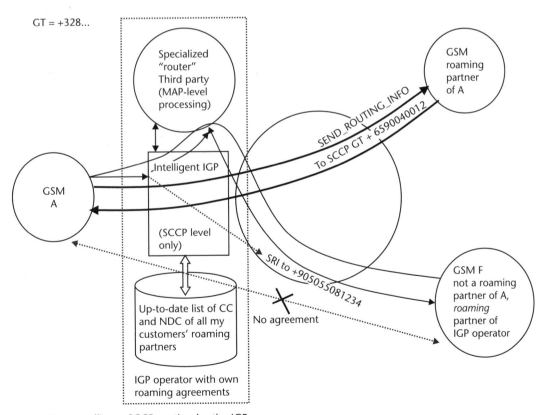

Figure 7.5   Intelligent SCCP routing by the IGP.

ing party address, e.g., Proximus) whether GSM A has roaming with Aycell (Turkey). If not, it translates the GT Aycell (+905055081234 to +3204 905055081234) so that the SRI is sent to the router function (the 338 method becomes a 3204 method). Then the rest is obvious. The SRI is relayed using the roaming agreement that exists between the IGP and GSM F. No setup at all is required in GSM A.

7.4 Creating Third-Party SCCP Routing When a GT Translation Is Unavailable

If the connected mobile operator can do an address translation, the question is trivial. They just open an SCCP route according to the roaming procedure involving their respective IGPs. If the operator cannot do address translation, the process is more complicated.

This was the case as of 2002 for networks that have a CMG SMSC (no GT translation) and an Ericsson GMSC with the R8 software: Nowhere is there a GT translation facility. A setup can be provided without GT translation if the interworking SMS network has an international point code.

7.4.1 Case in Which Connected Operator Acts as Its Own SCCP Gateway

When an operator acts as its own SCCP gateway, it means that their SS7 carrier provides them with MTP transit service. To address the virtual HLR of SMS interworking network C (Figure 7.6), consider these two cases:

1. *First case:* You will set up your STP to route to the IPC of the SMS network any GT with which A does not have roaming. So this is simple because *any*

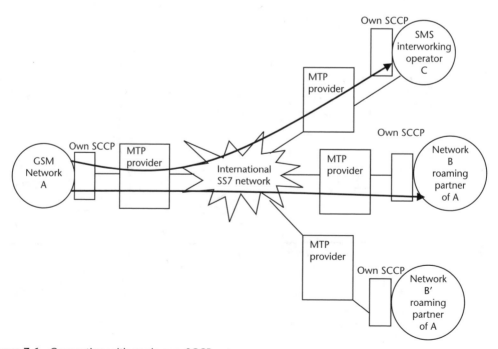

Figure 7.6   Connecting with one's own SCCP gateway.

GMSC has this function (no GT translation involved) of PC routing (choice of the distant SCCP gateway A). So the interworking network will be in the same situation as any partner of A: A will route to the IPC of C any GT with which it does not have roaming agreements. This is a nonstandard routing: It routes to France GT for the United States!

2. *Second case:* In these SMSCs, the only addressing provider that can really act is in the PC of the IGP (usually the operator's GMSC). If you set it to the SMS interworking network, all the traffic directed toward a foreign GT will reach it.

7.4.2 Case in Which Connected Operator Uses an International SCCP Gateway Service: No Solution

This case is the same as the preceding one; however, Operator A cannot do the routing itself (Figure 7.7). It should instruct (give a list) the IGP to route to C a list of GTs. However, the SCCP provider has several customers such as A' in addition to A. For A' it implements the standard routing on GT; a number starting with +65 is routed to Singapore. And most of the SCCP gateway (let's say all of it) cannot provide selective routing based on the calling address. So it will not be possible to route the +65 to C just when A is originating.

7.4.3 Case in Which GT Translation Is Not Possible and the Operator Is Not Its Own SCCP Provider: Use a Conversion Unit

In this case, the only thing that A can do is to route all traffic to GTs because it does not have roaming with an internal PC. The conversion unit is a private SCCP gateway, capable of GT translation, which its main GMSC does not have. The process is as follows: The conversion unit is a simple SS7 box with a GT translation capability. As an alternative, A may compare the price of the conversion unit and that of buying the software from its GMSC vendor.

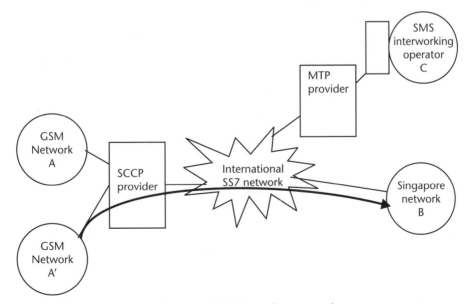

Figure 7.7 No connection possible to an SMS interworking network.

7.4.4 Transmission of Signaling Between a GSM and an IS-41 Network

The routing must be established at two levels:

1. *MTP:* The mobile operator and the third-party SMSC must use either ITU or ANSI or connect through an ANSI/ITU converter. This is the case, for example, between Europe and Southeast Asia where the CDMA networks are built on the ITU MTP.
2. *At the SCCP level:* The IS-41 networks do not use GT addressing inside their networks; however, they can agree with a third party to assign GTs to their equipment and *perform the routing on GTs to routing on PCs in their STP.* In this case the third party can easily connect through its ordinary SCCP provider.

An alternative if the third party has its own international PC is as follows: The IS-41 network could give the third party the private PC of all its equipment, HLR, and MSC. Then the third party would address them directly using routing on PCs at the SCCP level while using a direct MTP route between its international PC 6021 and the PC of the STP, that is, 4-022-1.

7.5　Conclusion

Table 7.4 sums up the different cases discussed in this chapter [1].

Table 7.4　Various Solutions for Rerouting to SMS Interworking Network C

A	Solution for Rerouting to the SMS Interworking Network C
1. A can do a GT translation (in the SMSC or the MSC); no investment.	A translates the GT, inserts +338, and then sends to its SCCP provider.
2. A cannot do a GT translation but is its own SCCP gateway; no investment.	Route any GT it does not have roaming with to C's international point code.
3. A has no GT translation and uses an SCCP gateway provider.	Practically, not possible without a conversion unit.
4. Conversion unit; investment.	There is an investment in equipment installed in A's network.
5. Intelligent SCCP routing by your IGP.	Nothing to do!

Reference

[1]　*Système de routage optimal de messages courts d'un center de messages vers un autre, avec traduction globale d'adresses,* French Patent FR-0209667.

## CHAPTER 8

Connecting ASPs and ISPs with SMPP

Wine is the healthiest and most hygienic beverage.
—Louis Pasteur, 1890

8.1 Introduction

This chapter provides an overview of the most commonly used protocol for connecting *external short message entities* (ESMEs) to SMSCs. ESMEs cover any kind of IP-based client who may need to connect to an SMSC in a manner other than through the signaling network. An ESME is typically used by *Internet service providers* (ISPs) who need to send or receive SMS to or from mobile phones for various Internet-based services, such as advertising, and do not have an SS7 connection.

At first, each SMSC vendor offered a proprietary interface to connect to its SMSCs, but this lack of standardization led to the definition of a common interface for ESMEs so that they would have the ability to use SMS services. The SMPP protocol is based on a TCP/IP client/server connection or X.25 connection between an ESME and the SMSC. The ESME is typically a voice mail server, an e-mail gateway, and so on. The SMPP protocol has been designed to offer SMS services for various cellular networks such as GSM, CDMA, and TDMA.

This chapter does not cover all aspects of the SMPP protocol, but does discuss how to use an SMPP as an interworking facility to connect IS41-based networks to GSM-based networks.

8.2 SMPP Sessions

To use the SMS services offered by the SMPP protocol, an ESME must establish a session with an SMSC. There are three kinds of SMPP sessions:

1. *Transmitter (TX):* This kind of session allows the ESME to submit short messages to an SMSC for onward delivery to a MS. This session also offers the possibility of querying the status of a previously submitted short message, replacing it, or canceling it.
2. *Receiver (RX):* This session allows an ESME to receive SMSs from the SMSC.
3. *Transceiver (TRX):* This session is a combination of transmitter and receiver sessions.

8.3   SMPP Commands

The SMPP protocol is a query/answer designed protocol. That is to say, each message sent from the SMPP client toward the SMPP server is acknowledged by the server side, indicating a status on the previously submitted request. Tables 8.1, 8.2, 8.3, and 8.4 describe, respectively, the commands for session management operations, message submission operations, message delivery operations, and ancillary submission operations.

8.4   Example of SMPP Sessions

Figures 8.1, 8.2, and 8.3 show, respectively, examples of a transmitter session, receiver session, and transceiver session with PDU sequencing.

8.5   Example of Message Operations

8.5.1   Session Management: Transceiver PDUs

8.5.1.1   bind_transceiver PDU

This PDU is used by an IP-based ESME to initiate an SMPP connection toward an SMSC. It includes, in particular, authentication information (system_id, password) used by the SMSC to filter unauthorized ESMEs. See Table 8.5.

Table 8.1   Session Management Operations Commands

SMPP PDU Name	Description
bind_transmitter	Authentication protocol data unit (PDU) used by a transmitter ESME to bind to the MC. This PDU includes identification and authentication information to protect access to the SMSC.
bind_transmitter_resp	SMSC response to the bind_transmitter PDU. It indicates the success or failure of the attempt to bind as a transmitter.
bind_receiver	Authentication PDU used by a receiver ESME to bind to the MC. This PDU includes identification and authentication information to protect the access to the SMSC.
bind_receiver_resp	SMSC response to the bind_receiver PDU. It indicates the success or failure of the attempt to bind as a receiver.
bind_transceiver	Authentication PDU used by a transceiver ESME to bind to the MC. This PDU includes identification and authentication information to protect access to the SMSC.
bind_transceiver_resp	SMSC response to the bind_transceiver PDU. It indicates the success or failure of the attempt to bind as a transceiver
outbind	Authentication PDU used by a MC to outbind to an ESME to inform it that messages are present in the MC. If the ESME authenticates the result, it will respond with a bind_receiver or bind_transceiver to begin the process of binding to the MC.
unbind	This PDU can be sent by the ESME or MC as a means of initiating the termination of an SMPP session.
unbind_resp	This PDU is sent to acknowledge the receipt of an unbind request. After sending this PDU, the network connection will typically be closed.
enquire_link	Test network connection.
enquire_link_resp	Acknowledgment of enquire_link PDU.
alert_notification	The SMSC sends an alert_notification to an ESME as a means of alerting it to the availability of an ESME.
generic_nack	This PDU is used as a means of indicating the receipt of an invalid PDU.

Table 8.2   Message Submission Operations Commands

SMPP PDU Name	Description
submit_sm	A transmitter or transceiver ESME that desires to submit a short message uses this PDU to specify all information related to the submitter SMS (sender, receiver, text, priority, and so on).
submit_sm_resp	The MC response to a submit_sm PDU, indicating the success or failure of the request. It also includes a message_id allowing the ESME to identify the submitted SMS in subsequent operations (cancel, replace, query).
submit_multi	A variation of the submit_sm that supports up to 255 recipients' addresses or use of delivery lists.
submit_multi_resp	This PDU is similar to the submit_resp. It can also include the list of failed recipients. It also includes the message_id for subsequent operations.
data_sm	A streamlined version of the submit_sm operations designed for packet-based applications (WAP).
data_sm_resp	Response to a data_sm PDU. It includes the message_id for subsequent operations.

Table 8.3   Message Delivery Operations Commands

SMPP PDU Name	Description
deliver_sm	This PDU is used by an MC to deliver an SMS to an ESME.
deliver_sm_resp	This PDU indicates to the SMSC the acceptance or not of the deliver_sm command.
data_sm	Same as deliver_sm [Wireless Application Protocol (WAP)].

Table 8.4   Ancillary Submission Operations Commands

SMPP PDU Name	Description
cancel_sm	This PDU is used to cancel a previously submitted short message. It identifies the message through the source address of the original message and the message_id returned by the original submit_sm_resp or submit_multi_resp.
cancel_sm_resp	Indicates the success or failure of the cancel_sm command.
query_sm	This PDU is used to query the SMSC about the status of a previously submitted short message.
query_sm_resp	PDU returning the status of the submission operations.
replace_sm	This PDU is used to replace a previously short message identified through the source_address and the message_id.
replace_sm_resp	PDU indicating the success or failure of the replace_sm request.

8.5.1.2   bind_tranceiver_resp PDU

This PDU is the response from the SMPP server located on the SMSC function to the bind_transceiver command. It indicates to the SMPP client that its demand for connection and authentication has been approved and that it may submit commands to the SMSC. See Table 8.6.

8.5.2   Message Submission Operation

8.5.2.1   submit_sm PDU

This PDU illustrates the equivalent in the IP SMPP protocol of a SUBMIT short message of the GSM 03.40 recommendation (Table 8.7). (Note that it is also equivalent to the SUBMIT of the IS-637 ANSI standard since the SMPP protocol has been

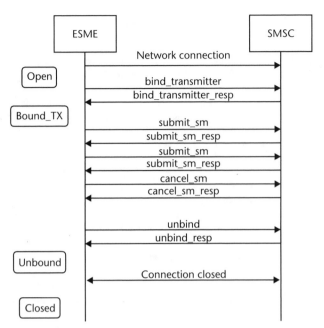

Figure 8.1   Example of a transmitter session.

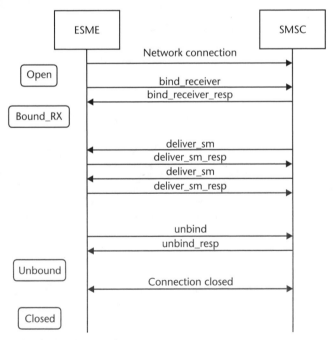

Figure 8.2   Example of a receiver session.

designed to work either with a GSM SMSC or ANSI TIA/EIA-41-D-based message center.) The most significant fields are of course the source_address, destination_address, and short_message defining the originator of the submitted short message, the destination, and the content of the SMS, respectively.

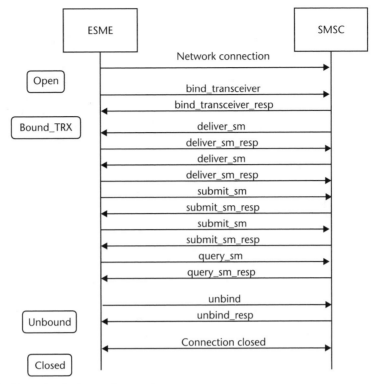

Figure 8.3   Example of a transceiver session.

Table 8.5   bind_transceiver PDU

Field Name	Size	Type	Description
command_length	4	Integer	Overall length of the PDU
command_id	4	Integer	0x00000009
command_status	4	Integer	0x00000000
sequence_number	4	Integer	Set to a unique sequence number. The bind_transceiver_resp_PDU will echo the same sequence number
system_id	Var. max 16	C-octet string	Identifies the ESME system.
password	Var. max 9	C-octet string	Used to authenticate the ESME.
system_type	Var. max 13	C-octet string	Identifies the type of ESME.
interface_version	1	Integer	Identifies the version of the SMPP protocol supported by the ESME.
addr_ton	1	Integer	Type of number for ESME address(es) served via this SMPP transceiver session.
addr_npi	1	Integer	Numbering plan indicator for ESME address(es).
address_range	Var. max 41	C-octet string	ESME address or range of address served via this SMPP transceiver session.

8.5.2.2   submit_sm_resp PDU

This PDU is the response from the SMPP server to a previously submitted short message (Table 8.8). It indicates in the field command_status the success or failure of

Table 8.6   bind_transceiver_resp PDU

Field Name	Size	Type	Description
command_length	4	Integer	Overall length of the PDU
command_id	4	Integer	0x80000009
command_status	4	Integer	Indicates status of original bind_transceiver request.
sequence_number	4	Integer	Set to sequence number of original bind_transceiver request.
system_id	Var. max 16	C-octet string	Identifies the SMSC system.
Optional Type, Length, and Values (TLVs)			
TLV Name	Type	Description	

Table 8.7   submit_sm PDU

Field Name	Size	Type	Description
command_length	4	Integer	Overall length of the PDU
command_id	4	Integer	0x00000009
command_status	4	Integer	0x00000000
sequence_number	4	Integer	Set to a unique sequence number. The bind_transceiver_resp_PDU will echo the same sequence number.
service_type	Var. max 6	C-octet string	It can be used to indicate the SMS application service associated with the message. It allows the ESME to control the teleservice used on the air interface.
source_addr_ton	1	Integer	Type of number for the source address.
source_addr_npi	1	Integer	Numbering plan indicator for the source address.
source_addr	Var. max 21	C-octet string	Address of the SME that originated the message.
dest_addr_ton	1	Integer	TON for destination.
dest_addr_npi	1	Integer	NPI for destination.
dest_addr	Var. max 21	C-octet string	Destination address of this short message.
esm_class	1	Integer	Message mode and message type.
protocol_id	1	Integer	Protocol identifier.
priority_flag	1	Integer	Priority level.
schedule_delivery_time	1 or 17	C-octet string	Time by which the message is to be scheduled for delivery.
validity_period	1 or 17	C-octet string	Validity period of this message.
registered_delivery	1	Integer	Indicator to signify if an MC delivery receipt, manual ACK, delivery ACK, or an intermediate notification is required.
replace_if_present_flag	1	Integer	Flag indicating if the submitted message should replace an existing one.
data_coding	1	Integer	Defines the encoding scheme of short_message user data.
sm_default_msg_id	1	Integer	Indicates the short message to send from a list of predefined short messages stored on the MC.
sm_length	1	Integer	Length in octets of the short_message user data.
short_message	Var. 0-255	Octet string	Short message user data.
Message Submission TLVs	Var.	TLV	

Table 8.8   submit_sm_resp PDU

Field Name	Size	Type	Description
command_length	4	Integer	Overall length of the PDU
command_id	4	Integer	0x80000004
command_status	4	Integer	Indicates outcome of submit_sm request.
sequence_number	4	Integer	Set to sequence number of original submit_sm PDU.
message_id	Var. max 65	C-octet string	This field contains the SMSC message ID of the submitted message. It may be used at a later stage to query the status of the message or to cancel or replace it.
Message Submission TLVs	Var.	TLV	

the submit_sm and gives to the SMPP client a unique identifier for the submitted short message in the field message_id. This message_id may be used in further commands such as cancel_sm and query_sm, for instance, to cancel the short message submission or to query the SMSC function the status of the delivery of the SMS.

8.5.3   Other SMPP Operations

For a complete description of all SMPP operations, refer to [1].

8.6   GSM IS-41 Interworking Through SMPP

The SMPP protocol has been designed to connect clients either to GSM MCs or to CDMA/TDMA MCs. It is then possible to open an interface between networks relying on different wireless technologies.

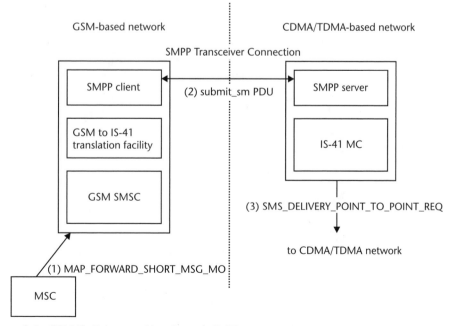

Figure 8.4   GSM IS-41 interworking through SMPP.

A GSM SMSC may open a transceiver connection to an IS-41-based MC and then offer its subscribers the ability to reach CDMA/TDMA-based MSs. To do that, the GSM SMSC must have a translation capability to translate a MAP_FORWARD_SHORT_MSG Indication SMS-MO into a submit_sm command compatible with the IS-41 specification. We then reach a possible architecture for a GSM/IS-41 interworking facility, as depicted in Figure 8.4.

Reference

[1]   *Short Message Peer-to-Peer Protocol Specification Version 5.0*, http://www.smpp.org.

## CHAPTER 9

MMS Interworking

The one who falls in a fight for liberty never dies.
—Christo Botev

9.1 Introduction

MMS technology has been introduced with the development of GPRS networks and handsets (Figure 9.1). GPRS allows any handset to exchange data with a network entity through a packet-designed protocol on a multiple *traffic channel* (TCH) circuit. Then, depending on the network and handset capabilities, the throughput of data exchange can reach up to 64 Kbps. Typically, on most deployed GSM networks providing GPRS technology and also on most GPRS handsets, the bit rate is approximately 40 Kbps. (Handheld phones use only 4 TCH among all those available.)

With this new technology it became possible for an MS to send more sophisticated data than SMSs to the network entities. MMS technology was introduced to provide GSM subscribers with the ability to send messages that are a combination of text, images or video, and sounds.

9.2 Standard Model for MMS Sending and Receiving

MMS technology introduces the entities discussed in the following sections in the *multimedia messaging service environment* (MMSE), as depicted in Figure 9.1.

9.2.1 MMS Relay/Server

The MMS relay/server (Figure 9.2) is a network entity or application controlled by the MMS provider. It brings to the network the ability to transfer messages, store them, or provide any specific operations needed by the mobile environment. This entity transcodes and delivers messages to mobile subscribers. The server provides the store and store-and-forward architecture.

9.2.2 MMS User Databases

The user databases that the MMS may have access to contain records such as subscription data and user profiles.

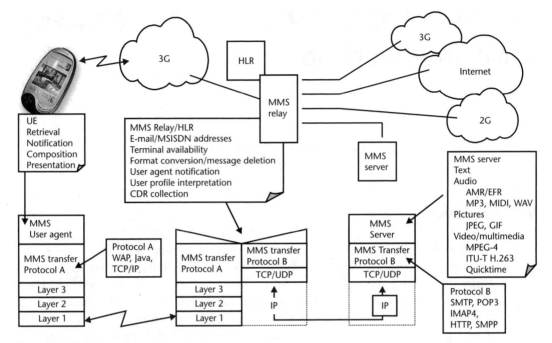

Figure 9.1   MMS service environment.

9.2.3   MMS User Agent

The MMS user agent provides the following application layer functionalities:

- Retrieval of MMs (initiate MM delivery to the MMS user agent);
- The MMS user agent may provide additional application layer functionalities such as MM composition, submission, presentation;
- Presentation of notifications to the user;
- Signing of an MM on an end user-to-end user basis;
- Decryption and encryption of an MM on an end user-to-end user basis;
- All aspects of storing MMs on the terminal;
- Handling of external devices;
- User profile management.

This optional list of additional functionalities of the MMS user agent is not exhaustive.

9.2.4   MMS VAS Applications

MMS *valued-added services* (VAS) applications provides MMS technology with some specific applications such as these:

- Prepaid;
- Voice mail alert;
- Call detail record mediation;

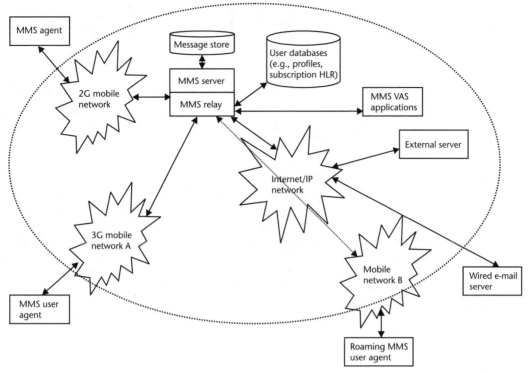

Figure 9.2   MMS network architecture.

- Mobile Internet access;
- Ringtone download;
- Mobile antispam filtering;
- Mobile e-mail;
- Over-the-air provisioning.

9.3   Standard Protocols for MMS

MMS technology has introduced various protocols by which the involved network entities can exchange information, as depicted in Figure 9.3. This architecture introduces the MM1 through MM8 protocols used in the MMSE.

9.3.1   MM1 Protocol over WAP

The MM1 protocol is used by any MMS user agent (e.g., MMS compatible cell phones) to submit or retrieve an MMS to or from the MMS relay/server. This protocol is based either on the WAP protocol for mobile stations or on HTTP for Internet-based user agents. The MM1 protocol is based on the use of MMS PDUs, which are passed in the content section of WSP or HTTP messages, with the content type of these messages being set to application/vnd.wap.mms-message. The MMS PDU consists of a header section and optionally a message body (depending on the PDU being sent). Here are several MM1 PDUs used by the protocol:

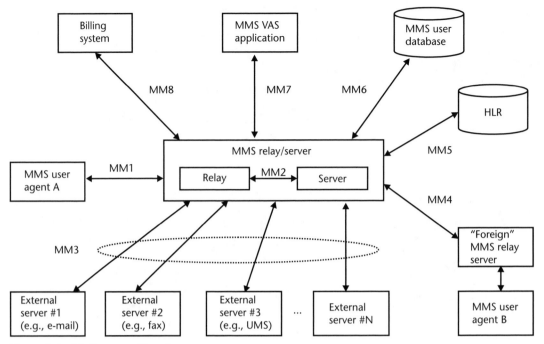

Figure 9.3 MMS protocol interfaces.

- M-Send.req (Originator → MMSC)
- M-Send.cnf (Originator ← MMSC)
- M-Notification.ind (MMSC → Receiver)

Immediate Retrieval	**Delayed Retrieval**
WSP GET.req (MMSC ← Receiver)	M-NotifyResp.ind (MMSC ← Receiver)
M-Retrieve.conf (MMSC → Receiver)	WSP GET.req (MMSC ← Receiver)
M-NotifyResp.ind (MMSC ← Receiver)	M-Retrieve.conf (MMSC → Receiver)
	M-Acknowledge.req (MMSC ← Receiver)

M-Delivery.ind (Originator MMSC)

9.3.1.1 Building an MMS PDU

To build a MMS PUD, we must first must build the MMS PDU header. Here is an example of a header:

    X-Mms-Message-Type : m-send-req
    X-Mms-Transaction-ID : 0123456789
    X-Mms-Version : 1.0
    From : +1234567/TYPE PLMN

To : +4567890/TYPE PLMN

Subject : Test Message

Content-Type : application/vnd.wap.multipart.related ;

type= "application/smil";

start= "<0000>"

All fields will be encoded. For instance, X-Mms-Message-Type is encoded 8Ch, m-send-req is encoded 80h, and so on.

After the header has been built, one must encode the PDU body with the information requested. This will lead to a binary data field that will be encapsulated into a WAP request that is forwarded to the MMSC through a WAP gateway.

9.3.1.2 Examples of Transaction Flows

Our first example depicted in Figure 9.4 shows the sending of an MMS message between an originating MS to its MMS proxy relay. The second example shows an immediate retrieval on the receiver side and is depicted in Figure 9.5. The last example shows the message flow of a deferred retrieval of MMS content on the receiver side, as depicted in Figure 9.6.

9.3.2 MM1 over M-IMAP

The *Mobile Internet Message Access Protocol* (M-IMAP) can be used as a transport layer for the MM1 interface through which an MMS user agent interfaces with the MMS relay/server entity. A typical MM1 over M-IMAP transaction flow is depicted in Figure 9.7.

Tables 9.1 through 9.4 show examples of various MAP information elements for MM1_Submit.REQ and MM1_Submit.RES. (For a complete description of M-IMAP MM1 commands, refer to [1].)

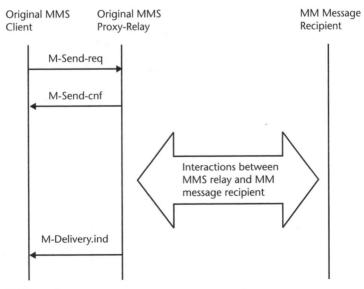

Figure 9.4 MMS sending.

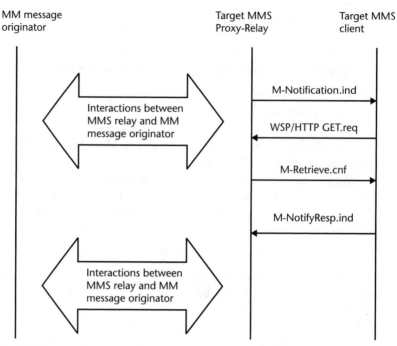

Figure 9.5 MMS immediate retrieval.

9.3.3   MM4 Protocol

The MM4 protocol is the interface between different operators' MMSCs. This protocol is based on the well-known internet protocol SMTP (standard e-mail protocol)

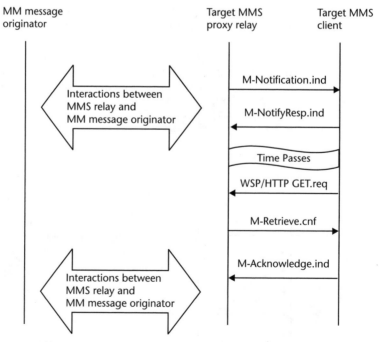

Figure 9.6 MMS deferred retrieval.

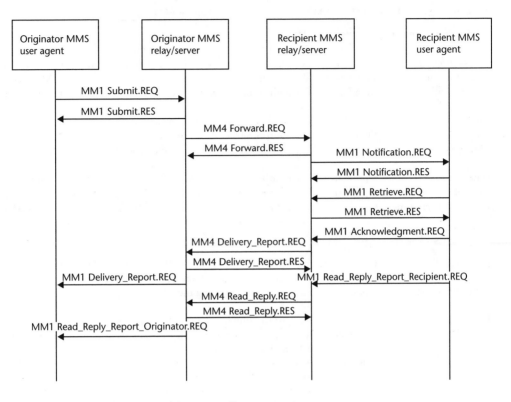

Figure 9.7 MM1 over M-IMAP transaction flow.

and provides MMSC with the capabilities of sending, retrieving, or forwarding MMS between network MMSC entities through the Internet network. Tables 9.5 through 9.8 show examples of various MM4 information elements for MM4_Forward.REQ and MM4_Forward.RES. (For a complete description of the MM4 protocol, refer to [2].)

9.3.4 MM7 Protocol

The MM7 protocol is used between the MMS relay/server and a *VAS applications provider* (VASP). An example of a standard MM7 transaction flow is depicted in Figure 9.8, and Tables 9.9 and 9.10 give examples of MM7_Submit.REQ and MM7_Submit.RES information elements, respectively. (For a complete description of the MM7 protocol, refer to [3].)

9.4 MMS Interworking Architectures Using a Third Party

The standard MMSCs requires anybody who desires to send and receive an MMS to register his or her phone number on his or her own MMSC. The MMS interworking procedure requires the sending MMSC (i.e., the originating MMSC) to provide a table of partner MMSCs. After an analysis of the destination number, the MMS interworking procedure will transfer the MMS to the terminating MMSC

Table 9.1   Example of MM1 M-IMAP Command: MM1_Submit.REQ

Information Element	Presence	Description
Message type	Mandatory	Identifies this message as MM1_Submit.REQ.
Transaction ID	Mandatory	The identification of the MM1_submit.REQ/MM1_submit.RES pair.
MMS Version	Mandatory	Identifies the version of the interface supported by the MMS UA.
Recipient address	Mandatory	The address of the recipient(s) of the MM; multiple addresses are possible.
Content type	Mandatory	The content type of the MM's content.
Sender address	Optional	The address of the MM originator.
Message class	Optional	The class of the MM (e.g., personal, advertisement, information service).
Date and time	Optional	The time and date of the submission of the MM (time stamp).
Time of expiry	Optional	The desired time of expiry for the MM or reply-MM (time stamp).
Earliest delivery time	Optional	The earliest desired time of delivery of the MM to the recipient (time stamp).
Delivery report	Optional	A request for delivery report.
Reply-Charging	Optional	A request for reply-charging.
Reply-Deadline	Optional	In case of reply-charging, the latest time of submission of replies granted to the recipient(s) (time stamp).
Reply-Charging-Size	Optional	In case of reply-charging the maximum size for reply-MM(s) granted to the recipient(s).
Priority	Optional	The priority (importance) of the message.
Sender visibility	Optional	A request to show or hide the sender's identity when the message is delivered to the recipient.
Store	Optional	A request to store a copy of the MM in the user's MMBox, in addition to the normal delivery of the MM.
MM State	Optional	The value to set in the MM State information element of the stored MM, if the store option is present.
MM Flags	Optional	One or more MM Flag keywords to set in the MM Flags information element of the stored MM, if the store option is present.
Read reply	Optional	A request for read reply report.
Subject	Optional	The title of the whole multimedia message.
Reply-Charging-ID	Optional	In case of reply-charging, when the reply-MM is submitted within the MM1_submit.REQ this is the identification of the original MM that is replied to.
Content	Optional	The content of the multimedia message.

(Figure 9.9). This process is quite different from that of SMS, in which the SMSC may deliver SMSs directly to a destination even if it belongs to another network. The interface between the two MMSCs is the MM4 protocol that we discussed earlier and it uses SMTP/IP, the standard e-mail transfer protocol.

MNP is a key issue in this process. For a U.K., Belgium, Swede, Hong Kong, Australian, and so on number, we cannot tell which is the destination operator with merely a table lookup—so your MMSC cannot determine the destination domain name to which to transfer the MMS!

One solution for this situation was introduced in Chapter 4. The MMSC can interrogate a *domain name server* (DNS) with the destination number to retrieve, through the MM5 protocol, the MMSC of the destination handset and then submit the MMS to it through the MM4 protocol. Refer to Section 4.6.1 for details.

Another solution is as follows: For any destination phone that is not registered in your MMSC [and you want to use a third party for this, *MMS interoperability operator* (MMS-IO)], you send the MMS to the third party's own MMSC, which

Table 9.2   Mapping of MM1_Submit.REQ to the RFC2822 Header

Information Element	Type	MM1_submit.REQ Headers
MMS MM1 Version	M	X-Mms-MM1-Version: MM1_Submit.REQ
Message type	M	X-Mms-MM1-Message-Type
Transaction ID	M	X-Mms-Transaction-ID
Message ID	M	X-Mms-Message-ID
Recipient address	M	One or more of To:, Cc:, Bcc:
Content type	M	Content-Type
Sender address	O	From:
Message class	O	X-Mms-Message-Class
Date and time	O	Date
Time of expiry	O	X-Mms-Expiry
Earliest delivery time	O	X-Mms-Delivery-Time
Delivery report	O	X-Mms-Delivery-Report
Reply-Charging	O	X-Mms-Reply-Charging
Reply-Charging size	O	X-Mms-Reply-Charging-Size
Reply-Deadline	O	X-Mms-Reply-Deadline
Priority	O	X-Mms-Priority
Sender visibility	O	X-Mms-Sender-Visibility
Read reply	O	X-Mms-Read-Reply
Subject	O	Subject:
Reply-Charging ID	O	X-Mms-Reply-Charging-ID
Content	O	E BODY

Table 9.3   MM1_Submit.RES

Information Element	Presence	Description
Message type	Mandatory	Identifies this message as MM1_Submit.RES.
Transaction ID	Mandatory	The identification of the MM1_Submit.REQ/MM1_Submit.RES pair.
MMS version	Mandatory	Identifies the version of the interface supported by the MMS relay/server.
Request status	Mandatory	The status of the MM submit request.
Request status text	Optional	Description that qualifies the status of the MM submit request.
Message ID	Conditional	The identification of the MM if it is accepted by the originating MMS relay/server.
Store status	Conditional	If the store request was present in MM1_Submit.REQ, the status of the store request.
Store status text	Optional	The explanatory text corresponding to the store status, if present.
Stored message reference	Conditional	If the store request was present in MM1_Submit.REQ, the message reference to the newly stored MM.

Table 9.4   Mapping of MM1_Submit.RES to the RFC2822 Header

Information Element	Type	MM1_Submit.RES Headers
Message type	M	X-Mms-MM1-Message-Type: MM1_Submit.RES
Transaction ID	M	X-Mms-Transaction-ID
MMS version	M	X-Mms-Version
Request status	M	X-Mms-Request-Status
Request status text	O	X-Mms-Request-Status-text
Message ID	M	X-Mms-Message-ID

integrates MNP handling (over an Internet connection) using the standard MM4 protocol.

Table 9.5   Example of MM4 Command: MM4_Forward.REQ

Information Element	Presence	Description
3GPP MMS Version	Mandatory	The MMS version of the originator MMS relay/server as defined by the present document.
Message type	Mandatory	The type of message used on reference point MM4: MM4_Forward.REQ.
Transaction ID	Mandatory	The identification of the MM4_Forward.REQ/ MM4_Forward.RES pair.
Message ID	Mandatory	The identification of the MM.
Recipient address	Mandatory	The address(es) of the MM recipient(s); multiple addresses are possible.
Sender address	Mandatory	The address of the MMS user agent that most recently handled the MM, that is, the one that either submitted or forwarded the MM. If the originator MMS user agent has requested her address to be hidden from the recipient, her address will not be provided to the recipient.
Content type	Mandatory	The content type of the MM's content.
Message class	Conditional	The class of the MM (e.g., personal, advertisement, information service) if specified by the originating MMS user agent.
Date and time	Mandatory	The time and date of the most recent handling (i.e., either submission or forwarding) of the MM by an MMS user agent (time stamp).
Time of expiry	Conditional	The desired time of expiry for the MM if specified by the originator MMS user agent (time stamp).
Delivery report	Conditional	A request for delivery report if the originator MMS user agent has requested a delivery report for the MM.
Originator R/S delivery report	Conditional	A request for delivery report that, when set to "Yes," means the originator MMS relay/server has requested a delivery report for the MM. Interpret as "No" in the absence of this information element.
Priority	Conditional	The priority (importance) of the message if specified by the originator MMS user agent.
Sender visibility	Conditional	A request to show or hide the sender's identity when the message is delivered to the MM recipient if the originator MMS user agent has requested her address to be hidden from the recipient.
Read reply	Conditional	A request for a read-reply report if the originator MMS user agent has requested a read-reply report for the MM.
Subject	Conditional	The title of the whole MM if specified by the originating MMS user agent.
Acknowledgment request	Optional	Request for MM4_Forward.RES.
Forward_counter	Conditional	A counter indicating the number of times the particular MM was forwarded.
Previously-sent-by	Optional	If forwarded, this information element contains one or more address(es) of MMS user agent(s) that handled (i.e., forwarded or submitted) the MM prior to the MMS user agent whose address is contained in the sender address information element. The order of the addresses provided is be marked. The address of the originating MMS user agent, if present, is be marked.
Previously-sent-date-and-time	Optional	The date(s) and time(s) associated with submission and forwarding event(s) prior to the last handling of the MM by an MMS user agent (time stamps).

If the destination phone has been previously used and recognized as an MMS-enabled phone, the third party sends to this cell phone (with an MMS notification) directly (no need to use a terminating MMSC, especially since there may not be one available) if it is known that the receiving network can connect to Internet using the standard WAP setup. (It is completely impractical in terms of business to ask the recipient to change its standard WAP setup.)

Table 9.6 MM4 Mapping to the RFC2822 Headers for MM4_Forward.REQ

Information Element	MM4_Forward.REQ Headers
MMS version	X-Mms-MMS-Version:
Message type	X-Mms-MM1-Message-Type
Transaction ID	X-Mms-Transaction-ID
Message ID	X-Mms-Message-ID
Recipient address	One or more of To:, Cc:, Bcc:
Content type	Content-Type
Sender address	From:
Message class	X-Mms-Message-Class
Sender address	From:
Message class	X-Mms-Message-Class
Date and time	Date
Time of expiry	X-Mms-Expiry
Delivery report	X-Mms-Delivery-Report
Priority	X-Mms-Priority
Sender visibility	X-Mms-Sender-Visibility
Read reply	X-Mms-Read-Reply
Subject	Subject:
Acknowledgment request	X-Mms-Ack-Request
Forward counter	X-Mms-Forward-Counter
Previously sent by	X-Mms-Previously-Sent-By
Previously-sent-date-and-time	X-Mms-Previously-Sent-Date-And-Time
Content	E BODY

Table 9.7 MM4_Forward.RES

Information Element	Presence	Description
3GPP MMS version	Mandatory	The MMS version of the recipient MMS relay/server as defined by the present document.
Message type	Mandatory	The type of message used on reference point MM4: MM4_Forward.RES.
Transaction ID	Mandatory	The identification of the MM4_Forward.REQ/ MM4_Forward.RES pair.
Message ID	Mandatory	The message ID of the MM, which has been forwarded within the corresponding MM4_Forward.REQ.
Request status	Mandatory	The status of the request to route forward the MM.
Request status text	Optional	Status text corresponding to the request status.

Table 9.8 MM4 Mapping to the RFC2822 Headers for MM4_Forward.RES

Information Element	MM4_Forward.RES Headers
MMS version	X-Mms-MMS-Version
Message type	X-Mms-MM1-Message-Type : MM4_Forward.RES
Transaction ID	X-Mms-Transaction-ID
Message ID	X-Mms-Message-ID
Request status	X-Mms-Request-Status-Code

Another method by which a third party can deliver MMS-MT is to use the MM4 protocol if it has an agreement with the receiving network. Alternatively, the agreement could allow direct sending merely by registering the third-party MMSC in the recipient network's WAP gateway! We consider this is a better method because it saves on resources.

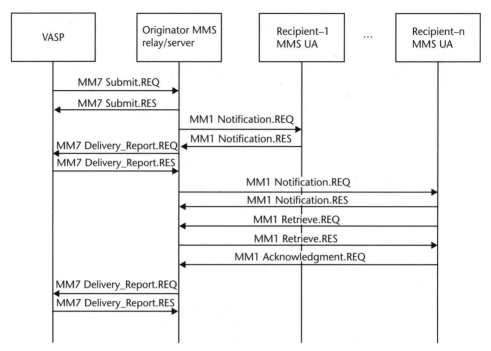

Figure 9.8   MM7 transaction flow.

If the third party does not know the characteristic of the destination cell phone, it will send a first normal notification by a text MMS giving a unique URL (with a cyclic code) to retrieve the MMS (in HTML form) with an ordinary browser (from a PC). It will also send it a second MMS notification with a different URL (MMS form). In MMSC, the MMS exists both in MMS/WAP form and also, after conversion, in ordinary HTML format.

When the user retrieves the MMS using an MMS phone, the third party recognizes it and the phone number is automatically registered for the next time. Regardless of the retrieval mode (MMS phone or PC) used, the two-form MMS is deleted. Also, the third party can obtain the terminal type (Ericsson, Nokia, and so on) and also has the choice of dynamically converting from one format to the other. If the third party cannot deliver the MMS-MT (i.e., if it is not retrieved after 24 hours), it will send a delivery failure back to the originating SMSC.

In the business model, the originating operator pays the third party a fixed fee for each MMS that is sent to it. The receiving operator has one (once the receiving cell phone has been MS registered) or two SMS-MT deposit charges from the third party. In most cases the MMS-IO will need to be connected to the other MMSCs through the GRX network (the Internet of the mobile operations used for GPRS roaming and MMS interconnection).

9.5   Setting Up the MMS Profile in the Cell Phone

To set up the MMS profile (Figure 9.10), you can use the OTA (the MMS profile is downloaded using SMS as a transport service) of your operator; otherwise, a good

Table 9.9   Example of MM7 Command: MM7_Submit.REQ

Information Element	Presence	Description
Transaction ID	Mandatory	The identification of the MM7_Submit.REQ/MM7_Submit.RES pair.
Message type	Mandatory	Identifies this message as a MM7_Submit request.
MM7 version	Mandatory	Identifies the version of the interface supported by the VASP.
VASP ID	Optional	Identifier of the VASP for this MMS relay/server.
VAS ID	Optional	Identifier of the originating application.
Sender address	Optional	The address of the MM originator.
Recipient address	Mandatory	The address of the recipient MM. Multiple addresses are possible as is the use of an alias that stands in for a distribution list. It is possible to mark an address to be used only for informational purposes. It is possible to mark that a recipient address be provided in encrypted or obfuscated format; for example, the address could have originally been provided in encrypted or obfuscated form in an associated MM7_Deliver.REQ.
Service code	Optional	Information supplied by the VASP that may be included in charging information. The syntax and semantics of the content of this information are out of the scope of this specification.
Linked ID	Optional	This identifies a correspondence to a previous valid message delivered to the VASP.
Message class	Optional	Class of the MM (e.g., advertisement, information service, accounting).
Date and time	Optional	The time and date of the submission of the MM (time stamp).
Time of expiry	Optional	The desired time of expiry for the MM (time stamp).
Earliest delivery time	Optional	The earliest desired time of delivery of the MM to the recipient (time stamp).
Delivery report	Optional	A request for delivery report.
Read reply	Optional	A request for confirmation via a read-report to be delivered as described in an earlier section.
Reply-Charging	Optional	A request for reply-charging.
Reply-Deadline	Optional	In case of reply-charging the latest time of submission of replies granted to the recipient(s) (time stamp).
Reply-Charging size	Optional	In case of reply-charging the maximum size for reply-MM(s) granted to the recipient(s).
Priority	Optional	The priority (importance) of the message.
Subject	Optional	The title of the whole multimedia message.
Adaptations	Optional	Indicates if VASP allows adaptation of the content (defaults to "True").[1]
Charged party	Optional	An indication of which party is expected to be charged for an MM submitted by the VASP, for example, the sending party, receiving party, both parties or neither, or a third party.
Content type	Mandatory	The content type of the MM's content.
Content	Optional	The content of the multimedia message
Message distribution indicator	Optional	If set to "False," the VASP has indicated that content of the MM is not intended for redistribution. If set to "True," the VASP has indicated that content of the MM can be redistributed.[2]
Charged party ID	Optional	The address of the third party that is expected to pay for the MM.

[1] From Release 6 onward, in case of misalignment between the value assigned to adaptations and DRM protection rules, the latter shall prevail.
[2] From Release 6 onward, in case of misalignment between the value assigned to MDI and DRM protection rules, the latter shall prevail.

level of understanding is required. Figure 9.10 represents the WAP profile in the cell phone and the various possible paths that are used to retrieve a MMS.

9.5.1   Data Access Profile

Even if you have GPRS you can use a GSM circuit mode for the data services. In such a case, you set up the telephone number of the circuit mode *network access sys-*

Table 9.10   MM7_Submit.RES

Information Element	Presence	Description
Transaction ID	Mandatory	The identification of the MM7_Submit.REQ/ MM7_Submit.RES pair.
Message type	Mandatory	Identifies this message as a MM7_Submit response.
MM7 version	Mandatory	Identifies the version of the interface supported by the MMS relay/server.
Message ID	Conditional	If status indicates success, then this contains the MMS relay/server-generated identification of the submitted message. This ID may be used in subsequent requests and reports relating to this message.
Request status	Mandatory	Status of the completion of the submission; no indication of delivery status is implied.

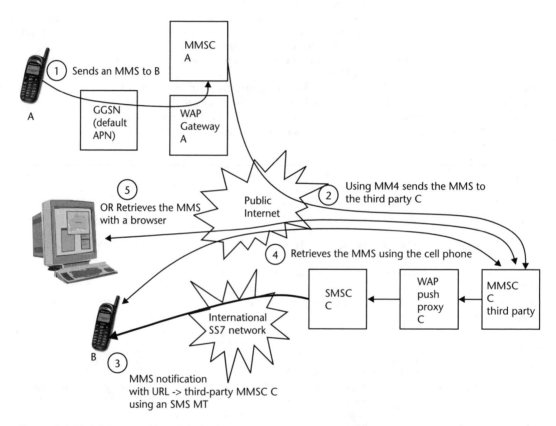

Figure 9.9   MMS interworking with third party.

tem (NAS) of your operator or the telephone number of a dial-up third-party Internet access provider (such as 08 36 01 93 01 in Figure 9.10).

You then set up the GPRS access by setting the *access provider name* (APN). This could again be a third-party WAP gateway (not that of your operator) provided that the GGSN (the access gateway) of your operators allows you to use a third-party WAP gateway. In Figure 9.10, we assume a very open system when the cell phone is free to use a third-party WAP gateway.

Next you set up the default access, GPRS most often, or circuit mode, which is limited to 14 Kbps. For all data services, Internet access, and MMS sending or

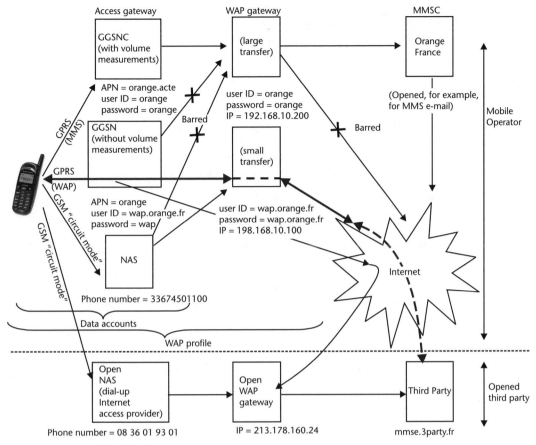

Figure 9.10   MMS profile.

retrieval, you use this default data access profile. Note that a MMSC service can be provided with only a circuit data mode, without GPRS, but it will be quite costly and slow if the user is roaming.

9.5.2   MMSC Profile

Lastly, one enters the name of the MMSC such as mmse.toto.fr. The mobile operators provide various restrictions for commercial reasons, as shown by crosses (+) in Figure 9.10:

- Their WAP gateway does not provide large-volume data transfer to the public Internet, only to their WAP portal or their own MMSC.
- The GGSN that is used for the fixed monthly charge data services cannot access the large transfer WAP gateway.

However, keep in mind that an MMSC service can be offered in a network

- Without GPRS, just the circuit data mode;
- Without a NAS (using a public Internet access);

- Without a WAP gateway;
- Using a third-party MMSC (an MMS-IO).

References

[1]   *X.S0016-311 v1.0.0: MMS MM1 Stage 3 Using M-IMAP for Message Submission and Retrieval.*
[2]   *X.S0016-340 v1.0: MMS MM4 Stage 3 Intercarrier Interworking.*
[3]   *X.S0016-370 v1.0: MMS MM7 VASP Interworking Stage 3 Specification.*

## CHAPTER 10

Optimal Routing Algorithms for an SMS Interworking Network

J'ai rêvé la nuit verte aux neiges éblouies
Baiser montant aux yeux des neiges
avec lenteurs
La circulation des sèves inouïes
Et l'éveil jaune et bleu des phosphores
chanteurs

On innocent nights, I'd dream of brilliant
snows
Slowly mounting kiss to the rolling seas
As though driven by rising sap
And the blue and yellow awakening of
the singing phosphorescence

—Arthur Rimbaud, *Le Bateau Ivre* (*The Drunken Ship*)

10.1 Maximizing the Margin of an SMS Interworking Network

In this chapter we want to discuss optimal routing of the traffic, so as to maximize the margin of an SMS interworking network. It is assumed that the reader knows only the vocabulary of graph theory and network flows [1, 2]. For Section 10.6, some background in mathematical programming theory [2, 3] is required.

10.2 Enumerating All Loopless Paths with the Latin Multiplication Algorithms

Let us consider the graph of Figure 10.1, where S is the set of customers who want to send SMS to all the mobile networks S'. As the basis of a crude or sophisticated optimization method, we want first to have an algorithm that computes all the paths from every origin (customers sending SMS) to every destination [various mobile networks (SMS-MT) or content providers]. In Figure 10.1, the indices i,j,k denote the various nodes, also called routers, of an SMS interworking network. It shows that several paths, using different nodes, can be used for the traffic demand from a customer $s \in S$ to a destination $s' \in S'$.

Remember that an oriented graph (the type we are concerned with in this chapter) is an application of a set X called *vertices* into itself. The elements U of the application are called *edges* of the graph. And the graph is noted $G = (X,U)$. A path is a list of vertices $a, .. i_k, i_{k+1}, .., c, ..d$ such that any of its edges $(i_k, i_{k+1}) \in U$. Here the sending customers (noted s_i), the destination networks (noted s'_j), and the routers of

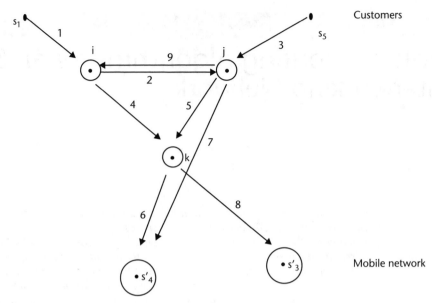

Figure 10.1 Graph model of a simple SMS interworking network.

the SMS network (noted $i, j, ..k$) are the vertices of the graph. An edge (a,b) exists between vertex a and b if an SS7 route exists from a to b.

Counting by hand because it is a small graph, we find that the list of all 10 paths in the graph is as follows: (s_1, i, k, s'_4) s_1 to s'_4, path 1 (from customer s_1 to destination s'_4 through routers i then k). The s_1 uses node i for his connection to the SS7 interworking network so (s_1, i) exists. There is an SCCP connection between router i and router k so (i,k) exists, and s_4 is a roaming partner of k so (k, s'_4) exists:

$$\left(s_1, i, j, k, s'_4\right) s_1 \text{ to } s'_4, \text{ path 2}$$

$$\left(s_1, i, j, s'_4\right) s_1 \text{ to } s'_4, \text{ path 3}$$

$$\left(s_5, j, s'_4\right) s_5 \text{ to } s'_4, \text{ path 4}$$

$$\left(s_5, j, k, s'_4\right) s_5 \text{ to } s'_4, \text{ path 5}$$

$$\left(s_1, i, j, k, s'_3\right) s_1 \text{ to } s'_3, \text{ path 6}$$

$$\left(s_1, i, k, s'_3\right) s_1 \text{ to } s'_3, \text{ path 7}$$

$$\left(s_5, j, i, k, s'_3\right) s_5 \text{ to } s'_3, \text{ path 8}$$

$$\left(s_5, j, i, k, s'_4\right) s_5 \text{ to } s'_4, \text{ path 9}$$

$$\left(s_5, j, k, s'_3\right) s_5 \text{ to } s'_3, \text{ path 10}$$

A systematic method is available that computes all loopless path in a network even if it has paths with loops as illustrated in Figure 10.1, called *Latin multiplication* [4].

Let A be the square matrix 7×7 (there are 7 nodes in the graph) that represents the edges of the graphs. Element A is an element of the set M of 7×7 matrices defined over elements, line u, column v that are a set of paths from u to v *without loops*:

$$
A = \begin{array}{c|ccccccc}
 & s_1 & s_5 & i & j & k & s_4' & s_3' \\
\hline
s_1 & 0 & 0 & (s_1,i) & 0 & 0 & 0 & 0 \\
s_5 & 0 & 0 & 0 & (s_5,j) & 0 & 0 & 0 \\
i & 0 & 0 & 0 & (i,j) & (i,k) & 0 & 0 \\
j & 0 & 0 & (j,i) & 0 & (j,k) & (j,s_4') & 0 \\
k & 0 & 0 & 0 & 0 & 0 & (k,s_4') & (k,s_3') \\
s_4' & 0 & 0 & 0 & 0 & 0 & 0 & 0 \\
s_3' & 0 & 0 & 0 & 0 & 0 & 0 & 0
\end{array}
$$

Define an operation in the set M, denoted $*$, which resembles the classical matrix product but is not a commutative operation. Take $B \in M$ ($A \in M$ because it has simple edge paths). Define $C = B * A$ by:

$$C_{uv} = \{\text{set of distinct paths from } u \text{ to } v \text{ of the form } B_{ul} \,|\, A_{lv} \text{ such that } v \text{ is not} \in B_{ul}\}$$

where "|" denotes the concatenation of path B_{ul} finishing at node l and path A_{lv} starting at l. So the path of C_{uv} may have one more edge than those of B_{ul} and there is no loop, so $C \in M$ and $*$ is a closed operation in the set M. We get

$$
A * A = A^2 =
$$

$$
\begin{array}{c|ccccccc}
s_1 & 0 & 0 & (s_1,i) & (s_1,i,j) & (s_1,i,k) & 0 & 0 \\
\\
s_5 & 0 & 0 & 0 & (s_5,j) & (s_5,j,k) & (s_5,j,s_4') & 0 \\
\\
i & 0 & 0 & 0 & (i,j) & \begin{array}{c}(i,k)\\(i,j,k)\end{array} & (i,j,s_4') & (i,k,s_3') \\
\\
j & 0 & 0 & (j,i) & 0 & (j,k) & \begin{array}{c}(j,s_4')\\(j,k,s_4')\end{array} & (j,k,s_3') \\
\\
k & 0 & 0 & 0 & 0 & 0 & (k,s_4') & (k,s_3') \\
\\
s_4' & 0 & 0 & 0 & 0 & 0 & 0 & 0 \\
\\
s_3' & 0 & 0 & 0 & 0 & 0 & 0 & 0
\end{array}
$$

For example, let us compute $A^3_{s_1,s_4'}$

$$\{0 \,|\, 0,0 \,|\, 0,(s_1,i)| \; 0,(s_1,i,j) \,|\, (j,s_4'),(s_1,i,k) \,|\, (k,s_4'),0 \,|\, 0,0 \,|\, 0\}, \text{ that is,}$$

$$\{(s_1,i,j,s_4'),(s_1,i,k,s_4')\}$$

We get

$A^2 * A = A^3 =$

$$
\begin{array}{c|ccccccc}
s_1 & 0 & 0 & (s_1,i) & (s_1,i,j) & (s_1,i,k) & (s_1,i,j,s_4') & (s_1,i,k,s_3') \\
 & & & & & & & (s_1,i,k,s_4') \\
s_5 & 0 & 0 & 0 & (s_5,j) & (s_5,j,k) & (s_5,j,s_4') & (s_5,j,k,s_3') \\
 & & & & & & & (s_5,j,k,s_4') \\
i & 0 & 0 & 0 & (i,j) & (i,k) & (i,k,s_4') & (i,k,s_3') \\
 & & & & (i,j,k) & (i,j,s_4') & (i,j,k,s_3') \\
 & & & & & (i,j,k,s_4') \\
j & 0 & 0 & (j,i) & 0 & (j,k) & (j,s_4') & (j,k,s_3') \\
 & & & & (j,i,k) & (j,k,s_4') \\
k & 0 & 0 & 0 & 0 & 0 & (k,s_4') & (k,s_3') \\
s_4' & 0 & 0 & 0 & 0 & 0 & 0 & 0 \\
s_3' & 0 & 0 & 0 & 0 & 0 & 0 & 0 \\
\end{array}
$$

which gives

$A^3 * A = A^4 =$

$$
\begin{array}{c|ccccccc}
s_1 & 0 & 0 & (s_1,i) & (s_1,i,j) & (s_1,i,k) & (s_1,i,j,s_4') & (s_1,i,k,s_3') \\
 & & & & (s_1,i,j,k) & (s_1,i,k,s_4') & (s_1,i,j,k,s_3') \\
 & & & & & (s_1,i,j,k,s_4') \\
s_5 & 0 & 0 & (s_5,j,i) & (s_5,j) & (s_5,j,k) & (s_5,j,s_4') & (s_5,j,k,s_3') \\
 & & & & (s_5,j,i,k) & (s_5,j,k,s_4') & (s_5,j,i,k,s_3') \\
 & & & & & (s_5,j,i,k,s_4') \\
i & 0 & 0 & 0 & (i,j) & (i,k) & (i,k,s_4') & (i,k,s_3') \\
 & & & & (i,j,k) & (i,j,s_4') & (i,j,k,s_3') \\
 & & & & & (i,j,k,s_4') \\
j & 0 & 0 & (j,i) & 0 & (j,k) & (j,s_4') & (j,k,s_3') \\
 & & & & (j,i,k) & (j,k,s_4') & (j,i,k,s_3') \\
 & & & & & (j,i,k,s_4') \\
k & 0 & 0 & 0 & 0 & 0 & (k,s_4') & (k,s_3') \\
s_4' & 0 & 0 & 0 & 0 & 0 & 0 & 0 \\
s_3' & 0 & 0 & 0 & 0 & 0 & 0 & 0 \\
\end{array}
$$

If we compute A^5, we find $A^5 = A^4$ (and in general $A^n = A^4$ for $n > 4$).

We then find all of the paths from s_1 to s'_4 in the element $A^4_{s_1 s'_4}$. All 10 paths from s to s' are in bold type in the preceding matrix. Note that the algorithm guarantees that there are no loops in the elementary paths that are found. However, nothing prevents us from finding the same path several times before they are made the same by the operation C_{uv}.

10.3 Shortest Path: Djsktra Algorithm

We find the shortest path from a given s to a given s. The Djsktra algorithm is efficient [5] when we look for all paths from one node of the graph to all others.

10.4 Least Cost Path

We find the least cost path from a given s to a given s'.

10.5 Least Trouble Path

This is the shortest path along those that terminate at s through a specific SMS termination agreement (this standard agreement is called AA 19 in the GSM association) and an agreed-on termination price for the SMS-MT. It will not be the cheapest but there is no risk of complaint.

10.6 The Best Flow Problem—Not a Classical Graph Problem

An SMS interworking operator will want to achieve global optimization:

- To satisfy customers by providing them with the agreed-upon bandwidth required to send SMS;
- To maximize the margin, the difference between what the SMS operator receives as payment from its customers and what it pays its suppliers, for example, the SS7 provider, transit partners, and termination costs;
- To avoid saturating any of its network bandwidths;
- To avoid sending any more SMS-MT traffic, which could create a complaint with operators with whom there is no agreement.

It is more general than a simple classical graph problems such as

- Finding the least cost route to deliver the SMS from one customer to all its destinations (shortest path type problems) [2].

- Finding the maximum flow of SMS from all customers to all their destination maximum multicommodity flows [1]. This is not applicable if the demand is less than the flow, we incur the traffic costs without benefit.

The best flow problem, which is nonclassical but is resolved as described later by nonlinear convex programming, is knowing

- The capacity of the network;
- The costs: traffic, transit, termination;
- The demand for SMS traffic;
- The price scheme agreed upon with all customers including penalties if the agreed-upon minimum bandwidth cannot be provided to find the optimal flows for each customer or each arc of the network.

10.6.1   Income Model for Customer Charges and Notations

For each customer $s \in S$ and each destination $s' \in S'$ (of all possible destinations), it is possible to have different prices. We will assume that the price charged is constant for each SMS sent to a given destination. If one sends more than the instant demand $d_{ss'}$ (by filling with void SMS), there will not be any additional revenue.

It is also possible to have agreed-upon penalties. That is, if the network cannot provide a minimum capacity $c_{ss'}$, it will pay a penalty proportional to the unsatisfied demand.

10.6.2   Noncontinuous Price Function Paid to the Interworking Network for an Unsatisfied Demand

Assume that the contract between customer s and the SMS interworking network is as follows: The price per SMS is only 0.01 euro if the throughput provided is less than

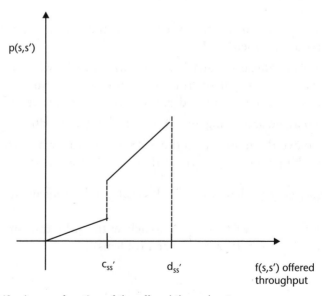

Figure 10.2   SMS price as a function of the offered throughput.

the minimum requested capacity $c_{ss'}$. The price becomes 0.03 euros when the minimum capacity is provided. This case yields a price function $p(s,s')$ as in Figure 10.2.

This function is very badly suited for finding an optimum, because it is not continuous and the gradient is 0 for $f(s,s') > d_{ss'}$. The global optimization would need to involve many integer variables (as many as $\{s,s'\}$ pairs) and is rather computationally difficult [6].

10.6.3   Continuous Concave Price Function

To compute a simple global routing solution, we must have a continuous functional with nonconstant derivatives over an interval. So, for customers who have a guaranteed level of service, the agreement is that, if the throughput is inferior to the agreed minimum $c_{ss'}$, a back payment of 0.05 euros per unsent SMS is made (Figure 10.3).

For $f(s,s') > d_{ss'}$ we smooth the function with a decreasing slope linear function. In fact, the optimum flow from s,s' will never be greater than $d_{ss'}$ because there are costs associated with that, but the algorithm needs to have nonzero derivatives. As a result, $p(s,s')$:

- Is continuous;
- Has nonzero derivatives except at $f = c_{ss'}$ and $f = d_{ss'}$ (no derivative defined at these points);
- Is a concave function of $f(s,s')$.

It is possible to compute effectively the maximum of a sum of such functions because their sum is also a concave function [2].

10.6.4   Network Model

To make the model more general, there could be different edges (i,j), corresponding to SS7 or IP connections (e.g., using SMPP; see Chapter 8), with different costs and

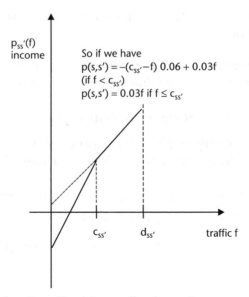

Figure 10.3   SMS price function with minimum offered capacity.

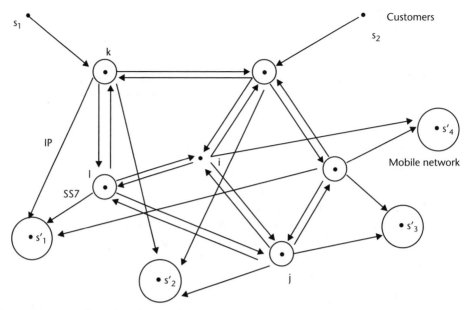

Figure 10.4 Graph model of a complex SMS interworking network.

performances. In Figure 10.4, (k,s'_1) is an edge corresponding to a router k sending SMS to the SMSC of network s'_1 using an IP-based protocol. While (l, s'_1) represents an SS7 edge, allowing the router k to send SMS-MT directly to the subscribers of network s'_1 using the procedures of Chapter 1:

- $s'_1, s_2, ..., s_p$ are customers sending SMS to
- $s'_1, s'_2, ..., s'_p$ (the mobile networks providing the SMS-MT termination)
- and b_{ij} is the capacity of edge (i,j), where S is the set of all customers sending SMS through the SMS interworking network with $s \in S$ one of them.
- S' is the set of all destinations receiving SMS from the SMS network with $s' \in S'$ one of them.

The optimization will be a multicommodity flow problem [1], in which each arc carries different flows from different origins to different destinations.

10.6.5 Mathematical Model for Optimization

For all pairs (s,s'), the graph has many paths. We want to select the best paths so that the arc capacities are not violated. We use the following process:

- Let us assume n edges in the networks and m (oriented) paths.
- Let A be the incidence matrix between the arcs and the paths, that is,

$$a_{ij} = \begin{cases} 1 \text{ if edge } i \text{ belongs to path } j \\ 0 \text{ if it does not} \end{cases}$$

- Let x_j be the flow of SMS in path j (several different paths could be used to provide the service $s \to s'$), such as in Figure 10.1, (s_1, i, k, s'_4) and (s_1, i, j, k, s'_4) for the SMS from s_1 to s'_4.
- We must have $Ax \le b$ so that the capacity of each edge i is not violated. This constraint becomes clear with this example.

10.6.5.1   Example of an Incidence Matrix Edge to Paths

In Figure 10.1, the four paths that use the edge (k, s'_4) (edge number 6) are the paths 1, 2, 5, and 9: Path 1 and 2 carry the SMS traffic from s_1 to s'_4 and path 5 carries the SMS from s_5 to s'_4. So if x_1, x_2, x_5, and x_9 represent the SMS traffic on these paths, we must have:

$$x_1 + x_2 + x_5 + x_9 \le b_6 \ (b_6 \text{ is the SMS capacity of edge number 6})$$

because line 6 of incidence matrix A is 1 1 0 0 1 0 0 0 1 0.

The traffic of edge number 6 is the sum of traffic from sender s_1 to the mobile network s'_4 (along two different paths, paths 1 and 2) and of the traffic (paths 5 and 9) from sender s_5 to the mobile network s'_4. The complete matrix A is 9×10 (9 edges and 10 paths):

$$A_{9 \times 10} =$$

	x_1	x_2	x_3	x_4	x_5	x_6	x_7	x_8	x_9	x_{10}
(s_1,i)	1	1	1	0	0	1	1	0	0	0
(i,j)	0	1	1	0	0	1	0	0	0	0
(s_5,j)	0	0	0	1	1	0	0	1	1	1
(i,k)	1	0	0	0	0	0	1	0	1	1
(j,k)	0	1	0	0	1	1	0	1	0	0
(k,s'_4)	1	1	0	0	1	0	0	0	1	0
(j,s'_4)	0	0	1	1	0	0	0	0	0	0
(k,s'_3)	0	0	0	0	0	1	1	1	0	1
(j,i)	0	0	0	0	0	0	0	0	1	1

10.6.5.2   Incidence Matrix Source × Destination to Paths

In the maximization of the margin, the flow of commodity from s to s' must be limited by the demand $d_{ss'}$. So the sum of the flows on all paths from s to s' must not be greater than $d_{ss'}$. We define the incidence matrix T such that $t_{ij} = 1$ if the path j is one of the paths between the ith pair (s,s') and 0 if it does not; and we must have

$$Tx \le d_{ss'} \quad \text{for all } (s,s')$$

This is a very general model. If, for example, one estimates that there is a free SMS-MT termination cost to destination s' until the traffic sent by a given node reaches a certain value, it may be included, by making b_s equal to this value, in order that the SMS small volume remains free and unchecked by the visited network.

Let c_j be the cost of path j (SS7 traffic + transit + termination). The traffic sent for customer s to destination s' is then

$$f(s,s') = \sum_{j \in (s,s')} x_j$$

The cost for this traffic $f(s,s')$ is

$$\sum_{j \in (s,s')} c_j x_j$$

so the net margin $f(x)$ of the SMS interworking network for this traffic is the difference between the price paid by the customers to the SMS network and the cost of the traffic, which is paid by the SMS network. This margin for customers s and destination s' is

$$f(x) = p_{ss'} \left(\sum_{j \in (s,s')} x_j \right) - \sum_{j \in (s,s')} c_j x_j$$

and we look for the solution x that maximizes this net margin (a concave function) for all customers and destinations:

$$\max f(x) = \sum_{s,s' \in (S,S')} \left(p_{ss'} \left(\sum_{j \in (s,s')} x_j \right) - \sum_{j \in (s,s')} c_j x_j \right)$$

$$X \begin{cases} Ax \leq b \\ Tx \leq d_{ss'} \end{cases}$$

$$x \geq 0$$

Note that the functions $p_{ss'}(x)$ (where x is a scalar) are concave, as shown in the following equation.

We introduce the new (equality) constraints with $f_{s,s'}$ scalar:

$$f_{s,s'} = \sum_{i \in (s,s')} x_i$$

where $f_{s,s'}$ is the total number of SMS sent from customer s along all possible paths to the destination s'. So the problem becomes

$$\max f(x) = \sum_{s,s' \in (S,S')} p_{ss'} \left(f_{s,s'} \right) - \sum_{j \in (S,S')} c_j x_j$$

$$X \begin{cases} Ax \leq b \\ Tx \leq d_{ss'} \end{cases}$$

$$x \geq 0$$

$$f_{s,s'} = \sum_{i \in (s,s')} x_i$$

which is a convex problem (max of a concave function), because the functions p_{ss} of the scalar variables $f_{s,s'}$ are concave, and the functions $c_j x_j$ are linear.

10.6.5.3   Linear Case (Simplified Revenue Function)

If we assume a linear revenue function (no penalty for unmet minimum capacity), the functional will be

$$\max \sum p_j x_j - \sum c_j x_j$$

If d_i is the cost per SMS for an edge i. We have $c = A^t d$ for the cost of the path which uses these edges, so the linear multiflow problem is

$$\max(p^t - d^t A)x$$
$$X \begin{cases} Ax \leq b \\ Tx \leq d_{ss'} \end{cases}$$
$$x \geq 0$$

Note A^j, the jth column of A, if $p_j^t - d^t A^j = 0 \rightarrow x_j = 0$. This means that if we lose money on a path j, we will not use that particular path.

10.6.6   Algorithm to Find the Global Optimum

Note $\nabla f(x)$, the gradient of the concave function $f(x)$. At each extreme point x_k of (Px), we compute this gradient to yield the *linearized tangential problem:*

$$\max \nabla f(x_k) \cdot x$$
$$X \begin{cases} Ax \leq b \\ Tx \leq d_{ss'} \end{cases}$$
$$x \geq 0$$

It is possible to obtain the global maximum of the concave function $f(x)$ over the convex set (polyhedral). *The optimum is not in general an extreme point of X.* So that even if the matrix A is unimodular and the capacities (b) are integers, the optimal x (that is, the x that gives optimal numbers for SMSs and their routing) is not necessarily an integer. Because these are large values, it really does not make any practical difference if we round the components of x to the nearest integer.

One can easily make counterexamples which show that the stepwise procedure consisting at each x_k to take a linear approximation of $f(x)$ by $\nabla f(x_k)x$ will not converge [3]. The Frank-Wolfe algorithm [3] is well suited and works with a sequence of tangential linear programming problems followed by a one-dimensional approximation.

10.6.7   Centralized Network Traffic Regulation Principle

Centralized network regulation is performed by equipment that receives traffic measurements every 10 minutes from the gateway nodes to the network, so that it can compute all instantaneous demands d_{ss} of the set of customers S. The equipment distributes this measurement of demand to all nodes on the network, including the gateway nodes such as node k in the graph, as well as updates (which may change over time) of c_{ss}.

Each node then recomputes the optimum of $f(x)$ at this time and thus determines all the flows x_j, including the paths that do not use this node. Then they will use the result differently.

The transit nodes will compute for each outgoing arc the optimal flow from each path, which, by definition, corresponds to different customers. They will distribute their traffic according to the optimum computed values. In the system, each node is able to identify the origin customer and the destination network (information that is used to create billing records) from the data received. They will never need to drop traffic because the optimal solution is feasible at each node.

The gateway nodes will also do this. In addition, they will bar traffic when demand $d_{ss'}$ is greater that the optimum computed throughput f_{ss}. This barred traffic will be stored (it is called a *spooler*) and resent later (adding to the demand). This deferred traffic could be accounted for by having a different price function. The simplified model assumes that the deferred traffic is lost. The system is fully coherent because each node computes the same optimum.

10.7   Example: Detailed Modeling of a Real SMS Interworking Network

This real SMS interworking network includes its own SS7 routers on the SS7 network and also uses relays, which are the SMSCs of the operators (with IP or SS7 direct connection agreements).

10.7.1   Modeling a Simple SS7 Router or a Relay

To account for transit costs (which are paid by the networking operator), each time the roaming of a hosting partner is used, and also to model the SS7 connection capacity and the SS7 traffic price, it is necessary to model a physical router or relay

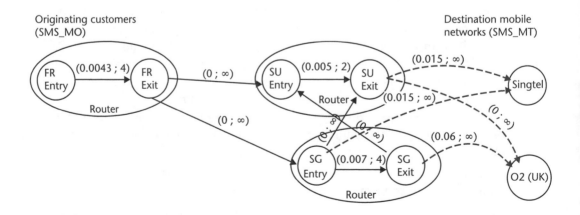

Figure 10.5   Graph model of an SS7 router on an SMS interworking network.

as two abstract nodes of the graph (Figure 10.5), and an edge has two valuations: the transit cost per SMS and the capacity, taking as a unit a 64-Kbps SS7 signaling channel (17 SMS per second; see Chapter 1).

For example, the arc (SG entry, SG exit) has a transit cost of 0.007 and a capacity of four SS7 links. Any path of SMS using this transit will then have a cost of 0.007 per SMS plus the termination cost (0.06 for the GSM operator O2 in the United Kingdom). The sum of the traffic from FR and SU to SG cannot exceed 4.

10.7.2   Modeling Traffic to Subscribers of a Network Hosting an SS7 Router

We see in Figure 10.5 that there is an edge from SG Entry to Singtel, which has the two valuations (0.015; ∞) as well as the paths from SU Exit to Singtel. So whether we go through the SG SS7 router or the SU SS7 router, the termination price will be 0.015. If we go through SU, there will be an additional transit cost, but if we go through SG, we do not add the path (SG Entry, SG Exit) that corresponds to the transit cost (we pay only the deposit cost, as agreed). Also, the arc capacity is 4 because there are 4 SS7 signaling channels for this router.

10.7.3   Modeling a Virtual SS7 Router with Several IGPs and Transit Agreements

As explained in Chapter 6, the SS7 routers belonging to an SMS interworking network could be virtually in several networks (that is, have several addresses or GTs) although the router is physically one machine. Figure 10.6 shows a simple SMSC but with two sets of different GTs that it can use as originating addresses, when using two different international gateways. One has a cost of approximately 0.0043 euros per SMS with a capacity of four SS7 links. The other uses two leased SS7 links, so there is no variable cost.

When we use the FREM GT (see Figure 10.7), we pay the operator, which lends us its GT for a transit fee of 0.009 euros or 0.012 euros when we use the FRSC GT.

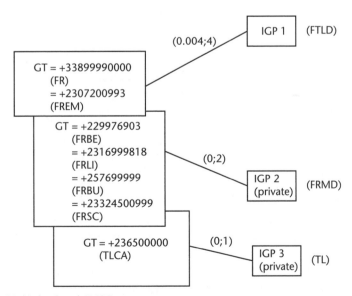

Figure 10.6   Multiple virtual SMSCs.

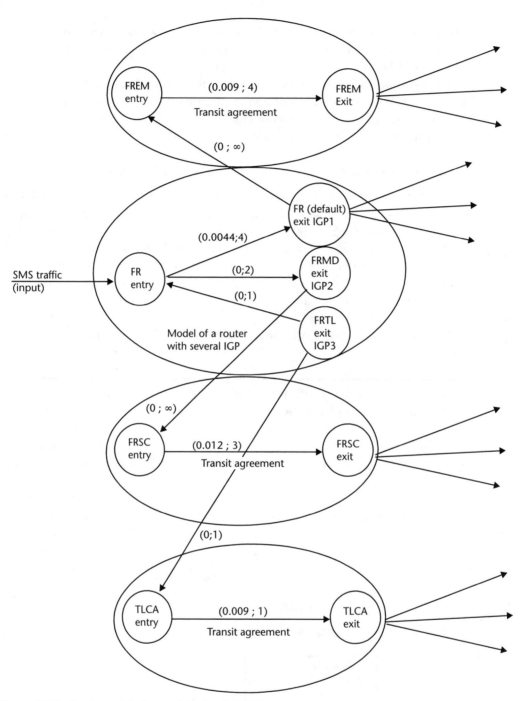

Figure 10.7 Graph model of a multiple virtual SMSC.

For the other GT, we do not have transit agreements. So we add another FREM node that has a lot of roaming agreements, in particular, with O2 (U.K.). (Note that this is just an example—the costs pertain only to these explanatory examples.)

As you can see in Figure 10.7, if an SMS is sent through the FRSC transit agreement, the cost is 0 (leased line) + 0.012 (SC transit cost) = 0.012 euros, with a maximum capacity of 2. When we send direct to our own roaming agreement S, we pay 0.0044 euros plus termination. When we use the FREM transit agreement of FREM, we pay 0.0044 (because FREM uses the IGP1) + 0.009 (EM transit cost) = 0.0134. In addition, there is a third IGP3 (FRTL) that has one (64-Kbps) leased line with another real node, TLCA, so the model of cost (no variable cost per SMS) and capacity is (0; 1) for the (FR, FRTL) edge, and (0.009; 1) for the transit agreement

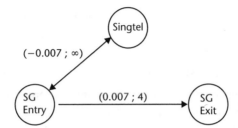

Figure 10.8   Graph model of a hosting partner.

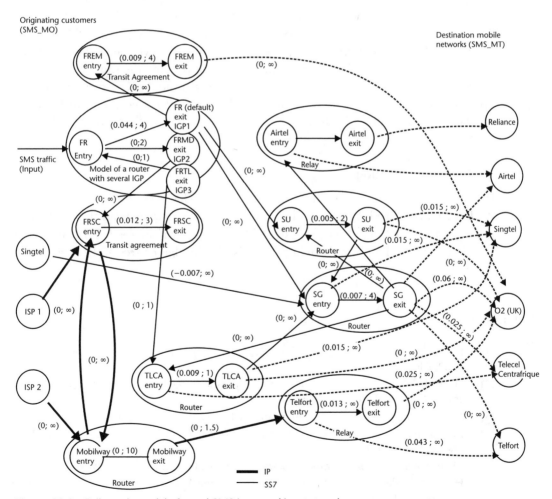

Figure 10.9   Full graph model of a real SMS interworking network.

Table 10.1   Five Possible Paths from ISP1 to O2

Paths	Cost	Capacity	QoS	Number of Hops
FR_{EN}, FR_{EX1}, SG, O2	$0.0043 + 0.007 + 0.06$	$4 = \min(4, \infty, 4, \infty)$	Router	2
FR_{EN}, FR_{EX1}, EM, O2	$0.0043 + 0.009$	$4 = \min(\infty, 4, \infty)$	Router	1
FR_{EN}, FR_{EX1}, SU, O2	$0.0043 + 0.005$	$2 = \min(4, \infty, 2, \infty)$	Router	2
FR_{EN}, IP hub, Telfort, O2	0.013	$1.5 = \min(1.5, \infty, \infty)$	Relay	3
FR_{EN}, TLCA, O2	0.009	$1 = \min(1, 1, \infty)$	Router	2

with the operator CA, whom we will pay 0.009 euros per SMS sent using its roaming agreements.

10.7.4   Connection of Hosting Partners

When a hosting partner, an operator which sends its roaming agreements, sends SMS-MO, we do not pay a transit fee, so it is appropriate to value the arc with a negative of the transit cost, so that the sum will cancel (Figure 10.8).

When we put all of this together, we obtain the graph shown in Figure 10.9. In reality, there are more nodes and hundreds of destinations. In Figure 10.9, you can see that the router SG, which is an SMSC, can send SMSs to Telfort (Netherlands). There is another path with an SMS hub using IP connections to the SMSCs that sends the SMS to the Telfort relay, when then sends the SMS on to its subscriber. The second path would be cheaper but it does not provide the same QoS. With a relay, the quality of service is not as good (no real time ACK, setting of OA not always possible, and so on). The commercial agreement is that the transit charges are paid when roaming agreements are used and termination charges are paid for the SMS-MT.

10.7.5   Path Valuations

Consider the five possible paths from ISP1 to O2 (U.K.), as shown in Table 10.1. ISP1 is connected to FR by an SMPP connection. However, what is the total maximum capacity? Because three paths use (FR_{EN}, FR_{EX}), it will be 4 + 1.5 (IP hub) paths + 1 (TLCA path). The number of hops, that is, the number of arcs of the path, is also a measure of quality.

References

[1]   Hu, T. C., *Integer Programming and Network Flows*, Reading, MA: Addison-Wesley, 1969, Chap. 11.

[2]   Henry-Labordère, A., *Cours de Recherche Opérationnelle*, Paris, France: Presses des Ponts et Chaussées, 1995, Chaps. 4 and 11.

[3]   Minoux, M., *Programmation Mathématique,* Tome I, Paris, France: Dunod, 1983, Chap. 5.

[4]   Kaufmann, A., *Méthodes et Modèles de la Recherche Opérationnelle*, Tome II, Paris, France: Dunod, 1968, Sec. 39.

[5]   Sakarovitch, M., *Optimisation Combinatoire*, Paris, France: Hermann, 1984.

[6]   Henry-Labordère, A., and A. Kaufmann, *Integer and Mixed Programming*, Reading, MA: Addison-Wesley, 1975.

## CHAPTER 11

INAP and CAMEL Overview and Other Solutions for Prepaid SMS

The gates of Heaven shall not be opened to them,
nor shall they enter Paradise until the camel passes
through the eye of the needle
—Koran 7.38

Intelligent Network Application Part (INAP) is a protocol that was first developed in the 1980s to bring intelligence into the PSTNs. In the first implementations of INAP, the protocol was operator or manufacturer specific and was only available for the main switching nodes. Its purpose was to allow an operator to develop new telephony services without having to update all of the software in its switching nodes. This meant that the installation of a new service in the network could be done in several months rather than in several years. The purpose of an intelligent network is to perform the routing of a telephone call in a far more sophisticated way than the routing that performs a simple switch based on a routing table (e.g., location-based services, free phone services, and so on).

Then in the mid-1990s, the development of new international services [such as *international virtual private network* (IVPN) or international free phone numbers] were crucial in the definition of a new standard, based on existing proprietary INAP protocols, to make international interworking possible. The *European Telecommunications Standards Institute* (ETSI) then specified ETSI CS-1 and ETSI CS-2 protocols to be used in the INAP-based architecture. The GSM then implemented its own INAP protocol based on ETSI CS-2, to be used in the PLMN, *customized application for mobile network enhanced logic* (CAMEL) [1]. This protocol layer is at the same level as MAP.

Phase 1 of CAMEL was released in 1997 without the specification of a service resource function interface. This lack of intelligent peripheral has been corrected in phase 2 of CAMEL, released in 1998 (thus allowing the mobile IN to interface with voice response units). The protocol is named *CAMEL Application Part* (CAP).

The architecture of an intelligent network is based on the *service switching function* (SSF); *service control function* (SCF), which runs on an IN machine; *service resource function* (SRF), also called *intelligent peripherals* (IPs); and *service management function* (SMF). The principal entity is the SCF, which controls the process of the invoked service. The IP is a sophisticated resource that may include test-to-speech synthesis for voice announcements or speech recognition for user interaction.

11.1   Use of CAMEL for SMS Prepaid Services

One of the most common applications of CAMEL services in a PLMN is to manage prepaid services, especially for SMS. CAMEL is used in many PLMNs for SMS payment from prepaid customers and credit reloading. Note that this also works if CAMEL roaming agreements are in place while the prepaid subscriber is in another country.

The address of the SCP as well as the service key are transferred by the HLRs in the INSERT_SUBSCRIBER_DATA response when the subscriber registers in a visited network, so that, as explained later, the visited MSC can address the SCP to invoke a CAMEL service that checks to see if the credit is sufficient.

11.1.1   SMS Payment from Prepaid Customers

A key issue with the value-added services is charging the prepaid customers. This is the primary reason why prepaid subscribers are barred from roaming and sometimes from sending SMS-MO in their own network. CAMEL offers a way to implement this service for a PLMN as shown in Figure 11.1.

In this network, a prepaid customer initiates an SMS-MO through its MSC toward an SMSC. The SMSC recognizes from the profile that the originating MS is a prepaid contract from the profile in the VLR and then initiates a CAMEL INITIAL_DP toward an SCP. The INITIAL_DP fields will contain the GSM subscriber number, such as the MSISDN. The SCP will interrogate the prepaid subscriber database to see if the initiating MS has enough credit to pay for the SMS. If so, it responds to the SMSC with a release call (with a cause = accepted). The SMSC then sends the SMS.

This service can also be used to provide a charging method for an ISP to charge special value-added SMS services, such as logo or ringtone sending, by charging them at a specific price in the account before sending them to the destination MS.

The credit database is the *service data point* (SDP), which has a proprietary interface with the HLR, so it is addressed only indirectly through the HLR and the CAMEL services.

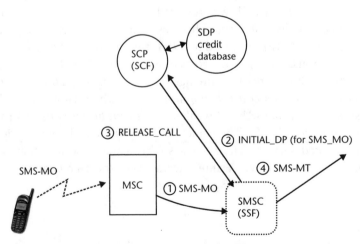

Figure 11.1   SMS charging for prepaid customers with CAMEL.

11.1.2   Credit Reloading for Prepaid Customers

This service is used by an ISP to credit a prepaid account. The ISP uses an IP connection on the SSF that initiates an INITIAL_DP request to the SCP that contains information on the account to be credited (MSISDN and a voucher number identifying the transaction). The SCP computes the INITIAL_DP, verifies the deposit voucher, and replies to the SSP with a RELEASE_CALL Invoke, which indicates the result of the operation. Then as a result of the credit reloading of the prepaid customer, the ISP can send an SMS to the MS indicating the result of the operation.

11.2   Useful Subset of CAMEL Services for Prepaid Customers

For prepaid services, we implement six CAMEL services: INITIAL_DP, REQUEST_ REPORT_EVENT_BCSM, REPORT_EVENT_BCSM, APPLY_CHARGING, CONTINUE, APPLY_CHARGING_REPORT, and RELEASE_CALL.

The total number of services in version 2 of CAMEL is about 25 and includes the PLAY_ANNOUNCEMENT to play voice messages and CONNECT to dial a number [1]. The following examples show some procedures that use these services.

11.2.1   Example 1: Prepaid SMS

Our first example is for prepaid SMS. Note that the following example contains a "cause" such as "0," which is interpreted as successful charging of an account.

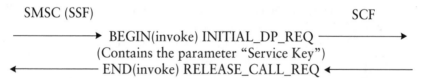

11.2.2   Example 2: Simple Prepaid Voice Call

Figure 11.2 shows the principle behind an IN-based voice prepaid service. We also describe the procedure in full detail. The network equipment with which we are concerned is the MSC (SSF function)/VLR, the HLR, and the SCP (SCF function).

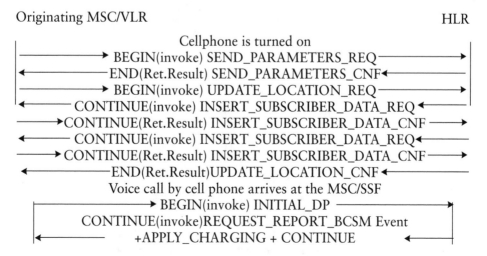

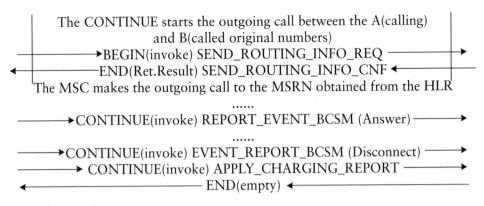

The CONTINUE starts the outgoing call between the A(calling)
and B(called original numbers)

→BEGIN(invoke) SEND_ROUTING_INFO_REQ ────────→

←────────END(Ret.Result) SEND_ROUTING_INFO_CNF ◄───────

The MSC makes the outgoing call to the MSRN obtained from the HLR

......

────────→CONTINUE(invoke) REPORT_EVENT_BCSM (Answer)────────→

......

────────→CONTINUE(invoke) EVENT_REPORT_BCSM (Disconnect)────→

────────→ CONTINUE(invoke) APPLY_CHARGING_REPORT ──────→

◄──────────────── END(empty) ◄────────────────

The first part of the signaling (between the VLR and the HLR) will load the CAMEL subscriber information in the VLR (see details in Chapter 1). In the second part, the SCF sends an INITIAL_DP to the SSF, which will respond to the SCF. When it responds, it determines which events to report (the Answer and the Disconnect) with a REQUEST_REPORT_EVENT_BCSM and it starts charging (APPLY_CHARGING). Then the MSC makes the outgoing call. It interrogates the HLR to obtain the MSRN (see Chapter 1).

The SSF reports the Answer of the called number, the Disconnection, and returns to the SSF in an APPLY_CHARGING_REPORT the call duration in seconds, so that the caller's account can be charged.

Some network vendors provide the SMS-MO prepaid support at the MSC level. The MSC recognizes that it is a prepaid subscriber (from the HLR profile) and checks and charges the credit by issuing an INITIAL_DP to the SCP. They may be several SCPs in a network. So when a CAMEL subscriber roams in a foreign network, when he registers in a VLR, this VLR will receive from the concerned HLR an INSERT_SUBSCRIBER_DATA message, which contains the VLR CAMEL subscription information parameter, which itself contains the service key and the SCF GT. So the MSC (in tandem with its VLR) will be able to address the INITIAL_DP whenever an outgoing call is initiated.

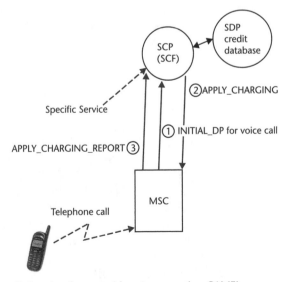

Figure 11.2 Voice call charging for prepaid customers using CAMEL.

11.2.3   Example 3: Voice Call Rerouted to an Announcement Machine

This is the most complicated example and will be explained with the help of a simplified figure, Figure 11.3, to aid in understanding the principles. The SRF is somewhat expensive and the assist procedure allows sharing between the various switches.

It starts with a voice call from a prepaid subscriber, so an INITIAL_DP is sent by the SSF of the visited MSC to the SCF. This IN machine finds that there is not enough credit and will reroute the call to a voice announcement machine (the SRF) to help the subscriber reload her credit.

So the SCF will send an ESTABLISH_TEMPORARY_CONNECTION message [(2) in Figure 11.3] to the SSF with a Correlation ID that allows the different equipment to coordinate. We assume that the SRF is not directly connected to the visited SSF, so this SSF will make a voice call (using the ISUP procedure) to an assisting SSF to which the SRF is connected. The Correlation ID will be used in the call, so the assisting SSF is able to include it, when it receives the call, in the ASSIST_REQUEST_INSTRUCTIONS message that it sends to the SCF (4). It will respond with a CONNECT_TO_RESOURCE (5); then the SSF will make a voice call to its attached SRF (6). The SRF will be fully driven by the SSF, through the assisting SSF (7) acting transparently (in relay mode).

Here are full details of the procedure with all the CAMEL services required to satisfy this example, including interactive services such as PROMPT_AND_COLLECT [to obtain the *dual-tone multiple-frequency* (DTMF) digits keyed by the mobile].

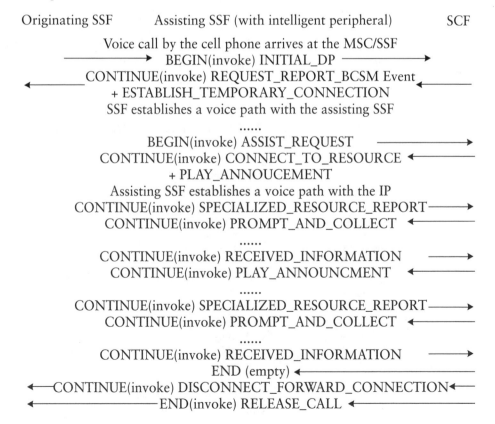

Originating SSF   Assisting SSF (with intelligent peripheral)   SCF

Voice call by the cell phone arrives at the MSC/SSF
———————▶ BEGIN(invoke) INITIAL_DP ———————▶
◀——————— CONTINUE(invoke) REQUEST_REPORT_BCSM Event ◀———————
+ ESTABLISH_TEMPORARY_CONNECTION
SSF establishes a voice path with the assisting SSF

......

BEGIN(invoke) ASSIST_REQUEST   ———————▶
CONTINUE(invoke) CONNECT_TO_RESOURCE ◀———————
+ PLAY_ANNOUCEMENT
Assisting SSF establishes a voice path with the IP
CONTINUE(invoke) SPECIALIZED_RESOURCE_REPORT ———————▶
CONTINUE(invoke) PROMPT_AND_COLLECT ◀———————

......

CONTINUE(invoke) RECEIVED_INFORMATION ———————▶
CONTINUE(invoke) PLAY_ANNOUNCMENT ◀———————

......

CONTINUE(invoke) SPECIALIZED_RESOURCE_REPORT ———————▶
CONTINUE(invoke) PROMPT_AND_COLLECT ◀———————

......

CONTINUE(invoke) RECEIVED_INFORMATION ———————▶
END (empty) ◀———————
◀—— CONTINUE(invoke) DISCONNECT_FORWARD_CONNECTION ◀——
◀——————— END(invoke) RELEASE_CALL ◀———————

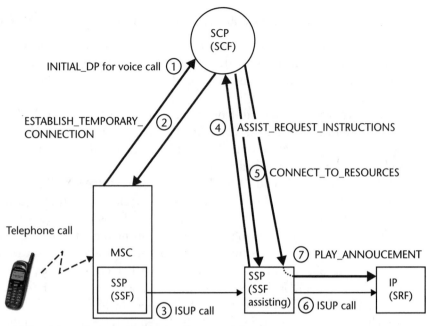

Figure 11.3 Voice call rerouted to an intelligent peripheral.

11.2.4   Details of Applicable CAMEL Services

The INITIAL_DP (destination point) is the initial service sent by a requesting SSF to the SCP (Table 11.1). It contains all parameters necessary for the service

The CAMEL service CONTINUE (Table 11.2) is sent by the SCF to request the establishment of a call between originating number A and terminating number B. It is different from the CONTINUE in the TCAP protocol.

Table 11.1   CAMEL Primitive Used for Prepaid Service: INITIAL_DP

Initial_DP

Parameter

InvokeID	HighLayerCompability	Vlr_Number
ServiceKey	ServiceInteractionIndicators	LocationNumber
DialedDigits	AdditionalCallingPartyNumber	CellIdOrLAlb
CalledPartyNumber	ForwardCallIndicators	CellIdFixedLength
CallingPartyNumber	BearerCapability	LAIFixedLength
CallingPartyBusinessGroupID	EventTypeBCSM	ExtensionContainer
CallingPartysCategory	RedirectingPartyID	LocationInfoEllipsis
CallingPartySubaddress	RedirectionInformation	Ext_BasicSrvCode
CGEncountered	IMSI	BearerServiceCode
IPSSPCapabilities	SubscriberState	TeleserviceCode
IPAvailable	AssumeIdle	CallReferenceNumber
LocationNumber	CamelBusy	MSCAddress
OriginalCalledPartyID	NetDetNotReachable	CalledPartyBCDNumber
Extensions	NotProvidedFromVLR	Ellipsis
ServiceProfileIdentifier	LocationInfo	
TerminalType	AgeOfLocationInformation	
TriggerType	GeographicalInformation	

Table 11.2   CAMEL Primitive Used for Prepaid Service: INITIAL_DP

Continue
Parameter
InvokeID

Table 11.3   CAMEL Primitives Used for Prepaid Services

REQUEST_REPORT_EVENT_BCSM	*REPORT_EVENT_BCSM*
Parameter	*Parameter*
InvokeID	InvokeID
Event-type BCSM (answer, disconnect, and so on)	Event type BSCM (e.g., answer)
Monitor Mode (e.g., notify and continue)	Leg ID
Leg ID	
SendingSideID	

The REQUEST_REPORT_EVENT_BCSM (Table 11.3) is sent by the SCP to the SSP to give a list of events to be monitored and reported back about (answer of the called number, disconnect of one of the legs, and so on). They are reported by an SSP in a REPORT_EVENT_BCSM. The Monitor Mode parameters tell the SSP what to do when the event occurs (such as notify the SCP and continue). The APPLY_CHARGING service (Table 11.4) from the SCP asks the SSP to report back charging information such as call duration.

11.2.5   Specificity of the CAMEL Services

To use existing CAMEL services (such as prepaid voice calls) for another application, such as SMS prepaid, we need to understand that although the CAMEL protocol is standard, the meaning of the parameters is completely specific to the script that was written by the SCP vendor. For example, the calling party BCD number may be the destination number or the originating number. It is decided by the SSF

Table 11.4   CAMEL Primitive Used for Prepaid Service: APPLY CHARGING

APPLY_CHARGING	*APPLY_CHARGING_REPORT*
Parameter	*Parameter*
InvokeID	InvokeID
AchBillingChargingCharacteristics	CallResult (PartyToCharge, TimeIfNoTariffSwitch: the call duration)
SendCalculationToSCPIndication	
Default=false	
PartyToCharge	
SendingSideID	
ReceivingSideID	
Extensions	
Ellipsis	
RELEASE_CALL	
Parameter	
InvokeID	
Cause	

(the originating call MSC) in a function of the service key. So the SSF may have a service-dependent specific INITIAL_DP, and of course the execution and the result (release call or apply charging) is completely service dependent.

For economic reasons, when deciding to use an existing CAMEL service for prepaid SMS, we will need to do some reverse engineering to determine the meaning of the parameters of the best suitable CAMEL service so that the implementation is simple.

Many parameters, such as cell ID, location number, must be supplied although, clearly. they are not useful. So the sending equipment, to properly emulate the INITIAL_DP, should, in general, have the MAP services that allow us to obtain these mandatory (but not useful) parameters.

Look at the CAMEL traces of Section 11.2.4, which correspond to an INITIAL_DP used to reload a credit, the value of a scratch card having the number 22213872438544963 in the called party BCD parameter, which is normally used for the destination number.

11.3   Implementation: Multiple-Protocol Services-Oriented Platform: CAMEL Gateways

Those two examples of CAMEL-based services show that for prepaid customer and SMS services, the SMSC and SSP must be bound together to implement the service. Then it becomes obvious that the best functional aritecture for developing this kind of service is the implementation of a multiple-protocol services-oriented platform based on the SS7 signaling network (MTP through SCCP), plus TCAP and, depending on the functions involved (SCP, HLR, MSC, SMSC), INAP, CAMEL, or MAP.

As we have seen before, this is made possible through the very flexible design of SCCP and TCAP. The multiple-protocol platform only implements several services bound to a single TCAP and SCCP layer, allowing the routing based on service access points linked with SSNs. As a consequence, it is possible for each layer over TCAP to connect to the TCAP and SCCP layer using its own SSN (for instance SSN = 6 identifies an HLR function; SSN = 146 identifies an IN SCP function, and so on). TCAP and SCCP will then route all invocations to these entities to the appropriate one upon analysis of the SSN involved. This architecture provides a very powerful way to implement complex services architecture that may involve several services (location based services + prepaid + ...). This concept of a multiple-protocol services-oriented platform is totally independent from the hardware on which it is processing. The term *platform* refers to a global title or point code that is shared among all these functionalities, including a CAMEL gateway.

Some vendors provide CAMEL gateways that can interface with IT servers, so they can easily use CAMEL through an IP protocol, such as SMPP. *Note, however, that the CAMEL gateway must also have the MAP protocol* because certain parameters for CAMEL, if needed by certain IN services (e.g., location-dependent tariffs for example), need MAP primitives (see Chapter 13). Consider the following example for SMPP (in which one makes use of the submit_sm):

SMPP Parameters   Corresponding CAMEL
Port_Address = 9274   INITIAL_DP

Protocol_ID (PID)	Service key
dest_address	Calling party number
source_address	Called party BCD number

11.4   Example of Analyzer Traces of a CAMEL Transaction

To gain a practical understanding of CAMEL, we now present the detailed trace for an INITIAL_DP service, Credit Recharging. The IN service number (13) is in INPN_ServiceKey, the account number is contained in INPN_CallingPartyNumber, and the credit number is contained in INPN_CalledPartyBCDNumber. So you can see that there are lots of parameters, but very few are really useful for this application. For the LAC and Cell ID, set it to 0 because it is not used.

```
INAP-INITIAL-DP
 INPN_InvokeID(1)
  L = 001
  Data: 128
 INPN_ServiceKey(3)
  L = 001
  Data: 13
 INPN_CallingPartyNumber(Q763)(6)
  L = 008
  Data: Nature of Address = International number
     Number incomplete Indicator = complete
     Numbering plan Indicator = ISDN/Telephony(E164)
     Presentation Indicator = presentation allowed
     Screening Indicator = network provided
     Address = 40744460013
 INPN_CallingPartysCategory(8)
  L = 001
  Data: 10
 INPN_LocationNumber(Q763)(13)
  L = 008
  Data: Nature of Address = International number
     Number incomplete Indicator = incomplete
     Numbering plan Indicator = ISDN/Telephon
      CalledPy(E164)
     Presentation Indicator = presentation restricted
     Screening Indicator = network provided
     Address = 40744007530
 INPN_BearerCapability(23)
  L = 003
  Data: (Hex) 8090A3
 INPN_EventTypeBCSM(0)(192)
  L = 001
```

```
        Data: 2
    INPN_IMSI(130)
     L = 008
     Data: Address = 226100000000000
    INPN_AgeOfLocationInformation(124)
     L = 001
     Data: age of location info 0 minutes
    INPN_Vlr_Number(126)
     L = 007
     Data: Ext = No extension
        Ton = International
        Npi = ISDN
        Address = 40744007530
    INPN_CellIdFixedLength(122)
     L = 007
     Data: MCC = 226 MNC = 10 LAC = 0
        Cell_ID = 0
    INPN_Ext_TeleserviceCode(121)
     L = 001
     Data: 17
    INPN_CallReferenceNumber(131)
     L = 008
     Data: (Hex) 3124952080000000
    INPN_MSCAddress(132)
     L = 007
     Data: Ext = No extension
        Ton = International
        Npi = ISDN
        Address = 40744007530
    INPN_CalledPartyBCDNumber(129)
     L = 010
     Data: Ext = No extension
        Ton = Unknown
        Npi = ISDN
        Address = 22213872438544963
- - - - - - - - - - - - - - - -   - - - - - - - - - - - - - -

 INAP-OPEN-CNF
     INDP_result(5)
      L = 001
      Data: (0):dialogue accepted
     INDP_applic_context_index(27)
      L = 001
      Data: (Hex) 20
```

```
-  -  -  -  -  -  -  -  -  -  -  -  -  -  -     -  -  -  -  -  -  -  -  -  -  -  -  -  -  -


INAP-RELEASE-CALL
    INPN_InvokeID(1)
     L = 001
     Data: 128
    INPN_Cause (41)
     L = 002
     Data:
      Coding/Location 8A
        Coding standard = CCITT standard
        Location = network beyond interworking point(BI)
      Cause: 94
        Class = normal event
        Value = FAIL:PREPAID ACCOUNT NOT FOUND
-  -  -  -  -  -  -  -  -  -  -  -  -  -     -  -  -  -  -  -  -  -  -  -  -  -  -  -


INAP-CLOSE-REQ
-  -  -  -  -  -  -  -  -  -  -  -  -  -  -     -  -  -  -  -  -  -  -  -  -  -  -  -  -  -
```

In the above trace, only the bold literal parameters are really useful; the others are mandatory, but only need to be syntactically correct. For example, the IMSI is 22610xxxxxxxxxx, the MCC and MNC are correct, and the rest does not matter.

11.5   Other Solutions for Prepaid SMS

11.5.1   Prepaid SMS with Service Nodes

Service nodes (SNs) are a third-party-provided solution for the provision of voice services to prepaid customers (Figure 11.4). They are an alternative to a full IN system (with CAMEL or a proprietary INAP protocol) and to the *Advice_of_Charge* (AoC) solution, which are in general provided by the main network vendors with whom they are competing. A SN is made of a switch connected in at the ISUP to one of the MSCs and a credit database.

In general, it is necessary for the prepaid customers to have a special range of numbers, so that the MSC can have a call routing table that directs all calls originated by prepaid subscribers to the SN. The subscriber profile in the HLR tells us whether or not it is a prepaid subscriber, but the SN cannot access it, because it does not have a VLR function. The SN computes the cost of the call and decrements the credit or releases the call when the credit has been used. They also provide a recharging system through an *interactive voice response* (IVR) system (with credit vouchers scratch cards) or USSD.

The provision of prepaid SMS is straightforward if there is an open interface to the credit database, which the SMSC can use to interrogate the credit and decrease it by the price of an SMS-MO. The difficulty is that the SN vendor will not release the interface to a competitor, and the SMSC vendor must use reverse engineering to establish the specifications.

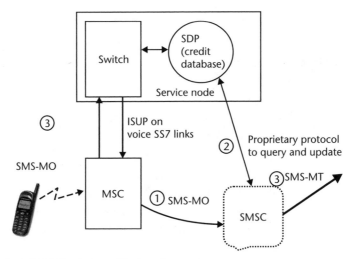

Figure 11.4 Prepaid SMS solution with a service node.

11.5.2 Prepaid SMS with AoC-Enabled Networks

This is certainly the cheapest solution [2] for small operators: No IN, no SN. It uses the two standard counters in the SIM card called the *accumulated call meter* (ACM) and *maximum value of the call meter* (ACMmax).

The credit reloading is very simplified: The customer goes with some cash to the operator's premises, which has a system (provided by the SIM card vendor) that is capable of updating directly the ACMmax once it is inserted in a SIM card reader: no scratch cards and no IVRs are required, but obviously the commercial distribution network is quite limited (Figure 11.5).

Whenever a call is initiated by the cell phone or a call is received while roaming (that is the call is initiated by the GMSC; see Chapter 1 on mobility), it receives a *charge advice information* (CAI) [3], which increments the ACM. When this value

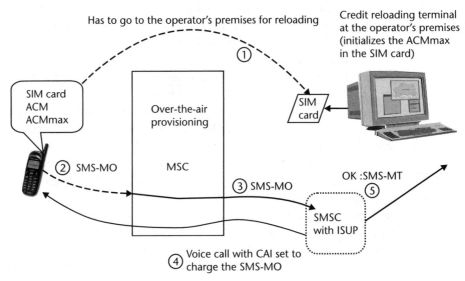

Figure 11.5 Prepaid SMS with an AoC system

reaches ACMmax, no call can be initiated, and if it receives a call with CAI (while roaming), it releases the call immediately. So in this prepaid solution, all the charging control is inside the cell phone. The value of the CAI, which sets the price depending on the call destination, is set in the MSC.

To charge an SMS, the SMSC must have a voice interface with ISUP signaling, so it can make a call to the cell phone, as if it were roaming, while setting the proper CAI. If the ACMmax is reached, it will block the voice calls from this cell phone, not the SMS because the ACMmax counter has no effect on the blocking of the SMS-MO sent by the cell phone. So this is not a satisfactory solution for prepaid SMS.

Note that the initial idea of AoC was to provide the cost of each call when completed. In most phones you can activate it in the Settings mode (Summary Call setting). However, most operators do not activate the feature and no CAI is received; instead you only see the duration of the call. Note also, that due to the risk of fraud, no operator allows AoC subscribers to have roaming service.

References

[1]   GSM 03.78 (ETS 101 441): Digital Cellular Telecommunication System (Phase 2+), Customized Applications for Mobile Network Enhanced Logic(CAMEL) Phase 2.

[2]   GSM 02.86: Digital Cellular Telecommunication System, Advice of Charge (AoC) Supplementary Services.

[3]   GSM 02.24 (ETS 300 923): Digital Cellular Telecommunication System, Description of Charge Advice Information (CAI).

## CHAPTER 12

USSD: A Still-Relevant Conversational Application Service

But why, father?
Not expensive enough, my son.
—Renault publicity for the R5

12.1 USSD Advantages over SMS

Unstructured supplementary service data (USSD) services allow operators to offer conversational text services to 2G (or more) subscribers. In contrast, SMS is not conversational: It takes a few seconds to send an SMS-MO and 5 to 7 seconds to receive an answer. "Conversational" means that the cell phone can initiate a session, get back and answer with a possible choice, reply to that answer, and so on. Most GSM networks can offer this service but have chosen not to because it was not as glamorous as WAP, although for many things, such as customer care, it is faster and simpler. It also has some very particular advantages:

- Unlike SMS-MO, there are no charges for roaming from roaming partners when one of your subscribers is using your USSD services in another country. This is the call-back application described later in Section 12.4.
- Your USSD services are accessible by your prepaid subscribers when roaming (receiving calls only) in another country (unlike with SMS-MO), which means that they can initiate USSD sessions even if the do not have SMS-MO service.
- The bandwidth requirement is much lower than that required for WAP services, but much higher than for SMS service. It is like a voice connection that is maintained for the duration of a session on the air interface. However, in the PLMN, it does not occupy a circuit, it only uses MAP messages.

12.2 How Does Mobile-Initiated USSD Service Work?

Mobile-initiated USSD service initiates a service by making a call to * service # by typing, for example, **113*33608123456*DOCTORS #** and pressing the call key of the cell phone. These services have a HLR service code, which means they cannot be delivered by the USSD applications in the visited MSC or by the VLR. The service uses MAP services (see GSM 09.02). Remember that the MSCs, VLRs, and HLRs have a USSD platform incorporated into them, which, depending on the vendors, has various functionalities. In the following, we assume that, because it is more flexible, the network has a specialized USSD platform.

A MAP_PROCESS_UNSTRUCTURED_SS_REQ is sent to the HLR from any visited MSC because the service has a HLR service code. The HLR has also been configured to route the MAP_PROCESS_UNSTRUCTURED_SS_REQ (Table 12.1) to the USSD platform to which the content provider interfaces through the SMPP, as illustrated in Figure 12.1.

The MAP_OPEN_IND message (Table 12.2) uses an ellipsis (additional parameters of the MAP protocol) with the MSISDN of the cell phone, which initiates this USSD session. This ellipsis is necessary to provide customer-specific applications and charging (because the MSISDN of the calling cell phone is not otherwise included in the MAP_PROCESS_UNSTRUCTURED_SS_REQUEST_REQ. The final BYE BYE is generally conveyed in the USSD string in the CNF shown in Table 12.3.

An interactive session takes place using the MAP_UNSTRUCTURED_SS_NOTIFY_REQ or MAP_UNSTRUCTURED_SS_REQUEST_REQ message between the cell phone and the USSD platform until the USSD session is released.

Table 12.1 MAP_PROCESS_UNSTRUCTURED_SS_REQUEST_REQ Message Details

MAP_PROCESS_UNSTRUCTURED_SS_REQUEST_REQ		
Parameter	Class	Context
Primitive type octet	M	V1,V2
Timeout	O	V1,V2
Invoke ID	M	V1,V2
USSD data coding scheme	M	V2
USSD string	M	V1,V2

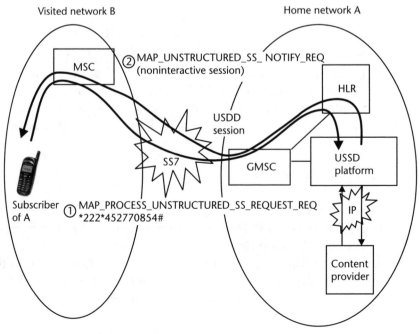

Figure 12.1 Mobile-initiated USSD services.

Table 12.2 MAP_OPEN_IND Message Details

MAP_OPEN_IND		
Parameter	*Class*	*Content*
Called party address	M	USSD platform
Calling party address	M	HLR
Origination reference	O	HLR
Ellipsis	M	OA of cell phone

Table 12.3 MAP_PROCESS_UNSTRUCTURED_SS_REQUEST_ CNF Message Details

MAP_PROCESS_UNSTRUCTURED_SS_REQUEST_CNF		
Parameter	*Class*	*Context*
Primitive type octet	M	V1,V2
Invoke ID	M	V1,V2
Where user error not included:		
USSD data coding scheme	M	V2
USSD string	O	
V1,V2		
Where user error included:		
User error	M	V1,V2
Network resource	O	V1,V2

The MAP_UNSTRUCTURED_SS_REQUEST_REQ is conversational (Table 12.4). It prompts the cell phone with a text (with a variable timeout set by the application) and receives the response in the CNF. Note that some old phones do not implement this properly and cut the session without waiting for the phone to answer. The MAP_UNSTRUCTURED_SS_NOTIFY_REQ is not conversational (does not ask for a response) so there is no timeout parameter (Table 12.5).

Figure 12.2 shows the exchange of data between a cell phone and a content provider. This exchange would use the SMPP protocol with the USSD platform:

- Exchange of data between the USSD platform and the content provider occurs by using the sm-data PDUs from the SMPP protocol (IP connection).

Table 12.4 MAP_UNSTRUCTURED_SS_REQUEST_ REQ and CNF Message Details

MAP_UNSTRUCTURED_SS_REQUEST_REQ		
Parameter	*Class*	*Context*
Primitive type octet	M	V2
Timeout	O	V2
Invoke ID	M	V2
USSD data coding scheme	M	V2
USSD string	M	V2
MAP_UNSTRUCTURED_SS_REQUEST_CNF		
Parameter	*Class*	*Context*
Primitive type octet	M	V2
Invoke ID	M	V2
USSD data coding scheme	C	V2
USSD string	C	V2
User error	C	V2

Table 12.5   MAP_UNSTRUCTURED_SS_NOTIFY_
REQ and CNF Message Details

MAP_UNSTRUCTURED_SS_NOTIFY_REQ		
Parameter	Class	Context
Primitive type octet	M	V2
Timeout	O	V2
Invoke ID	M	V2
USSD data coding scheme	M	V2
USSD string	M	V2
MAP_UNSTRUCTURED_SS_NOTIFY_CNF		
Parameter	Class	Context
Primitive type octet	M	V2
Invoke ID	M	V2

- The content provider uses its own server connected to the USSD platform by these protocols (SMPP/IP).
- The interactive applications can be written in any language suitable for Internet applications (PERL, Java, C, and so on).

12.3   Example of USSD Service

In this example, we look at a prepaid customer account that is recharged with a scratch card. Scratch cards are vouchers that a person buys that have a certain value and are protected by a hidden secret number. They are a very popular way of distributing subscriptions for prepaid customers both with large and medium mobile operators.

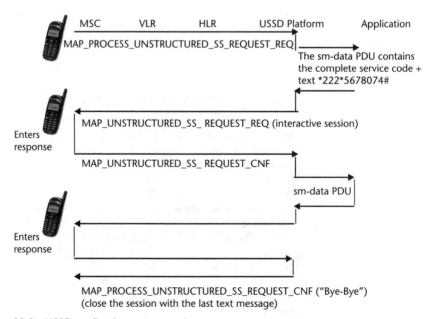

Figure 12.2   USSD application programming.

A USSD platform is much simpler to use to recharge a scratch card than is an interactive voice system (Figure 12.3). The big advantage is that if the prepaid subscriber might roam (with CAMEL), he can recharge his account from abroad with a scratch card that he has taken with him, and he can do so without paying for an expensive voice call. The USSD platform must be able to interrogate and update the scratch card database and the customer accounts, regardless of whether the prepaid implementation uses an IN (with CAMEL) or an SN.

12.4   USSD Is Free: A Call-Back Application

The interesting thing about USSD is that it is free. As of 2002, the visited network did not charge the home network for several reasons, including the billing issue (one would need to measure the traffic). A big difference with SMS is that only the USSD platform in the home network can dialogue with its (own) subscribers.

Figure 12.4 explains the call-back system for prepaid subscribers that has been implemented by some specialized *mobile virtual network operators* (MVNOs) that use the roaming agreements of a partner (and the partner's SMS cards as well). In (1) of Figure 12.4, the roaming subscriber wants to call a Hong Kong number. The subscriber cannot make a voice call because he or she is roaming in another country and is therefore barred from making calls and also sending SMS-MO (because we assume there is no CAMEL setup). But the subscriber is able to establish a USSD session with his or her USSD platform, passing the service code Call Back = 211 and the

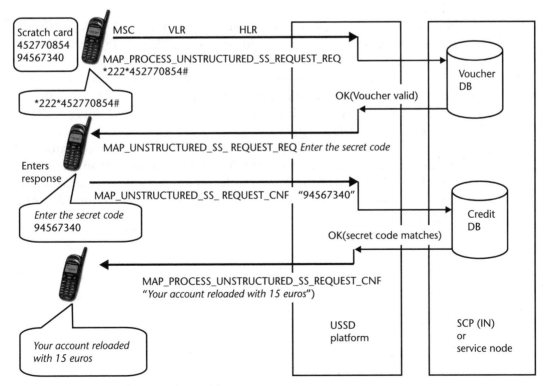

Figure 12.3   USSD recharging of prepaid accounts.

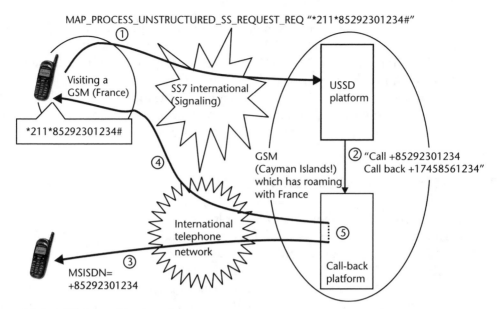

Figure 12.4   USSD free call-back service.

number that the subscriber wants to call. (He or she just types 211*85292301234#; the cell phone recognizes from the * and # symbols that this is not a voice call but a USSD session.) The USSD platform passes it to a call-back system (a switch) (2), which calls the destination (3). Then if the call is answered, it calls back (4) the originating roaming mobile subscriber and (5) connects the two legs of the call. Then the call-back system will charge the call.

A few vendors make such systems. Obviously, the visited GSM networks that are used without getting any revenue would not be very happy if there was a lot of this call-back traffic.

We have seen in details how mobile-originated USSD sessions and their applications work. It is also possible for the network to make a USSD call to a mobile. One application is AoC for prepaid subscribers. They receive their remaining credit and are prompted to reload their account.

USSD is a very flexible and cheap solution for small or medium operators who want to offer cheap but efficient services to their subscribers more than they want to offer fancy but slow portals. Most USSD platforms can be application programmed using standard languages, so that the in-house IT department resources can be used—this is also an advantage.

## CHAPTER 13

Location-Based Services

Wherever I wonder, wherever I rove,
My heart is in the Highlands wherever I go.
                    —Robert Burns (1759–1796)

13.1 Location-Based Services: Examples and Revenue Possibilities

Even if they do not have a *universal mobile telephony system* (UMTS), the second generation GSM operators can provide *location-based services* (LBS) to their subscribers as well as to their roaming visitors. This technology will still be valid for 2.5G or 3G systems. Some examples of applications include the following:

- Make a call to a short number (a special number such as 66123) and obtain an SMS with the address of the closest *automatic teller machine* (ATM) (money retrieval).

- Send an SMS to an special number with the text BANK, DOCTOR, and so on and receive by SMS a list of the closest addresses for those services. Such a service could be offered to roaming visitors and it is a way to attract them to one's network.

- Locate missing people or track stolen vehicles.

Revenues are obtained from roaming charges, from SMS-MO, and from sponsorships by restaurants and entertainment places that pay to be in the database.

In this chapter, we explain the detailed mathematical methods for accomplishing these LBS offerings. If this is not your cup of tea, you may want to stop reading after Section 13.5.

13.2 Mobile-Originated LBS

The principle behind LBS is illustrated in Figure 13.1. The LBS system must be connected to an operator SS7 network, and it integrates three functions:

1. A *gateway mobile location center* (GMLC), which receives requests from external (third) parties, obtains the location of the inquiring subscriber (using only its MSISDN), and transmits the result (using the SMPP protocol with additional parameters) to the mobile location center to accomplish the required localization;

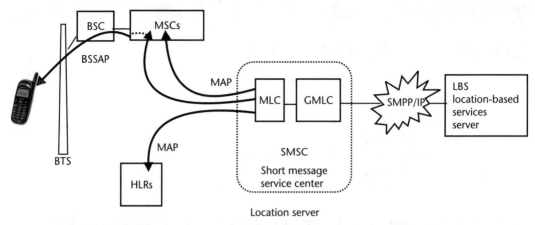

MLC: Mobile location center
GMLC: Gateway mobile location center

Figure 13.1   Principle of mobile-originated LBS.

   2. A *mobile location center* (MLC);

   3. An optional SMSC, which can send the message received from the LBS
      servers or transmit the SMS-MO received.

13.3   Methods

Various LBS methods are explained in [1]. These methods require that certain features be available in the network before they can offer LBS.

13.3.1   MSC Location Method

The MSC location method works in all cases. The MLC sends a SEND_ROUTING_
INFO_FOR_SM message to the HLR of a subscriber with its MSISDN (mobile
number) and gets back the address, such as + 6596100125, of the visited MSC as
well as the IMSI for later use.

    A table located at the operator's MSC gives the site name, for example, Singa-
pore North.

    The localization area is crude, from 5 km (large cities, such as London, that have
many MSCs) to 100 km (areas in the countryside), but is quite sufficient for services
such as weather forecasts.

13.3.2   Cell ID Method

If the visited network has MAP V3 capability, the cell ID method is used. In this
method, the MLC sends a *PROVIDE_SUBSCRIBER_INFO* (PSI) message [2] to
the visited MSC with the IMSI (which we know from the preceding method) (Table
13.1). This method works only if the MSC and MLC are part of the same network.

    There is another MAP primitive, *ANY_TIME_INTERROGATION* (ATI), that
allows foreign networks to interrogate the HLR of a roaming partner to obtain the

Table 13.1   PROVIDE_SUBCRIBER_INFO_REQ and
CNF Message Details

PROVIDE_SUBSCRIBER_INFO_REQ

Parameter	Class	Context
Primitive type octet	M	V3
Timeout	O	V3
Invoke ID	M	V3
Requested info	M	V3
IMSI	M	V3
MSISDN	O	V3

PROVIDE_SUBSCRIBER_INFO_CNF

Parameter	Class	Context
Primitive type octet	M	V3
Invoke ID	M	V3
Where user error not included:		
Age of location information	C	V3
Geographical information	C	V3
VLR number	C	V3
Location number	C	V3
Cell ID	C	V3
Subscriber state	C	V3
Not reachable reason	C	V3
Where user error included:		
User error	M	V3

same information as above (Table 13.2). To do so, the HLR of the home network
sends the PSI to the VLR. The ATI was designed so that a location number could be

Table 13.2   ANY_TIME_INTERROGATION_REQ and
CNF Message Details

ANY_TIME_INTERROGATION_REQ

Parameter	Class	Context
Primitive type octet	M	V3
Timeout	O	V3
Invoke ID	M	V3
Requested info	M	V3
GSM SCF address	M	V3
IMSI	C	V3
MSISDN	C	V3

ANY_TIME_INTERROGATION_CNF

Parameter	Class	Context
Primitive type octet	M	V3
Invoke ID	M	V3
Where user error not included:		
Age of location information	C	V3
Geographical information	C	V3
VLR number	C	V3
Location number	C	V3
Cell ID	C	V3
Subscriber state	C	V3
Not reachable reason	C	V3
Where user error included:		
User error	M	V3
Network resource	O	V3

provided to a CAMEL-based network (see Chapter 11) so that zone-dependent rates could be applied to voice calls or other services.

These two methods, PSI and ATI, give the visited location area the cell ID, information about the age of the localization (how many minutes since the cell phone was used), and whether the mobile is currently reachable. This information is meaningful only for the visited GSM operator, which has the corresponding database. So its IT department can build LBS while keeping confidential the cell database and its coverage. Note that very few operators allow ATI access by their roaming partners.

To estimate the number of different MSCs, LACs, and cell IDs in a given network, you can use these estimates: 1 MSC per 200,000 visitors, 5 LACs per MSC, and (in cities) 200 cells per MSC. The accuracy of this method goes from a few hundred meters (cities or picocells) to 15 km (countryside).

13.3.3 Extended Cell ID Method

If the network has the BSSAP-*location extension* (LE) extension of the *base station system application part* (BSSAP), the extended cell ID method is used. In this method, the MLC sends an information request [3, 4] to the MSC, which knows the cell ID from the IMSI. The MSC relays it to the BSC, using the SCCP connection oriented class 2 with time advance request as a parameter.

In addition to the cell ID, it provides the *time advance* (TA), a measure of the BTS-to-mobile distance, performed by the serving *base transceiver station* (BTS), which allows the system to know that the mobile is within a certain distance from the BTS. If the BTS is a 120° sector, it is a partial circle (Figure 13.2). The accuracy of one TA measurement is 48/13 (period of a timing advance click) μs × 300,000 km/sec = 1,107m, that is, ±554m.

13.3.4 Mobile Location Units and BSSAP-LE

A network that has *mobile location units* (MLUs) and BSSAP-LE represents the most complex method and a mathematically interesting case. The visited cell ID is first obtained as discussed earlier. From the cell ID database, one determines the three or four closest cells' IDs. The MLUs refer to special hardware installed in all BTS. Then the MLC sends proper commands to the visited BTS and to the neighboring BTS so that they simultaneously get measurements of the *time of arrival* (TOA) of a burst sent by the mobile first identified. The *difference in the TOAs* is measured

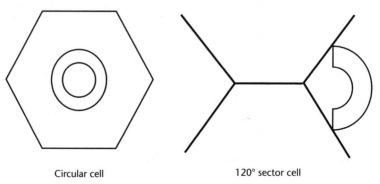

Circular cell   120° sector cell

Figure 13.2 Cell shapes and timing advance.

for each pair of BTS, which gives the difference of distance with $d = c \times$ TOA and propagation delay corrections (where $c =$ speed of light). If we use n hyperbolas for these TOA measurements, we can expect an average error of $\pm 554/\sqrt{n}$.

Knowing that the possible localizations for a mobile that has a given difference of distance (delay) from two BTS is represented by a hyperbola (Figure 13.3), an estimate can be built by the MLC based at the intersection of the hyperbolas.

As you can see from the example, this method refers to least squares distance estimates. The localization time depends on the number of BTS used, because it implies an increase in the number of hyperbolas that are computed to the closest point. This algorithm is integrated in the MLC and involves a method for an unconstrained optimization problem that is nonconvex. For further information, see [5] and Section 13.5.

As illustrated by the example, ambiguity remains between the possible locations A and B and a fourth station measurement is recommended.

13.4   Other Methods: Mobile Measured Power Level

The methods we discuss next are based on mobile power level measures. For the handover procedure, the mobile measures the power received from the visited cell, as well as that of its neighbors. These measurements must receive the proper (secret) [6] commands from your cell phone to be seen and stored:

- Carrier number;
- Reception level (Rx) in dBm;
- Received signal quality,
- Transmission power level;
- Radio link timeout;
- Time slot;
- Identification of the transmitter status;
- Information on the network parameters;

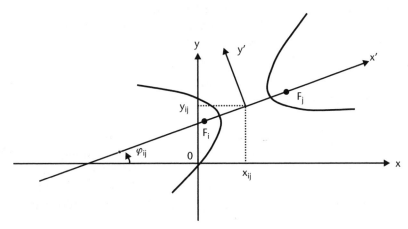

Figure 13.3   Hyperbolic triangulation.

- *Temporary mobile subscriber identity* (TMSI);
- Cell ID;
- Number of cells being used;
- MCC;
- MNC,
- LAC;
- Ciphering (on/off);
- Hopping (on/off);
- Discard cell barred information.

A particular SIM toolkit application in the SIM card may return an SMS-MO with these measurements on receipt of a special SMS-MT. However, this method is not satisfactory for these two reasons:

- *Marketing issue:* This solution requires a SIM toolkit and a special application, both of which depend on the cell phone manufacturer.
- *Technical issue:* Power level measurements with propagation models do not give a reliable distance measurement.

Our opinion is that this estimation is not much better than just having the cell ID.

What about GPS assistance? GPS assistance is quite impractical in many cities. In Hong Kong it was consistently impossible to access even a single stable satellite when at least three are necessary. We can get the cell ID using the cell ID method but that does not mean that we can map this to a geographical location. Only GSM networks that have the MLC and the up-to-date database are able to use this abstract information to obtain a geographical location.

How are we to assess the *n*-triangulation methods that do not make use of GPS assistance in a cell phone? In cities, the lack of precision of the TA [4] will not give a much better estimate than just the cell ID of a cell, which is already crude. In the countryside, with large cells, it will often not be possible to make TOA measurements from more than one BTS. For medium-sized areas, the TOA is quite useful and makes use of the sophisticated algebraic equations resolver of Section 13.6.

Our conclusion is that in most cases, it is not useful to look, as an estimate, at more information than the cell ID.

13.5   3G UMTS Networks

The 3G UMTS networks are designed to easily obtain a location estimate. In these types of networks (although all the computations are done as above by a MLC), it is quite easy for a service provider to obtain a location estimate using only one MAP service, *PROVIDE_SUBSCRIBER_LOCATION* (PSL) (Table 13.3). The location estimates need the full geographical coordinates and information to compute the accuracy.

Table 13.3 PROVIDE_SUBCRIBER_LOCATION_REQ and CNF Message Details

PROVIDE_SUBSCRIBER_LOCATION_REQ

Parameter	Class	Context
Primitive type octet	M	V5
Invoke ID	M	V5
MLC number	M	V5
IMSI or	C	V5
MSISDN (one is mandatory)	C	V5
LCS client ID	M	V5

PROVIDE_SUBSCRIBER_LOCATION_CNF

Parameter	Class	Context
Primitive type octet	M	V5
Invoke ID	M	V5
Where user error not included:		
Location estimate	M	V5
Age of location information	C	V5
Additional location estimate	C	V5
Where user error included:		
User error	M	V5

13.6   Best Estimate of a Location Using Hyperbolic *n*-Triangulation

This method works with more than three BTS so we use the term *n-triangulation*. The mathematics, discussed next, are quite elementary.

13.6.1   Algebraic Equation of a Hyperbola

In an ordinary coordinate system (the tilted $x' \perp y'$ coordinates in the following equation), a hyperbola has the equation:

$$\frac{x'^2}{a_{ij}^2} - \frac{y'^2}{b_{ij}^2} \tag{13.1}$$

with $a^2 + b^2 = c^2$ and where $2c$ is the distance between the two focus points (F_i and F_j, the two base stations of the hyperbola).

13.6.1.1   Notation

Equation (13.2) shows the coordinates of the middle point of the two focus points of a rotated, translated hyperbola and the angle between the main axis that joins the two focus points, φ_{ij}, of the rotated hyperbola and the original Ox axis:

$$X' = \begin{pmatrix} x' \\ y' \end{pmatrix}, \quad X_{ij} = \begin{pmatrix} x_{ij} \\ y_{ij} \end{pmatrix} \tag{13.2}$$

The rotation matrix is

$$R_{ij} = \begin{pmatrix} \cos \varphi_{ij} & -\sin \varphi_{ij} \\ \sin \varphi_{ij} & \cos \varphi_{ij} \end{pmatrix} \tag{13.3}$$

where $R_{ij}^{-1} = R_{ij}^T$, in which T refers to the transposed matrix.

A negative semi-definite matrix [5] is shown in (13.4) that makes our optimization problem difficult:

$$Q_{ij} = \begin{pmatrix} 1/a_{ij}^{2} & 0 \\ 0 & -1/b_{ij}^{2} \end{pmatrix} \tag{13.4}$$

In the original coordinate system, the hyperbola equation is written as follows:

$$X^T Q_{ij} X = 1 \tag{13.5}$$

13.6.1.2   General Rotated and Translated Hyperbola

We now look at the equation for a general rotated and translated hyperbola, such as that shown in Figure 13.4. Replacing X by $R_{ij}^T(X - X_{ij})$ in (13.5), we can see immediately that this hyperbola has the following equation in the $x \perp y$ coordinate system:

$$\left(X - X_{ij}\right)^T R_{ij} Q_{ij} R_{ij}^T \left(X - X_{ij}\right) - 1 = 0 \tag{13.6}$$

13.6.1.3   Computing the Hyperbola from the TOA Difference

The position of the two BTS used in the measurement of the TOA delay difference allows us to compute the angle φ_{ij} (assuming a flat earth). Their distance gives $2c_i$. The measured difference between the TOA is $2a_i$. If you assume $y = 0$, you get point A, and $AF_1 - AF_2 = 2a_i$ as illustrated in the example. This allows us to then compute b_i.

13.6.2   Finding the Best Localization Estimate

13.6.2.1   Optimization Criterion

Let

$$P_{ij} = \left[\left(X - X_{ij}\right)^T R_{ij} Q_{ij} R_{ij} T\left(X - X_{ij}\right) - 1\right] \tag{13.7}$$

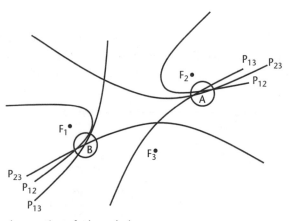

Figure 13.4   Algebraic equation of a hyperbola.

If there are several pairs of BTS measurements, the best estimate is arrived at by trying to find a point close to all the hyperbolas (in this case $P_{ij}=0$). Because all of the hyperbolas will not intersect exactly, a least squares estimate for the mobile location is arrived at by minimizing the square, so that negative and positive values are as important as all of the P_{ij}. We write the equation for this as follows:

$$\min_{X} Q(X) = \Sigma_{ij} P_{ij}^2 (X) \tag{13.8}$$

where *i* and *j* represent the BTS index. This is an unconstrained optimization problem. We look for the point that minimizes (13.8).

13.6.2.2   Hessian of $Q(X)$ and Gradient of $Q(X)$

The Hessian $H(X)$ of $Q(X)$ is the matrix of second derivatives of $Q(X)$. When $X^T H(X)X \geq 0$, we have a local minimum if $\nabla Q(X) = 0$ [the gradient of $Q(X)$]; see [5].

If $X^T H(X)X \geq 0$, $\forall X$, which is not the case here, we have a rather easy convex problem with a single local minimum, because even with three BTS, the local minima to $Q(X)$, such as A and B on Figure 13.2, give large estimates of the location differences.

13.6.3   Exact Solution (True Optimum)

Assuming that the TOA measurements have been performed using four BTS, we have $(4 \times 3)/2 = 6$ hyperbolas. We arrive at an exact solution by transforming the optimization of $Q(X)$ into the solution of the algebraic, two unknown, third-degree equations $\nabla Q(X) = 0$.

Notice that the third-degree unknown equation was partially resolved in radicals by Cartano [7], who could only find the real root because he did not know complex numbers, and eventually fully resolved by Lagrange.

We then evaluate $Q(X)$ for each real number and use that to arrive at the least value:

$$\nabla Q(X) = 2\Sigma_{ij} P_{ij}(X)\Delta P_{ij}(X) = 0 \tag{13.9}$$

System $\nabla Q(X)$ is a system of two third-degree algebraic equations in *x* and *y*, because $Q(X)$ is a fourth-degree equation. It has an odd number of real roots: 1, 3, 5, 7, or 9. To resolve it, we use the resultant method of Macaulay [8] to eliminate one equation and one variable. See details in Section 13.7 for the implementation of the resultant method by means of formal computation software that works with symbols and not numbers, so you have explicitly a ninth-degree equation.

As just mentioned, the method yields a ninth-degree equation in *y* (which cannot be resolved in radicals as shown by Galois [9]), but can easily be solved numerically based on Sturm's theorem [10]. This solution can also be used for GPS computations with *n* satellites, but because we must consider the altitude *z*, the single equation becomes the twenty-seventh degree from third-degree equations in *x, y, z*. The resolving algorithm, which gives the real roots, is explained in Section 13.7.

In Figure 13.2, having found A and B as local minima, it is obvious that there is another extrema (local maxima) at the center of the hyperbolas. It is a necessary condition for an extrema to have $\nabla Q(X) = 0$. The method is more computationally efficient than that offered by a gradient (which could not yield the true global optimum), because it is independent in its computation of the number of hyperbolas.

13.7   Main Results in the Theory of Resultants and Sturm's Theorem

13.7.1   Purpose of the Theory of Resultants

We want to transform the resolution of several algebraic equations of $m, n, ..., t$ degrees with unknown $x, y, ..., z$ into a single algebraic equation of degree $m + n + ... + t$ with coefficients in Q, if the original coefficients are in Q. We use it to find the optimum estimate of the intersection of p hyperbolas of R^2, which yields two third-degree homogeneous equations in x and y. This is the problem we looked at in the preceding section on the location method.

13.7.2   Main Result for Two Algebraic Equations

From Macaulay [8], we get

$$f(x) = a^n x^n + a^{n-1} x^{n-1} + ... + a^0$$

$$g(x) = b^m x^m + b^{m-1} x^{m-1} + ... + b^0$$

A necessary and sufficient condition for f and g to have a common root is that the following $m + n$ determinant must equal 0:

$$
m = \begin{cases}
\begin{vmatrix}
a^n & & a^{n-1\dots} & & & a^0 & & \\
& a^n & & a^{n-1\dots} & & & & \\
0 & & & & & & a^0 & 0 \\
& & a^n & & & a^{n-1\dots} & & a^0 \\
\end{vmatrix}
\end{cases}
$$

$$
n = \begin{cases}
\begin{vmatrix}
b^m & b^{m-1\dots} & & & b^0 & 0 & & 0 \\
& b^m & b^{m-1\dots} & & & b^0 & \dots & 0 \\
0 & & & & & & & \\
& & b^m & b^{m-1\dots} & b^{0\dots} & & & 0 \\
\end{vmatrix}
\end{cases}
$$

This determinant is called the resultant of $f(x)$ and $g(x)$ and is denoted Res(f,g,x).

Proof.   The proof is in two parts:

1. *Lemma:* A necessary and sufficient condition for two polynomials $f(x)$ and $g(x)$ to have a (nonconstant) common root is that there exist an equation of the form

$$h(x)f(x) = k(x)g(x)$$

where $h(x)$ is a polynomial of degree $m-1$ at most and $k(x)$ of degree $n-1$ at most.

Proof: The condition is *necessary:* If $f(x)$ and $g(x)$ have a common root $x = x_a$, they must write

$$f(x) = (x - x_a)k(x)$$

$$g(x) = (x - x_a)h(x)$$

and we can check that

$$h(x)f(x) = h(x)(x - x_a)k(x) = k(x)(x - x_a)h(x) - k(x)g(x)$$

The condition is *sufficient:* Every prime factor of $f(x)$, which is of degree n, corresponding to a root (there are n roots) must divide $k(x)g(x)$. It is not possible for them to divide only the factors of $k(x)$ [because $k(x)$ has at most $n-1$ factors], so they must divide at least one factor $(x - x_a)$ of $g(x)$. We conclude that $f(x)$ and $g(x)$ have at least one common factor $(x - x_a)$ corresponding to a common root $x = x_a$.

2. *Proposition:* If $h(x)\, f(x) = j(x)g(x)$ as above, Res$(f,g,x) = 0$.
    Proof: Let

$$h(x) = c^{m-1}x^{m-1} + c^{m-2}x^{m-2} + \ldots + c^0$$

$$k(x) = d^{n-1}x^{n-1} - d^{n-2}x^{n-2} \ldots - d^0$$

We perform the multiplication of $h(x)$ by $f(x)$, $k(x)$ by $g(x)$ and write that all the coefficients of $h(x)f(x)$ and $k(x)g(x)$ must be equal for the powers $m+n-1$, $m+n-1$, ..., 1, 0 of x.

$c^{m-1}a^n$	$= -d^{n-1}b^m$	(degree $m+n-1$)
$c^{m-1}a^{n-1} + c^{m-2}a^n$	$= -d^{n-1}b^{m-1} - d^{n-2}b^m$	(degree $m+n-2$)
$c^{m-1}a^{n-2} + c^{m-2}a^{n-1} + c^{m-3}a^n$	$= -d^{n-1}b^{m-2} - d^{n-2}b^{m-1}\ldots - d^{m-3}b^n$	(degree $m+n-3$)

$c^{m-1}a^{n-m+1} + c^{m-2}a^{n-m+2} + \ldots + c^0 a^n =$	$-d^{n-1}b^{m-m+1} - d^{n-2}b^{m-m+2} - \ldots - d^0 b^n$	(degree n)
$c^1 a^0 + c^0 a^1 =$	$-d^1 b^0 - d^0 b^1$	(degree 1)
$c^0 a^0 =$	$-d^0 b^0$	(degree 0)

Put the right-hand sides by the left-hand sides of the equation [this was the reason for the minus signs in the coefficients of $k(x)$] so that it yields homogeneous linear equations in $m+n$ unknowns c^{m-1}, c^{m-2}, ..., c_0, d^{n-1}, d^{n-2}, ..., d_0:

$$c^{m-1}a^n \qquad\qquad\qquad\qquad +d^{n-1}b^m \qquad\qquad\qquad\qquad\qquad = 0$$

$$c^{m-1}a^{n-1}+c^{m-2}a^n \qquad\qquad +d^{n-1}b^{m-1}+d^{n-2}b^m \qquad\qquad\qquad = 0$$

$$c^{m-1}a^{n-2}+c^{m-2}a^{n-1}+c^{m-3}a^n \quad +d^{n-1}b^{m-2}+d^{n-2}b^{m-1}+...+d^{m-3}b^n \quad = 0$$

$$c^{m-1}a^{n-m+1}+c^{m-2}a^{n-m+2}+...+c^0 a^n \quad +d^{n-1}b^{m-m+1}+d^{n-2}b^{m-m+2}+...+d^0 b^n \quad = 0$$

$$c^1 a^0 + c^0 a^1 \qquad\qquad\qquad +d^1 b^0 + d^0 b^1 \qquad\qquad\qquad\qquad = 0$$

$$c^0 a^0 \qquad\qquad\qquad\qquad +d^0 b^0 \qquad\qquad\qquad\qquad\qquad = 0$$

So that we arrive at a nontrivial 0 solution, the determinant must be 0. Taking the transpose of this determinant, we get the resultant $\mathrm{Res}(f,g,x)=0$.

Application 1   Double root for a second-degree equation $f = ax^2 + bx + c$; f and f must have a common root:

$$\mathrm{Res}(f,f',x)=\begin{vmatrix} a & b & c \\ 2a & 2b & 0 \\ 0 & 2a & b \end{vmatrix} = a(b^2 - 4ac) \Rightarrow \Delta = 0, \text{ the discriminant!}$$

Application 2   To transform an x,y algebraic equation of the second degree into an algebraic equation in x of the fourth degree:

$$f = a_{20}x^2 + a_{11}xy + a_{02}y^2 +...+ a_{10}x + a_{01}y + a_{00}$$
$$g = b_{20}x^2 + b_{11}xy + b_{02}y^2 +...+ b_{10}x + b_{01}y + b_{00}$$

we write on a two-equation form of x, with coefficients depending on y. We then have:

$$\mathrm{Res}(f,g,x)=\begin{vmatrix} a_{20} & a_{11}y+a_{10} & a_{02}y^2+a_{01}y+a_{00} & 0 \\ 0 & a_{20} & a_{11}y+a_{10} & a_{02}y^2+a_{01}y+a_{00} \\ b_{20} & b_{11}y+b_{10} & b_{02}y^2+b_{01}y+b_{00} & 0 \\ 0 & b_{20} & b_{11}y+b_{10} & b_{02}y^2+b_{01}y+b_{00} \end{vmatrix}$$

By setting $\mathrm{Res}(f,g,x) = 0$, we have a fourth-degree equation in y with coefficient in Q. This is the case when we look for the intersection of two ellipses. The same method gives a 6×6 determinant with an $m \times n$ = ninth-degree equation in y when we look for a common root of two third-degree equations in x and y (approximation of n hyperbola intersections).

The MLC software must supply a formal recursive computation of this determinant, which gives the resulting fourth-degree equation in y only.

13.7.3 Sturm's Theorem

Sturm's theorem [10] allows us to determine how many real roots of a polynomial on R are included in an $[a,b]$ interval.

13.7.3.1   Definition: Variation

The variation $\text{Var}(a)$ of a sequence of real numbers $a = \{a_0, \ldots, a_n\}$ is the number of pairs $(i, i + k)(k \geq 1)$ such that:

$$a_i a_{i+k} < 0$$
$$a_{i+r} = 0 \text{ for } 0 < r < k$$

For example, $\text{Var}(a) = 3$ for the sequence $a = \{-1,0,0,2,0,4,0,0,-5,6\}$ corresponding to the pairs $(0,3)$, $(5,8)$, $(8,9)$ of indices.

13.7.3.2   Definition of the Sturm Sequence

Let $P \in R[X]$. A Sturm sequence for P in $[a,b]$ is a sequence of polynomials $f_0(x)$, $f_1(x), \ldots, f_s(x)$ such that:

$$f_0 = P \tag{13.10}$$

$$fs \neq 0 \text{ in } [a,b] \tag{13.11}$$

$$\text{for } 0 < i < s \text{ and for } \alpha \in [a,b], \text{ if } f_i(\alpha) = 0, \text{ then } f_{i-1}(\alpha)f_{i+1}(\alpha) < 0 \tag{13.12}$$

$$\text{if } f_0(a) = 0 \text{ for } \alpha \in [a,b], \text{ then } f_0 f_1(\alpha - \varepsilon) < 0 \text{ and } f_0 f_1(\alpha + \varepsilon) > 0 \text{ for a small } \varepsilon \tag{13.13}$$

13.7.3.3   Sturm Sequence Theorem

Let $w(x)$ be the variation of a Sturm sequence. Then $w(a) - w(b)$ is the number of real roots of $f_0(x)$ in $[a,b]$.

Proof   Let's consider a root α of $f_0(x)$:

	$\alpha - \varepsilon$	α	$\alpha + \varepsilon$		$\alpha - \varepsilon$	α	$\alpha + \varepsilon$
$f_0(x)$	$-$	0	$+$		$+$	0	$-$
$f_1(x)$	$+$	$+$	$+$	OR	$-$	$-$	$-$

because of (13.13) and of the continuity of $P(x)$, so that in any of the two preceding cases $w(x)$ decreases by 1 when x goes from $\alpha - \varepsilon$ to $\alpha + \varepsilon$.

Let us now consider a root $f(x)$ of $f_i(x)$, $0 < i < s$, for which we have four cases:

	$\alpha - \varepsilon$	α	$\alpha + \varepsilon$		$\alpha - \varepsilon$	α	$\alpha + \varepsilon$
$f_{i-1}(x)$	$+$	$+$	$+$		$-$	$-$	$-$
$f_i(x)$	$-$	0	$+$		$+$	0	$-$
$f_{i+1}(x)$	$-$	$-$	$-$	OR	$+$	$+$	$+$

	$+$	$+$	$+$		$-$	$-$	$-$
OR	$+$	0	$-$	OR	$-$	0	$+$
	$-$	$-$	$-$		$+$	$+$	$+$

The contribution to $w(x)$ for $\alpha - \varepsilon \le x \le \alpha + \varepsilon$ of the subsequence $f_{i-1}(x)$, $f_i(x)$, $f_{i+1}(x)$ remains constant in all four cases. So $w(x)$ changes only (decreases by 1) when there is a root of $f_0(x)$. So $w(a) - w(b)$ will be equal to the number of roots in $[a,b]$.

13.7.3.4 Quotient Sturm Sequence

The algorithm in this section allows the building of a practical Sturm sequence. This sequence will then give a way of building a practical Sturm sequence. Let us then define the following sequence:

$$f_0(x) = P \in R(x), x \in [a,b]$$

Assume that P does not have multiple roots:

$$f_1(x) = f'(x)$$

$$f_{i-2}(x) = f_{i-1}(x)g_i(x) - f_i(x)$$

so that $g_i(x)$ is the Euclidean quotient of $f_{i-2}(x)$ by $f_{i-1}(x)$.

We have (13.12) by definition of the sequence and (13.11) because we assume that there are no multiple roots. We have (13.13) because P is a continuous function. Thus, the defined sequence is a Sturm sequence. For example:

$$\begin{aligned} f_0 &= x^3 + 3x^2 - 1 &= P \\ f_1 &= 3x^2 - 6x &= P' \\ f_2 &= 2x + 1 \\ f_3 &= 9/4 \end{aligned}$$

13.7.3.5 Theorem

The number of real roots of P is $w(-\infty) - w(\infty)$.

Proof. The proof is obvious because of Section 13.7.3.3. As a consequence of P being polynomial, the number of roots of P depends only on the sign of the leading coefficients (higher power) of the Sturm sequence. In the preceding example:

$$w(-\infty) = \text{var}(-\infty, +\infty, -\infty, 9/4) = 3$$
$$w(\infty) = \text{var}(\infty, \infty, \infty, 9/4) = 0$$

Thus, there are three real roots.

13.7.4 Bounds on the Value of Roots

Let

$$P(x) = x^n + a_{n-1}x^{n-1} + \ldots + a_0$$

13.7.4.1 First Bound

If is α root, we have $|\alpha| < 1 + \sup|a_i|$.

Proof   Let $A = \sup|a_i|(A > 0)$:

$$|P(x)| \geq |x|^n - A\left(|x|^{n-1} + \ldots + 1\right) = |x|^n - A\frac{\left(|x|^n - 1\right)}{|x| - 1}$$

If we had $|\alpha| \geq 1 + A > 0$, then by dividing by x we would have $1 \geq \dfrac{A}{|a| - 1}$, and

then $|\alpha|^n \geq \dfrac{A|\alpha|^n}{|\alpha| - 1}$, which implies:

$$P(\alpha) \geq \frac{A|\alpha|^n}{|\alpha| - 1} - \frac{A\left(|\alpha|^n - 1\right)}{\left(|\alpha| - 1\right)} = \frac{A}{|\alpha| - 1}$$

which is > 0, and α could not be a root. This theorem allows us to have the bounding interval $[-(1+A), 1+A]$ to look for the real roots. Reference [10] discusses better bounds from Lagrange and Cauchy.

13.7.4.2   Lagrange Bound

Let $m = \arg \sup\{a_i < 0\}$, that is, the largest index of an $a_i < 0$, and $B = \sup\{-a_i \mid a_i < 0\}$, with by definition $B = 0$ if there are no $a_i < 0$. Then $\alpha < 1 + B^{1/(n-m)}$.

Proof   Here is our proof:

$$0 = P(x) = \alpha^n + a_{n-1}\alpha^{n-1} + \ldots + a_0$$
$$\geq \alpha^n - B\left(\alpha^m + \alpha^{m-1} + \ldots + 1\right)$$
$$= \alpha^n - B\frac{\left(\alpha^{m+1} - 1\right)}{\alpha - 1}$$
$$0 > \alpha^{m+1}\left[\alpha^{n-m-1} - B/(\alpha - 1)\right], \text{ that is,}$$
$$\alpha^{m+1}\frac{B}{\alpha - 1} > \alpha^n, \text{ that is,}$$
$$B > \alpha^{n-m-1}(\alpha - 1) \text{ and as } \alpha > \alpha - 1$$
$$B > (\alpha - 1)^{n-m}, \text{ which gives } \alpha < 1 + B^{1/n-m}$$

In the preceding example, we have $B = 1$, $m = 0$, and $n = 3$, this gives $1 + 1^{1/3} = 2$, when Section 13.7.4.1 was giving $1 + 3 = 4$.

13.7.5   Application: Recursive Algorithm to Find All the Real Roots

What follows is a recursive algorithm that can be used to find all of the real roots.

```
    void SOLVE(int *nbRoot, float *Root, int *nbPoly, float **Sturm-
Poly, float A, float B, float Accuracy)
    /*- - - - - - - - - - - - - - - - - - - - - - - - - - -*/
    /* FUNCTION: solves f0 = 0 by a recursion on the interval A,B      */
```

```
/* the number of real roots in [A,B] is computed by V(A)-V(B).
      Where*/
/* V(X) is the number of variations of the Sturm polynomials */
/* f0(X),f (X),f2(X),...,fs (SturmPoly)                      */
/* ENTRY: A, B and the Accuracy                              */
/*          the Sturm Polynomials SturmPoly                  */
/*          the number of Sturm Polynomials nbPoly           */
/* RETURNS: nbRoot The number of real roots between [A,B]    */
/*      Root the table of values of the roots                */
/*    A.Henry-Labordere 12/2/2002                            */
/*                                                           */
{
int VA,VB;
    VA = VAR(nbPoly,SturmPoly,A);   // Variation(A)
    VB = VAR(nbPoly,SturmPoly,B);        // Variation(B)
    if ((VA-VB) == 1 && (B-A) Accuracy)// VA-VB is number of
          roots!
      {
      Root[*nbRoot] = (B+A)/2;   // if interval small and 1 root
      *nbRoot = *nbRoot + 1;       // the middle = the root !
      return;
      }
    else if ((VA-VB) == 0)      // no root, returns
      return;
    else
      {     // Halve the interval and resolves 2 problems
      SOLVE(nbRoot,Root,nbPoly,SturmPoly,A,A+(B-A)/2,Accuracy);
      SOLVE(nbRoot,Root,nbPoly,SturmPoly,A+(B-A)/2,B,Accuracy);
      return;
      }
}
 f0 = 1.000 X**5 2.000 X**4 -5.000 X**3 8.000 X**2 -7.000 X**1 -3.000
 X**0
```

From 13.6.4.1,the absolute value of all the roots are 1 + sup|ai|=
        1+8 = 9.000

Here is the Quotient Sturm sequence of 6 polynomial computed by
        13.6.3.4:
```
 1.000 X**5 2.000 X**4 -5.000 X**3 8.000 X**2 -7.000 X**1 -3.000 X**0

 5.000 X**4 8.000 X**3 -15.000 X**2 16.000 X**1 -7.000 X**0
 2.640 X**3 -6.000 X**2 6.880 X**1 2.440 X**0
-15.978 X**2 39.084 X**1 24.897 X**0
-12.113 X**1 -3.153 X**0
-13.640 X**0
```

```
w(-Max)= 4 w(Max)= 1
The number of roots of f0 is then 4 - 1 = 3 (theorem 13.6.3.5) which
     are:
root = -3.90780, the value of f0 = 0.00087
root = -0.30233, the value of f0 = -0.00005
root = 1.30682, the value of f0 = 0.00005
```

References

[1]   *Location Services (LCS)—Stage 2 (GSM 03.71)*, ETSI TS 101 974. v7.3.0

[2]   *Mobile Application Part (MAP) Specification: (GSM 09.02)*, ETSI TS 100 974. v7.5.0, pp. 137–138.

[3]   *Location Services (LCS) Base Station Application Point LCS Extension (BSSAP-LE): (GSM 09.31)*, ETSI TS 101 530, v8.1.0, 2000–2005.

[4]   *Location Services (LCS) Serving Mobile Location Center. Base Station System (SMLC-BSS) Interface, Layer 3 Specification (GSM 08.71)*, ETSI TS 101 726, v8.1.0, 2000–2005.

[5]   Henry-Labordère, A., *Recherche Opérationnelle*, Paris, France: Presses des Ponts et Chaussées, 1995.

[6]   http://www.astalavista.com/mobile/6110.shtml.

[7]   Cartano, G., *Artis Magnae Sive de regulis algebraicis liber unus*, 1545; translated by T. R. Witner, *Cartano: Great Art or Rules of Algebra*, Cambridge, MA: MIT Press, 1968.

[8]   Macaulay, F. S., *The Algebraic Theory of Modular Systems*, Cambridge, England: Cambridge University Press, 1916, reedited by Stechert-Hafner, 1964.

[9]   Tignol, J. P., *Galois' Theory of Algebraic Equations*, Singapore: World Scientific, 2001.

[10] Benedetti, R., and J. J. Risler, *Real Algebraic and Semi-Algebraic Sets*, Paris, France: Hermann, 1990.

SMS-MO Premium Number Services and Architectures

> La Flèche: As-tu quelque négoce avec le patron du logis?
> Frosine: Oui, je traite pour lui quelque petite affaire, dont j'espère une récompense.
> La Flèche: De lui? Tu sera bien fine si tu en tires quelque chose.
> Frosine: Il y a certains services qui touchent merveilleusement.
>
> Do you have some business with the boss?
> Yes, I have a deal which I hope a reward for.
> From him? You will be sharp if you get anything.
> There are services which bring a lot of cash.
>
> —Molière, *L'Avare*, 1668, Acte II, Scène 4, *The Miser*

14.1 The Premium SMS-MO Number Business

Today everyone's heard of or participated in SMS voting games on TV, in which the viewer casts an "instant" vote for or against a person or situation. In such a case, the viewer sends a premium SMS, which is much more expensive than a person-to-person SMS. The normal SMS-MO price is charged by the visited mobile operator, and a special additional charge is shared between the mobile operator, the content provider, and eventually a third-party SMS interworking network operator, which may also act an intermediate and redistribute this revenue to the content providers.

In this chapter, we discuss the various architectures that allows an SMS-MO of this type to be sent to content provider. Bu first we discuss the classical implementations and their limits. In Chapter 5, we discussed situations in which we wanted to bar the inbound SMS; now we want to attract them and, in doing so, make some revenue.

14.1.1 Use of a GSM Modem: Small Throughput

Consider a situation in which a cell phone (commonly called a modem, which is just the electronic core of a cell phone without a case and a keyboard) is connected to the server of content provider C. The cell phone receives the SMS that has been sent to its number and transmits it with the standard AT serial protocol to the content provider's server [1]. As explained in Chapter 1, because of the queuing to this number and the 5-second paging delay of the MAP_FORWARD_SM_MT, the throughput cannot exceed 7 to 10 SMSs per second. The modem can receive, albeit slowly, SMS from any subscriber of a mobile operator that has roaming agreements with the destination number (provided there is no SMS-MT barring). However, the setup is very clumsy (some content providers have racks with more than 100 modems, each of which must keep a valid subscription), which makes it difficult to provide good service.

14.1.2   Use of a Direct IP Connection to an SMSC: Negotiation and Setup Tasks

Content provider C is connected by a protocol such as SMPP (see Chapter 8) to an SMSC A with which it has a contract. Each SMS received by SMSC A for the code (short) associated with the content provider is routed over the IP connection. The accounting is performed by the SMSC and used by the SMSC to give a fee (sometimes more than $1) to the content provider for each SMS that it receives and handles. The throughput can be very important, but the SMS can only be received from A's subscribers, because they are the only ones whose SMS-MOs go through SMSC A.

At the time of this writing, this is the only practical setup available for receiving large payments from a mobile operator. (Total revenues in 2003 were about $4 billion.) But this situation requires a direct connection with every operator for revenue sharing. In addition, the IP interface may need to be customized for each connection. For these reasons, only local or very big content providers can participate in offering this service.

14.2   Virtual Roaming Subscriber Architecture

We now explain other SS7 architectures that combine (1) receiving SMS from all roaming partners; (2) large throughputs, although they are smaller than with a direct IP connection; and (3) the possibility of sharing small revenues (the AA19 SMS-MT charge) for each SMS received and sent to the content provider. These three choices represent quite good techniques for conducting a worldwide promotion campaign, such as allowing the user to send an SMS worldwide to a unique number to win a prize.

Assume that a mobile operator wants to promote an SMS-MO-based service, with a unique national destination number (one of its numbers, such as +39341234567). In such a case, SMS interworking must exist between all operators in the same country. Applicable examples are Italy, Germany, France, and the United Kingdom. For instance, if a Wind cell phone sends an SMS to a Omnitel cell phone (using the normal Wind SMSC), it will reach the Omnitel cell phone.

Obviously using modems simulating the cell phone +39311234567 is out of question because of the number of SMS per second that it must receive. (It could be at most 10 SMSs per minute and would occupy much of the radio spectrum in the cell in which it is installed, and the network operator would not like it.)

Using the preceding scenario about the Wind and Omnitel cell phones, we now look at two specific cases.

14.2.1   Case 1: Omnitel and Third-Party Operator

Here we consider the case in which Omnitel does not have any equipment in its network, so it has to use a third party, Operator C, in another network (with roaming, of course). After they have received the SMS-MO, Blu, Tim, Wind, and Omnitel's SMSCs will interrogate the Omnitel HLR that corresponds to +39341234567 in order to send an SMS-MT (Figure 14.1).

Looking at Figure 14.1, on this HLR, Omnitel (A) will have done the following setup: Declare, using a man/machine command (1), that this number is roaming on third-party C's virtual MSC, which gives them the GT address in the other country where the node is installed. The HLR returns (4) the address of MSC C to the interrogating SMSC to which the SMS-MT will be sent (5), that is, to the IMSI corresponding to +39341234567, because they have a roaming agreement with this virtual MSC. The third-party MSC, which holds a correspondence IMSI ↔ +39341234567, will then route (6) the message to an *application server* (ASP).

14.2.2   Case 2: Mobile Operator Has a Virtual MSC

This case is the same as Case 1, except that the mobile operator has purchased a virtual MSC and the equipment is installed in its own network. In this way, the operator can administer by themselves all the setup functions required by a new SMS-MO application.

This business case is the simplest one, because the negotiations on revenue sharing have to be conducted with only one operator. Remember that, more and more, the receiving network (Omnitel in the example) will receive a payment from the sending network (Wind, for example) and can thus share revenue with the ASP even for SMS received from its competitors. In Step 7 of Figure 14.1, the third party sends a CDR to Omnitel so that that it can claim payment on Wind (and then give some part of the revenue to the third party).

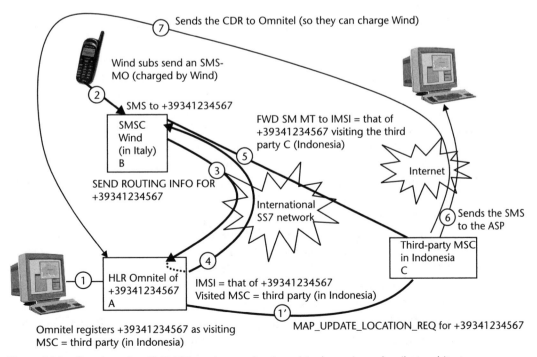

Figure 14.1   Creation of an SMS-MO service number in a virtual roaming subscriber architecture.

No radio paging is involved when third-party MSC C sends [(6) in Figure 14.1] the SMS to the ASP. So the throughput will be about 0.5 SMS per second for each SMSC that sends the SMS to the number +39341234567 because each is queuing until it has received an acknowledgment (5). So if it is a worldwide service, with hundreds of possible SMSCs, *the throughput will be limited only by the SS7 bandwidth available to MSC C.*

14.3 SMS-MO with a Real SIM Card

Now consider the case in which numbers are created by the SMS-MO using a real SIM card. This is a VLR procedure that is required to localize a mobile phone in third-party node C.

If C is not interested in receiving a share of the revenue from A (say, if it has revenues from the content provider, which is paying for the marketing campaign), it does not need A to do the setup. Third-party C merely buys the subscription for +39341234567 and gets the SIM card, which it stores, without physically using it, which is totally unlike the case with the modem in Section 14.1.1). Buying the subscription is done merely to make just so the transaction legal and does not disrupt an existing subscriber of A.

If third-party C has VLR functionality in its equipment, it can do a MAP_UPDATE_LOCATION_REQ to simulate the cell phone +39341234567 visiting its VLR as explained in Figure 1.4 in Chapter 1. This leads to the exact same result as if Operator A had done this using the man/machine interface to its HLR. This is the alternative (1′) instead of (1) in Figure 14.1.

14.4 Short Code: A Costly and Time-Consuming Setup

For marketing reasons, the number used for the SMS-triggered services should be a simple one, that is, a short code. To secure unique short code numbers, such as 25123, in at least one country, it is necessary to rent or reserve a list of such numbers with all operators in this country and, if one is in use, you must restart the negotiations. This is costly and adds to the cost of establishing direct links (with proprietary configurations) to all the SMSCs of this country. By not being part of the numbering plan, the SMS to these short codes cannot be routed naturally as is true of the preceding methods, which are vastly simpler and more cost-effective because they rely on normal operator mechanisms.

Use of the short code is completely unrealistic for worldwide service, because, from a marketing viewpoint, the SMS-MO numbers must be unique in a given country and preferably over a community (Europe). Only long numbers such as +628120109abc or +39341234567, which belong to the international numbering plan, are by definition unique.

New numbers such as +3383664abcd and +3383665abcd, could be introduced for the SMS-MO business in their international form and their national form (083664abcd and 083665abcd), but as part of a numbering plan that would guarantee their uniqueness. They should be decided by the various regulatory bodies. The

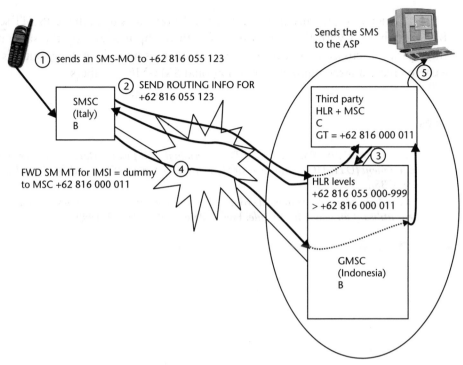

Figure 14.2 FSG architecture.

following discussion of the *foreign subscriber gateway* (FSG) architecture explains their setup, which is quite easy.

14.5 FSG Architecture

The setup of an international SMS-MO number requires the cooperation of A (to whom the number belongs) in Section 14.2, no cooperation in Section 14.3 (if C subscribes to a number and a SIM card from A). In the FSG architecture (Figure 14.2), C will require the cooperation of a mobile operator in the same country that hosts the MSC of C. So C benefits from its roaming agreement. The method used is almost identical to the connection method of Chapter 7. The third party has equipment that has HLR and MSC functions.

When a cell phone sends an SMS-MO [(1) in Figure 14.2] from Italy to +62816055123 (one of the MO numbers allocated to GSM C of Operator C) it will reach SMSC B. This will send over the international SS7 network a MAP_SEND_ROUTING_INFO_FOR_SM_REQ message to the called party SCCP +62811055123, which will reach the GMSC of the Indonesian GSM operator.

This one has created an additional HLR for the 1,000 numbers +62 816 000 xxx with HLR levels in its GMSC. So the MAP_SEND_ROUTING_INFO_FOR_SM_REQ message will reach C. As usual ([2] and Chapter 7), it will return its own GT (3) as the visited MSC and a dummy IMSI (it does not matter). SMSC B will send (4) the SMS to the called party number +62 816 000 011, which will reach third-party C, which will pass it with an IP connection (5) to the content provider.

If Partner C of the third party collects AA19 revenues from the SMS-MT sent by B, it can share it with the third party. The throughput is exactly the same as the throughput discussed in Section 14.2 because there is no radio paging, but it is quite easy to create a large number of international SMS-MO numbers.

References

[1]   ETSI, *GSM 07.05, V5.5.0, Use of Data Terminal Equipment, Data Circuit Terminating Equipment (DTE-DCE) Interface for Short Message Service (SMS) and Cell Broadcast Service (CBS)*.

[2]   Henry-Labordère, A., *Système de routage dynamique d'un message court émis par un émetteur utilisant un téléphone mobile,* French Patent FR-9906989, 1999.

## CHAPTER 15

Numbering Plan Creation and Maintenance Algorithms

Hello, said the little prince
Hello, said the points-man
What are you doing here?
I sort the passengers by bundles of one thousand,
I dispatch the trains once on the right, once on the left.
—Antoine de Saint-Exupery, *The Little Prince*

15.1   Purpose of Computing Numbering Plans for an SMS Interworking Network

In Chapter 1, we saw that the addressing of a HLR to obtain the information to send an SMS-MT uses the E164 routing tables in the GSMC of the destination network, which does not raise any adverse issues. Until 1999 this was not the case. Up to that point, many mobile operators had not commercially opened their SMS-MT inbound services so they had had no need for the tables. To send SMS to these networks, it was necessary for the third-party SMS network to build its own HLR numbering plan, which allows us to obtain the HLR GT from the MSISDN (GSM network) or MIN (IS-41 networks).

This is still important when one wants to send SMS-MT to certain countries, for example, Macau [1], that have unregulated MNP. That is, the regulator does not oblige each operator to terminate the calls or the SMS-MT sent to a ported-out subscriber. So we look in this chapter at how an accurate HLR numbering plan is computed to avoid the spanning of all HLRs, which mobile operators do not like to do because it loads their HLRs.

Another need is to build the full list of all MSCs of operators in multiple time zones (United States, Canada, Russia, Australia). Remember that a good SMS-MT service should be able to display on the cell phone the local time of reception, which depends on the time zone of the visited MSC. So we should know all the MSCs and enter their time zones in a table, so that the sending SMSC can compute the correct time from the GT of the visited MSC, which is obtained from the HLR interrogation. We will see in this chapter how the use of probability allows us to decide when a full list has been obtained based on the total number of SMS-MTs sent to such operators.

15.2   Entropy of a Numbering Plan as a Quality Indicator

15.2.1   Avoiding the Multiple Spanning of HLRs

The numbering plan (also called HLR levels) defines the address (GT) of the HLR containing the routing information of all the number intervals of a given mobile operator. This numbering plan becomes very complex even for a medium-sized operator (for instance, the main Moroccan GSM operator has more than 300 different intervals for five HLRs).

If a precise address is not available, it is necessary to try all the HLRs of this operator until the destination handset is found (we call it *HLR spanning*, not "spamming"), the maximum number of tries being h if h is the total number of HLRs of the destination home PLMN. The average number of tries will be (without a strategy better than a sequential search [2]):

$$(h+1)/2 (\text{for a fully random HLR numbering plan})$$

If the precise address of the HLR is known, only this HLR is interrogated using the MAP_SEND_ROUTING_INFORMATION_REQ for a GSM network or an SMS_REQUEST_REQ for an IS-41 network.

Successive attempts to send SMS to various numbers allow us to build sequentially a more precise numbering plan for each operator, so the average number s of HLR interrogations will be a decreasing function of the number of SMS samples, such as

$$1 \le s \le (h+1)/2 (\text{for a fully random HLR numbering plan})$$

When we send a lot of SMS and get back the HLR for each number, the algorithm of Section 15.3 computes step by step the HLR numbering plan. The experimental convergence rate is much better than linear because the interval numbers are chosen by the operators to start and finish on xxxx...000, xxxx...999 and the algorithm uses this knowledge, so that a small number of samples allows for accurate estimates of the intervals.

So when the numbering plan becomes perfect (when we have more samples to estimate it), we will have only one HLR interrogation, that is:

$$s = 1 (\text{for a fully random HLR numbering plan})$$

In this case, we minimize the load of the sending SMSC and avoid any unnecessary interrogations of HLRs, called *HLR spanning*. A quality indicator shows the accuracy of the numbering plan included in the SMS interworking network. To do this, we use as an analogy the statistical entropy of a set of elements (the sending of each SMS here) used in information theory or thermodynamics.

15.2.2   Average Entropy of the Numbering Plan

Information theory [3] defines

$$\log_2 s_i$$

as the information quantity necessary to specify the HLR of the subscriber i to which we send an SMS-MT, where s_i is the average number of HLRs interrogation for this subscriber number, which is expressed in bits. We define below the individual operator's and the global entropy-based quality indicators. This entropy-based indicator is good for showing small improvements in the overall quality once the numbering plan converges.

15.2.2.1   Entropy S_j of the Numbering Plan of a Given Operator j

In the following, the term card(U) stands for the number of elements of a finite set U. Noting O_j, the set of SMS-MT samples sent to operator j, we define S_j as follows:

$$S_j = \sum_{i \in O_j} (\log_2 s_i)/\mathrm{card}(O_j)$$

Note that $S_j = 0$ if the numbering plan is fully accurate.

15.2.2.2   Global Entropy S of the Numbering Plan

Instead of averaging for a given operator, we average the global set of all SMS-MT samples, O, over all of them:

$$S = \sum_{i \in O} (\log_2 s_i)/\mathrm{card}(O)$$

15.2.3   Resulting Global Entropy

The value of the global entropy S for all of the GSM networks was below 0.10 bit in 2000. For a particular operator such as Vodafone (United Kingdom), S_j had gone from 4.5 (more than 40 HLRs at the time) initially to 0.14 bit, as more samples were recorded.

With GMSCs now implementing the translation of MSISDN to HLR GT for almost all networks, S is close to 0 for the worldwide numbering plan, which means that the numbering plan is perfect (i.e., HLR spanning is not needed anymore).

15.3   "Little Prince" Algorithm to Compute an HLR Numbering Plan

Consider a Moroccan operator with the MSISDN range +21236000100 to +2123699999 and four HLRs with the GTs +21236000001 to 4. We assume that each interval of the HLR numbering plan is at least 1,000 numbers (to decrease the number of samples required to build the plan). Set an initial numbering plan that arbitrarily assigns all numbers to HLR 1 (see Table 15.1).

Table 15.1   Initial Numbering Plan (No Samples Yet)

MSISDN Interval		HLR
+21236000100	+21236999999	+21236000001

15.3.1 Numbering Plan After One Try

We send an SMS-MT to +21236897250. It is not found in the HLR +21236000001 (see Chapter 1, unknown subscriber error in the SEND_ROUTING_INFO_ FOR_SM_CNF. So the SMSC that implements this algorithm starts interrogating HLR +21236000002 then +21236000003, which returns the information for this number. We split the initial interval *in two disjoined and adjacent intervals*, using the hypothesis that every interval ends with 999 (see Table 15.2).

15.3.2 Numbering Plan After Two Tries

We send an SMS-MT to +21236923120, which is assumed to be found in HLR +2123600001, so we do not modify the numbering plan, then to +21236907123 found in HLR +21236000004 (see Table 15.3).

15.3.3 Numbering Plan After Three Tries

Then we find +21236145234 in HLR +21236000001 (see Table 15.4). The algorithm is thus quite simple to implement and shows that a third-party SMS network can compute the HLR numbering plan very efficiently, so there is no HLR spanning.

15.4 MSC Search Problem

As explained in Section 15.1, the problem is to find various estimates of the number of tries we must make to find the list of all MSCs of a given GSM operator by interrogating the HLRs with valid mobile numbers belonging to this operator and obtaining the visited MSC number. There are various practical cases where N (the number of MSCs of this network) is known or not known. This will be used to

Table 15.2 Numbering Plan After One Try

MSISDN Interval		HLR
+21236000100	+21236897999	+21236000003
+21236898000	+21236999999	+21236000001

Table 15.3 Numbering Plan After Two Tries

MSISDN Interval		HLR
+21236000100	+21236897999	+21236000003
+21236898000	+21236907999	+21236000004
+21236908000	+21236999999	+21236000001

Table 15.4 Numbering Plan After Three Tries

MSISDN Interval		HLR
+21236000100	+21236145999	+21236000001
+21236146000	+21236897999	+21236000003
+21236898000	+21236907999	+21236000004
+21236908000	+21236999999	+21236000001

decide how to stop a search for the list of MSCs. Due to the number of different mobile networks that exist, it is useful to have a procedure to search MSCs in order to keep their lists up to date for a given mobile operator. There are several practical applications for this. This problem is thus quite different from that of the HLR search problem. Let's take a look at some problems.

15.4.1   Problem 1

Knowing N, find the initial average number of searches required to obtain all of the different objects N. This problem is similar to that of children who need to buy many different packs of soccer players' cards to fill their keepsake albums. Near the end, they will need to buy thousands(!) of packs to get the last pictures. (Obviously, they will team with other children and exchange; otherwise, this activity would be very costly.) So Problem 1 gives the average cost of a completely filled album.

15.4.2   Problem 2

Find the probability that $N = j$ exactly, if we have found j different MSCs after i searches, so as to have a stopping rule if this probability is close to 1. It is like having a bag with marbles that are all different colors, but we do not know how many colors there are and we draw balls many times while replacing them. When can we estimate that we have drawn all the colors?

15.5   Definitions and Properties

Let $P_N(i,j)$ = Probability {that j MSC have been found after i samples, if the number of MSC = N}. We have

$$\text{Probability }\left\{\text{to find a new MSC if } j - 1 \text{ have been found already}\right\} = \frac{N+1-j}{N}$$

$$\text{Probability }\left\{\text{to find a previous MSC if } j \text{ have been found already}\right\} = \frac{j}{N}$$

We then have the following recursive formulas:

$$P_N(i, j) = P_N(i-1, j-1) \times \frac{N+1-j}{N} + P_N(i-1, j) \times \frac{j}{N} \tag{15.1}$$

with $j = [0, N], i = [0, \infty]$

$$P_N(0,0) = 1 \tag{15.2}$$

(because it is certain that we have found 0 MSCs when there are 0 samples).

$$P_N(i, j) = 0 \text{ for } j > i \tag{15.3}$$

(because we cannot find i MSC in less than i samples). Then let

$K_N(i,j) =$ probability {to find the jth MSC on the ith sample}

We have

$$K_N(i,j) = P_N(i-1, j-1) \times \frac{N+1-j}{N} \tag{15.4}$$

$K_N(i,N) =$ Probability {to find the Nth MSC on the ith sample}
and (15.4) simplifies to

$$K_N(i,N) = P_N(i-1, N-1) \times \frac{1}{N} \tag{15.5}$$

By using (15.1), (15.2), and (15.3), we can build Table 15.5 (case $N = 6$):
In Table 15.5 for P_N, we have the following property:

$$P_N(i,j) = \frac{N!}{N^i(N-i)!} \quad \text{for } i \le N; \text{ otherwise } = 0 \tag{15.6}$$

Table 15.5   Case $N = 6$

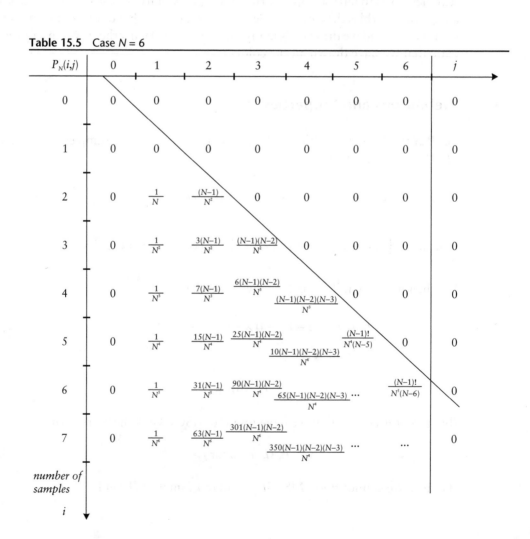

Proof   Because of (15.3) we can write, using the definition of P_N for $i \leq N$,

$$P_N(i,i) = P_N(i-1,i-1) \times \frac{N+1-i}{N}$$

and using (15.2),

$$P_N(i,i) = 1 \times \frac{N-1}{N} \times \frac{N-2}{N} \times \ldots \times \frac{N-(i-1)}{N}$$

By multiplying the denominator and numerator by N and $(N-i)!$, we have

$$P_N(i,i) = \frac{N(N-1)(N-2)\ldots(N-(i-1))(N-i)!}{N^i(N-i)!} = \frac{N!}{N^i(N-i)!}$$

Let us define the series $C_N(i,j)$ as follows:

$$C_N(i,j) = C_N(i-1,j-1) + C_N(i-1,j) \times j \tag{15.7}$$

with $C_N(i,j) = 0$ if $j > N$ (for $j \leq N$, they are independent of N), we have the property that generalizes (15.6).

We can also derive $C_N(i,j)$ from $C(i,j) = C(i-1, j-1) + C(i-1, j) \times j$ given in Table 15.6, which is *independent of N*.

$$P_N(i,j) = \frac{N!}{N^i(N-j)!} C_N(i,j) \tag{15.8}$$

(which makes it very easy to compute the elements of the table P_N).

Proof of (15.8)   Assuming that (15.8) holds, we use the definition of (15.1), which gives

$$\frac{N!}{N^i(N-j)!} C_N(i,j) = \frac{N-j+1}{N} \times \frac{N!}{N^{i-1}(N-(j-1))!} C_N(i-1,j-1)$$

$$+ \frac{j}{N} \times \frac{N!}{N^{i-1}(N-j)!} C_N(i-1,j)$$

This simplifies to:

$$C_N(i,j) = C_N(i-1,j-1) + jC_N(i-1,j)$$

so (15.7) is true.

We have the following property for the series C_N:

$$\sum_j \frac{C_N(i,j)}{(N-j)!} = \frac{N^i}{N!} \tag{15.9}$$

Table 15.6   Table of Coefficients C

C(i,j)	0	1	2	3	4	5	number of found MSC 6	7	j
0	1	0	0	0	0	0	0	0	
1	0	1	0	0	0	0	0	0	
2	0	1	1	0	0	0	0	0	
3	0	1	3	1	0	0	0	0	
4	0	1	7	6	1	0	0	0	
5	0	1	15	25	10	1	0	0	
6	0	1	31	90	65	15	1	0	
7	0	1	63	301	350	140	21	1	
8	0	1	127	966	1701	1050	266	28	1

i
number of
samples

Proof of (15.9)   Equation (15.9) is proved as a result of the definition (15.8) and the definition of $P_N(i,j)$ as a probability: $\sum_j P(i,j) = 1$.

15.6   Problem 1: Average Number of Searches for a Known *N*

This can be analytically computed for small N ($N = 2$ and 3); for larger values of N there is no simple analytical solution, and you may use Table 15.6 instead.

15.6.1   Case *N* = 2 MSCs

We need to compute $K_2(i, 2)$, the probability of finding the second MSC (last) at the *i*th sample, and then find the average of *i*. Without using (15.5), we find directly:

$$K_2(i,2) = \frac{1}{2^{i-1}}$$

(all the previous tests i have given the same MSC).

$$\text{Average } M_2 = \sum_{i=2} \frac{i}{2^{i-1}}$$

This is computed using continuous series (here $x = 1/2$), because $x < 1$ (which makes the series converge):

$$\sum_{i=2} ix^{i-1} = \left(\frac{d}{dx}\Sigma x^i\right) - 1 = \frac{d}{dx}\frac{1}{(1-x)} - 1$$
$$= \frac{1}{(1-x)^2} - 1$$

Thus, with $x = 1/2$, we find:

$$M_2 = 3 \text{ searches in average, for } N = 2 \text{ MSCs}$$

15.6.2   Case N = 3 MSCs

If we find the third MSC on search j, we can do it directly without using (15.5) and the sequence must be as follows:

$$a\,b\,a\,a\,b\,a\,c$$

which means that the $i - 1$ previous samples must not contain any c and cannot contain all a or all b (otherwise we would not find the third MSC on the ith sample). The number of such sequences is

$$2^{i-1} - C_i^1 - C_i^i = 2^{i-1} - 2$$

Thus, the probability is given by the total number of sequences corresponding to the result divided by the number of all possible sequences.

$$K_3(i,3) = \frac{(2^{i-1} - 2)\times 3}{3^i} = \left(\frac{2}{3}\right)^{i-1} - 2 \times \left(\frac{1}{3}\right)^{i-1}$$

$$\text{Average } M_3 = \sum_{i=3} iK_3(i,3)$$

Using the same method as for $N = 2$, we find easily:

$$\sum_{i=3} i\left(\frac{2}{3}\right)^{i-1} = \frac{1}{(1-2/3)^2} - 1 = 8$$

$$\sum_{i=3} i\left(\frac{1}{3}\right)^{i-1} = \frac{1}{(1-1/3)^2} - 1 = \frac{5}{4}$$

$$M_3 = \frac{27}{4} \approx 6.75 \text{ MSC searches}$$

15.6.3 Asymptotic Bound of M_N

The multinomial formula gives

$$\sum_l C_i^{a_e} C_i^{b_e} \ldots C_i^{k_e} = N^i, \text{ with } \sum_l a_l + b_l + \ldots + k_l = N$$

Thus, N^i is the total number of sequences. Let $K(N)$ be the number of sequences that do not have all N objects. Thus, the total number of sequences of i objects containing all N is as follows:

$$N^i - K(N)$$

As the earlier case of $N = 3$, we have

$$K_N(i, N) = \frac{\left((N-1)^{i-1} - K(N-1)\right) \times N}{N^i}$$

for the probability of finding all N on the ith sample [because $K(N) > 0$]:

$$M_N = \sum_{i=N} i K_N(i, N), \text{ which is}$$

$$\leq \sum_{i=N} i\left(\frac{N-1}{N}\right)^{i-1}, \text{ that is,}$$

$$\leq \frac{1}{\left(1 - \frac{N-1}{N}\right)^2} - 1 \leq N^2$$

N^2 gives an *upper bound estimate* of the average number of searches to find all N MSCs, so we know that this is a $0(N^2)$ problem. So for $N = 120$ MSCs (like Telecom Italia), you need about 24,000 SMS-MTs to be almost assured of having the full list of their MSCs. Going back to Problem 1 given earlier, with 20 teams of 15 players and packs of 18 players, you can estimate that the child must buy $(20 \times 15)^2/18 = 5,000$ packs. You are better off buying him a beautiful bicycle!

We can compute the following table with the real M^N based on the exact computation using (15.5):

N	M_N	N^2
1	1	1
2	3	4
3	6.75	9
4		16
5		25
6		36
7		49
8		64

15.7   Problem 2: Estimate of the Probability That the Number of MSCs $N = j$

We assume that all values of the unknown number m of MSCs are equiprobable after we have obtained j of them on the ith sample:

$$Prob(\text{number of MSCs } N = m) = \frac{1}{M + 1 - j}$$

where $j \leq m \leq M$ and M is a known upper bound on the number of MSCs (these are at least j and a maximum of M). We then have

Note $P(i,j) = Prob(j \text{ MSC found after } i \text{ samples})$

$$\sum_{m=j}^{M} P_m(i, j) Prob(N = m) = P(i, j)$$

$$= \sum_{m=j}^{M} P_m(i, j) \times \frac{1}{M + 1 - j} = \frac{1}{M + 1 - j} \sum_{m=j}^{M} P_m(i, j)$$

$Prob \{j \text{ MSC found after } i \text{ samples AND } N = j\}$

$$= Prob(N = j \mid i, j) \times P(i, j)$$

(according to conditional probability), which is also $P_j(i,j)$ from the definition in (15.5); so we get:

$$Prob(N = j \mid i, j) = \frac{P_j(i, j)}{\sum_{m=j}^{M} P_m(i, j) \times \dfrac{1}{M + 1 - j}} \tag{15.10}$$

Thus, if the $Prob(N = j \mid i,j)$ value is close to 1, we will stop the search.

References

[1]   Henry-Labordère, A., *Procédé d'envoi de messages courts à des mobiles avec portabilité du numéro ou plan de numérotation incomplet avec auto-apprentissage,* French Patent FR-0315, 2000.

[2]   Henry-Labordère, A., and C. M. Zerhouni, *Décision bayésienne avec information incomplète*, Paris, France: Revue METRA, 1972.

[3]   Shannon, C. E., and W. Weaver, *The Mathematical Theory of Communication,* Champaign-Urbana, IL: University of Illinois Press, 1949.

## CHAPTER 16

Worked-Out Examples

16.1   Example 1

Answer the following questions:

1. The following MAP traces are taken at the SMSC of Figure 1.1. What is the number of the destination cell phone?
2. What is the GT of the visited MSC?
3. What is the GT of the HLR of the destination cell phone?
4. What is the GT of the sending SMSC?
5. At what time GMT did the cell phone turn on again?
6. What is the IMSI of the receiving phone?

```
- - - - - - - - - HDR- - - - - - - - -[ Thu Nov 28 18:00:40 2002 ]
  MAPE-E Instance = 0
  MAPE-E Type = MAP_MSG_DLG_REQ (0000C7E2)
  MAPE-E Dialog_ID = 731
  MAPE-E Src  = 1D
  MAPE-E Dst  = 15
MAPE-E Rsp_req = 0  Class = 0  Status = 0  Err_info = 0  *Nxt = 0000
  - - - - - - - - - - PARAMETER - AREA - - - - - - - - - -
      PA_Len = 40
      MAP-OPEN-REQ
        MAPPN_dest_address(Q713)(1)
          L = 013
          Data:  Routing on GT, Global Title included(4),
                 Signalling Point Code (ITU) = 4-160-7 ( 9479)
                 Subsystem Number = HLR(6),
                 Global Title :
                     Translation Type = 0,
                     Numbering Plan = ISDN/Telephony(E164),
                 Nature Address Indicator = International number,
                     Address information = 85292395815
        MAPPN_orig_address(Q713)(3)
          L = 010
          Data:  Routing on GT, Global Title included(4),
                 No SPC in address
```

```
                        Subsystem Number = MSC(8),
                        Global Title :
                            Translation Type = 0,
                            Numbering Plan = ISDN/Telephony(E164),
                    Nature Address Indicator = International number,
                            Address information = 6596197979
            MAPPN_applic_context(11)
              L = 009
              Data: (Hex) 060704000001001402
                    ShortMsgGatewayPackage_v1_or_v2 MAP V2
- - - - - - - - -HDR- - - - - - - - - [Thu Nov 28 18:00:40 2002 ]
  MAPE-E Instance = 0
  MAPE-E Type = MAP_MSG_SRV_REQ (0000C7E0)
  MAPE-E Dialog_ID = 731
  MAPE-E Src  = 1D
  MAPE-E Dst  = 15
MAPE-E Rsp_req = 0  Class = 0  Status = 0  Err_info = 0  *Nxt = 0000
- - - - - - - - - - - PARAMETER - AREA - - - - - - - - - - -
      PA_Len = 25
      MAP-SEND-ROUTING-INFO-FOR-SM-REQ
        MAPPN_invoke_id(14)
          L = 001
          Data: 1
        MAPPN_msisdn(15)
          L = 007
          Data: Ext = No extension
                Ton = International
                Npi = ISDN
                Address = 85292395815
        MAPPN_sm_rp_pri(16)
          L = 001
          Data: (1):High Priority
        MAPPN_sc_addr(17)
          L = 006
          Data: Ext = No extension
                Ton = International
                Npi = ISDN
                Address = 6596197979
- - - - - - - - - HDR - - - - - - - - - -[ Thu Nov 28 18:00:40 2002 ]
  MAPE-E Instance = 0
  MAPE-E Type = MAP_MSG_DLG_REQ (0000C7E2)
  MAPE-E Dialog_ID = 731
  MAPE-E Src  = 1D
  MAPE-E Dst  = 15
MAPE-E Rsp_req = 0  Class = 0  Status = 0  Err_info = 0  *Nxt = 0000
- - - - - - - - - - PARAMETER - AREA - - - - - - - - - - -
      PA_Len = 2
```

```
          MAP-DELIMITER-REQ
  - - - - - - - - - HDR- - - - - - - - [ Thu Nov 28 18:00:40 2002 ]
    MAPE-R Instance = 0
    MAPE-R Type = MAP_MSG_DLG_IND (000087E3)
    MAPE-R Dialog_ID = 731
    MAPE-R Src  = 15
    MAPE-R Dst  = 1D
    MAPE-R Rsp_req = 0  Class = 0  Status = 0  Err_info = 0  *Nxt = 0000
    - - - - - - - - - PARAMETER - AREA - - - - - - - - - -
        PA_Len = 16
        MAP-OPEN-CNF
          MAPPN_result(5)
            L = 001
            Data: (0):Accept
          MAPPN_applic_context(11)
            L = 009
            Data: (Hex) 060704000001001402
                        ShortMsgGatewayPackage_v1_or_v2 MAP V2
  - - - - - - - - - HDR- - - - - - - - - - - [ Thu Nov 28 18:00:40 2002 ]
    MAPE-R Instance = 0
    MAPE-R Type = MAP_MSG_SRV_IND (000087E1)
    MAPE-R Dialog_ID = 731
    MAPE-R Src  = 15
    MAPE-R Dst  = 1D
    MAPE-R Rsp_req = 0  Class = 0  Status = 0  Err_info = 0  *Nxt = 0000
    - - - - - - - - - - PARAMETER - AREA - - - - - - - - - -
        PA_Len = 30
        MAP-SEND-ROUTING-INFO-FOR-SM-CNF
          MAPPN_invoke_id(14)
            L = 001
            Data: 1
          MAPPN_imsi(18)
            L = 008
            Data: Address = 454161002767389
          MAPPN_msc_num(19)
            L = 007
            Data: Ext = No extension
                  Ton = International
                  Npi = ISDN
                  Address = 33689000651
          MAPPN_lmsi(20)
            L = 004
            Data: Ext = Extension
                  Ton = Unknown
                  Npi = Unknown
                  Address = 003f78
    - - - - - - - - - - - - - - - - - - - - - - - - - - - - -
```

```
# ROUTER #   MTU_SEND_ROUTING_INFO_CNF :
             16 731 0 0 0 0 +454161002767389 +33689000651


- - - - - - - - - - -HDR- - - - - - - - [ Thu Nov 28 18:00:40 2002 ]
    MAPE-R Instance = 0
    MAPE-R Type = MAP_MSG_DLG_IND (000087E3)
    MAPE-R Dialog_ID = 731
    MAPE-R Src  = 15
    MAPE-R Dst  = 1D
   MAPE-R Rsp_req = 0  Class = 0  Status = 0  Err_info = 0  *Nxt = 0000
    - - - - - - - - - - PARAMETER - AREA - - - - - - - - - - -
        PA_Len = 2
        MAP-CLOSE-IND

- - - - - - - - - - - - - - - - - - - - - - - - - - - - - - - -
  # ROUTER # =  MTU_FORWARD_SHORT_MSG_MTA_REQ
                27  736 +6596197979 +6596197979 +33689000651  9479
+454161002767389  c0401276027fa91e  +0  +85292395815  0  0  0  1   254
20118202001400 0 3 0 18F4F29C0E6ABFC569761934BFA7E96374990C7A9BCD


- - - - - - - - - - - -HDR- - - - - - - - [ Thu Nov 28 16:48:33 2002 ]
    MAPE-E Instance = 0
    MAPE-E Type = MAP_MSG_DLG_REQ (0000C7E2)
    MAPE-E Dialog_ID = 291
    MAPE-E Src  = 1D
    MAPE-E Dst  = 15
   MAPE-E Rsp_req = 0  Class = 0  Status = 0  Err_info = 0  *Nxt = 0000
    - - - - - - - - - - - PARAMETER - AREA - - - - - - - - - - -
        PA_Len = 40
        MAP-OPEN-REQ
          MAPPN_dest_address(Q713)(1)
            L = 013
            Data:  Routing on GT, Global Title included(4),
                   Signalling Point Code (ITU) = 4-160-7 ( 9479)
                   Subsystem Number = MSC(8),
                   Global Title :
                       Translation Type = 0,
                       Numbering Plan = ISDN/Telephony(E164),
                  Nature Address Indicator = International number,
                       Address information = 33689000651
          MAPPN_orig_address(Q713)(3)
            L = 010
            Data:  Routing on GT, Global Title included(4),
                   No SPC in address
                   Subsystem Number = MSC(8),
                   Global Title :
                       Translation Type = 0,
                       Numbering Plan = ISDN/Telephony(E164),
```

```
                    Nature Address Indicator = International number,
                        Address information = 6596197979
            MAPPN_applic_context(11)
              L = 009
              Data: (Hex) 060704000001001501
                      ShortMsgRelayPackage_v1 MAP V1
     - - - - - - - - -HDR- - - - - - - - [ Thu Nov 28 16:48:33 2002 ]
      MAPE-E Instance = 0
      MAPE-E Type = MAP_MSG_SRV_REQ (0000C7E0)
      MAPE-E Dialog_ID = 291
      MAPE-E Src  = 1D
      MAPE-E Dst  = 15
     MAPE-E Rsp_req = 0  Class = 0  Status = 0  Err_info = 0  *Nxt = 0000
     - - - - - - - - - - PARAMETER - AREA - - - - - - - - - - -
          PA_Len = 63
          MAP-FORWARD-SHORT-MESSAGE-REQ
            MAPPN_invoke_id(14)
              L = 001
              Data: 1
            MAPPN_sm_rp_da(23)
              L = 010
              Data: TA_IMSI
                    Address = 454161002767389
            MAPPN_sm_rp_oa(24)
              L = 008
              Data: TA_SC_ADR
                    Ext = No extension
                    Ton = International
                    Npi = ISDN
                    Address = 6596197979
            MAPPN_sm_rp_ui(25)
              L = 034
              Data: Message Type = SMS_DELIVER(SMS-MT)
                    TP_RP  = No request for reply path
                    TP_UDHI= No Header in TP-User-Data
          TP_SRI = A status report will not be returned to the SME
                    TP_VPF = TP_VP field not present
          TP_MMS = No more messages are waiting for the MS in this SC
                    Originating mobile address                  =
                        Type of number = Unknown
                        Numbering Plan = Unknown
                        Address =
    TP-Protocol-Identifier = 00 (No Interworking: SME-to-SME protocol)
                    TP-Data-Coding-Scheme  =  00
                        Message Class = Default (Handset MEmory)
                    Alphabet     = Default alphabet(7 bit packed)
                TP_Service_Centre_Time_Stamp = 02.11.28 18:48:32 00
```

```
                    TP-User-Data-length = 24
                    TP-User-Data =
                      "test mobile switched off"
- - - - - - - - - -HDR- - - - - - -[ Thu Nov 28 16:48:33 2002 ]
 MAPE-E Instance = 0
 MAPE-E Type = MAP_MSG_DLG_REQ (0000C7E2)
 MAPE-E Dialog_ID = 291
 MAPE-E Src  = 1D
 MAPE-E Dst  = 15
MAPE-E Rsp_req = 0  Class = 0  Status = 0  Err_info = 0  *Nxt = 0000
- - - - - - - - - - PARAMETER - AREA - - - - - - - - - - -
     PA_Len = 2
     MAP-DELIMITER-REQ
- - - - - - - - - -HDR- - - - - - - [ Thu Nov 28 16:48:41 2002 ]
 MAPE-R Instance = 0
 MAPE-R Type = MAP_MSG_DLG_IND (000087E3)
 MAPE-R Dialog_ID = 291
 MAPE-R Src  = 15
 MAPE-R Dst  = 1D
MAPE-R Rsp_req = 0  Class = 0  Status = 0  Err_info = 0  *Nxt = 0000
     - - - - - - - - - - PARAMETER - AREA - - - - - - - - - - -
     PA_Len = 16
     MAP-OPEN-CNF
       MAPPN_result(5)
         L = 001
         Data: (0):Accept
       MAPPN_applic_context(11)
         L = 009
         Data: (Hex) 060704000001001501
                 ShortMsgRelayPackage_v1 MAP V1
- - - - - - - - - - - - HDR- - - - - - [ Thu Nov 28 16:48:41 2002 ]
 MAPE-R Instance = 0
 MAPE-R Type = MAP_MSG_SRV_IND (000087E1)
 MAPE-R Dialog_ID = 291
 MAPE-R Src  = 15
 MAPE-R Dst  = 1D
MAPE-R Rsp_req = 0  Class = 0  Status = 0  Err_info = 0  *Nxt = 0000
     - - - - - - - - - - PARAMETER - AREA - - - - - - - - - - -
     PA_Len = 8
     MAP-FORWARD-SHORT-MESSAGE-CNF
       MAPPN_invoke_id(14)
         L = 001
         Data: 1
       MAPPN_user_err(21)
         L = 001
         Data: (27):Absent Subscriber !
- - - - - - - - - - - - - - -   - - - - - - - - - - - - - -
```

```
    # ROUTER #   MTU_FORWARD_SHORT_MSG_CNF :
                   3 291 27 0 0 0


- - - - - - - - - - - HDR - - - - - - - [ Thu Nov 28 16:48:41 2002 ]
      MAPE-R Instance = 0
      MAPE-R Type = MAP_MSG_DLG_IND (000087E3)
      MAPE-R Dialog_ID = 291
      MAPE-R Src  = 15
      MAPE-R Dst  = 1D
    MAPE-R Rsp_req = 0  Class = 0  Status = 0  Err_info = 0  *Nxt = 0000
     - - - - - - - - - PARAMETER - AREA - - - - - - - - - -
         PA_Len = 2
         MAP-CLOSE-IND
- - - - - - - - - -HDR- - - - - - - - - - [ Thu Nov 28 16:48:41 2002 ]
     MAPE-E Instance = 0
     MAPE-E Type = MAP_MSG_DLG_REQ (0000C7E2)
     MAPE-E Dialog_ID = 418
     MAPE-E Src  = 1D
     MAPE-E Dst  = 15
    MAPE-E Rsp_req = 0  Class = 0  Status = 0  Err_info = 0  *Nxt = 0000
     - - - - - - - - - - PARAMETER - AREA - - - - - - - - - -
         PA_Len = 40
         MAP-OPEN-REQ
           MAPPN_dest_address(Q713)(1)
             L = 013
             Data:   Routing on GT, Global Title included(4),
                     Signalling Point Code (ITU) = 4-160-7 ( 9479)
                     Subsystem Number = HLR(6),
                     Global Title :
                         Translation Type = 0,
                         Numbering Plan = ISDN/Telephony(E164),
                     Nature Address Indicator = International number,
                         Address information = 85292395815
           MAPPN_orig_address(Q713)(3)
             L = 010
             Data:   Routing on GT, Global Title included(4),
                     No SPC in address
                     Subsystem Number = MSC(8),
                     Global Title :
                         Translation Type = 0,
                         Numbering Plan = ISDN/Telephony(E164),
                     Nature Address Indicator = International number,
                         Address information = 6596197979
           MAPPN_applic_context(11)
             L = 009
             Data: (Hex) 060704000001001402
                     ShortMsgGatewayPackage_v1_or_v2 MAP V2
```

```
- - - - - - - - - -HDR- - - - - - - - - [ Thu Nov 28 16:48:41 2002 ]
 MAPE-E Instance = 0
 MAPE-E Type = MAP_MSG_SRV_REQ (0000C7E0)
 MAPE-E Dialog_ID = 418
 MAPE-E Src  = 1D
 MAPE-E Dst  = 15
MAPE-E Rsp_req = 0  Class = 0  Status = 0  Err_info = 0  *Nxt = 0000
 - - - - - - - - - - PARAMETER - AREA - - - - - - - - - - -
     PA_Len = 25
     MAP-SHORT-MESSAGE-DELIVERY-STATUS-REQ
       MAPPN_invoke_id(14)
         L = 001
         Data: 1
       MAPPN_msisdn(15)
         L = 007
         Data: Ext = No extension
               Ton = International
               Npi = ISDN
               Address = 85292395815
       MAPPN_sc_addr(17)
         L = 006
         Data: Ext = No extension
               Ton = International
               Npi = ISDN
               Address = 6596197979
       MAPPN_sm_delv_outcome(27)
         L = 001
         Data: (1):absent subscriber !
- - - - - - - - - - - -HDR- - - - - - [ Thu Nov 28 16:48:41 2002 ]
 MAPE-E Instance = 0
 MAPE-E Type = MAP_MSG_DLG_REQ (0000C7E2)
 MAPE-E Dialog_ID = 418
 MAPE-E Src  = 1D
 MAPE-E Dst  = 15
MAPE-E Rsp_req = 0  Class = 0  Status = 0  Err_info = 0  *Nxt = 0000
 - - - - - - - - - - PARAMETER - AREA - - - - - - - - - - -
     PA_Len = 2
     MAP-DELIMITER-REQ
- - - - - - - - - - - - HDR - - - - - -[ Thu Nov 28 16:48:42 2002 ]
 MAPE-R Instance = 0
 MAPE-R Type = MAP_MSG_DLG_IND (000087E3)
 MAPE-R Dialog_ID = 418
 MAPE-R Src  = 15
 MAPE-R Dst  = 1D
MAPE-R Rsp_req = 0  Class = 0  Status = 0  Err_info = 0  *Nxt = 0000
 - - - - - - - - - - PARAMETER - AREA - - - - - - - - - - -
     PA_Len = 16
```

```
                    MAP-OPEN-CNF
                      MAPPN_result(5)
                        L = 001
                        Data: (0):Accept
                      MAPPN_applic_context(11)
                        L = 009
                        Data: (Hex) 060704000001001402
                                ShortMsgGatewayPackage_v1_or_v2 MAP V2
           - - - - - - - - - - -HDR- - - - - - [ Thu Nov 28 16:48:42 2002 ]
         MAPE-R Instance = 0
         MAPE-R Type = MAP_MSG_SRV_IND (000087E1)
         MAPE-R Dialog_ID = 418
         MAPE-R Src  = 15
         MAPE-R Dst  = 1D
        MAPE-R Rsp_req = 0  Class = 0  Status = 0  Err_info = 0  *Nxt = 0000
           - - - - - - - - - - PARAMETER - AREA - - - - - - - - - - -
             PA_Len = 5
             MAP-REPORT-SHORT-MESSAGE-DELIVERY-STATUS-CNF
               MAPPN_invoke_id(14)
                 L = 001
                 Data: 1
      - - - - - - - - - - - - - -    - - - - - - - - - - - - - - - -
# ROUTER #   MTU_REPORT_SM_DELIV_STATUS_CNF :
               17 418 0 0 0 0


- - - - - - - - - - HDR- - - - - - - - [ Thu Nov 28 16:48:42 2002 ]
       MAPE-R Instance = 0
       MAPE-R Type = MAP_MSG_DLG_IND (000087E3)
       MAPE-R Dialog_ID = 418
       MAPE-R Src  = 15
       MAPE-R Dst  = 1D
      MAPE-R Rsp_req = 0  Class = 0  Status = 0  Err_info = 0  *Nxt = 0000
         - - - - - - - - - - PARAMETER - AREA - - - - - - - - - - -
             PA_Len = 2
             MAP-CLOSE-IND
        - - - - - - - - - - - -HDR - - - - [ Thu Nov 28 18:00:39 2002 ]
       MAPE-R Instance = 0
       MAPE-R Type = MAP_MSG_DLG_IND (000087E3)
       MAPE-R Dialog_ID = 33219
       MAPE-R Src  = 15
       MAPE-R Dst  = 1D
      MAPE-R Rsp_req = 0  Class = 0  Status = 0  Err_info = 0  *Nxt = 0000
         - - - - - - - - - - PARAMETER - AREA - - - - - - - - - - -
             PA_Len = 42
             MAP-OPEN-IND
               MAPPN_dest_address(Q713)(1)
                 L = 012
```

```
             Data:    Routing on SPC, Global Title included(4),
                      Signalling Point Code (ITU) = 4-002-4 ( 8212)
                      Subsystem Number = MSC(8),
                      Global Title :
                         Translation Type = 0,
                         Numbering Plan = ISDN/Telephony(E164),
                   Nature Address Indicator = International number,
                         Address information = 6596197979
          MAPPN_orig_address(Q713)(3)
             L = 013
             Data:    Routing on GT, Global Title included(4),
                      Signalling Point Code (ITU) = 4-160-7 ( 9479)
                      Subsystem Number = HLR(6),
                      Global Title :
                         Translation Type = 0,
                         Numbering Plan = ISDN/Telephony(E164),
                   Nature Address Indicator = International number,
                         Address information = 85292347990
          MAPPN_applic_context(11)
             L = 009
             Data: (Hex) 060704000001001702
                      AlertPackage_v1_or_v2 MAP V2
- - - - - - - - - -HDR- - - - - - - [ Thu Nov 28 18:00:39 2002 ]
  MAPE-E Instance = 0
  MAPE-E Type = MAP_MSG_DLG_REQ (0000C7E2)
  MAPE-E Dialog_ID = 33219
  MAPE-E Src  = 1D
  MAPE-E Dst  = 15
MAPE-E Rsp_req = 0 Class = 0 Status = 0 Err_info = 0 *Nxt = 0000
- - - - - - - - - - PARAMETER - AREA - - - - - - - - - -
       PA_Len = 16
       MAP-OPEN-RSP
          MAPPN_result(5)
             L = 001
             Data: (0):Accept
          MAPPN_applic_context(11)
             L = 009
             Data: (Hex) 060704000001001702
                      AlertPackage_v1_or_v2 MAP V2
- - - - - - - - - -HDR - - - - - - - [ Thu Nov 28 18:00:39 2002 ]
  MAPE-R Instance = 0
  MAPE-R Type = MAP_MSG_SRV_IND (000087E1)
  MAPE-R Dialog_ID = 33219
  MAPE-R Src  = 15
  MAPE-R Dst  = 1D
MAPE-R Rsp_req = 0 Class = 0 Status = 0 Err_info = 0 *Nxt = 0000
- - - - - - - - - - PARAMETER - AREA - - - - - - - - - -
```

```
                    PA_Len = 22
                    MAP-ALERT-SERVICE-CENTRE-IND
                      MAPPN_invoke_id(14)
                        L = 001
                        Data: 159
                      MAPPN_msisdn(15)
                        L = 007
                        Data: Ext = No extension
                              Ton = International
                              Npi = ISDN
                              Address = 85292395815
                      MAPPN_sc_addr(17)
                        L = 006
                        Data: Ext = No extension
                              Ton = International
                              Npi = ISDN
                              Address = 6596197979
           - - - - - - - - - - - HDR- - - - [ Thu Nov 28 18:00:39 2002 ]
        MAPE-E Instance = 0
        MAPE-E Type = MAP_MSG_SRV_REQ (0000C7E0)
        MAPE-E Dialog_ID = 33219
        MAPE-E Src  = 1D
        MAPE-E Dst  = 15
       MAPE-E Rsp_req = 0  Class = 0  Status = 0  Err_info = 0  *Nxt = 0000
           - - - - - - - - - - PARAMETER - AREA - - - - - - - - - - -
                    PA_Len = 5
                    MAP-ALERT_SERVICE-CENTRE-RSP
                      MAPPN_invoke_id(14)
                        L = 001
                        Data: 159
           - - - - - - - - - - - HDR- - - - [ Thu Nov 28 18:00:39 2002 ]
        MAPE-E Instance = 0
        MAPE-E Type = MAP_MSG_DLG_REQ (0000C7E2)
        MAPE-E Dialog_ID = 33219
        MAPE-E Src  = 1D
        MAPE-E Dst  = 15
       MAPE-E Rsp_req = 0  Class = 0  Status = 0  Err_info = 0  *Nxt = 0000
           - - - - - - - - - - PARAMETER - AREA - - - - - - - - - - -
                    PA_Len = 5
                    MAP-CLOSE-REQ
                      MAPPN_release_method(7)
                        L = 001
                        Data: (0):Normal Release
           - - - - - - - - - - - - - - - -   - - - - - - - - - - - - - - -
       # ST/FWD #   MTU_ALERT_SERVICE_CENTRE_IND :
                        5 33219 +85292395815 +6596197979
       # ST/FWD #   MTU_ALERT_SERVICE_CENTRE_IND :
```

```
          5  33219  +85292395815  +6596197979


- - - - - - - - - -HDR- - - - - - - -[ Thu Nov 28 18:00:39 2002 ]
  MAPE-R Instance = 0
  MAPE-R Type = MAP_MSG_DLG_IND (000087E3)
  MAPE-R Dialog_ID = 33219
  MAPE-R Src  = 15
  MAPE-R Dst  = 1D
MAPE-R Rsp_req = 0  Class = 0  Status = 0  Err_info = 0  *Nxt = 0000
- - - - - - - - - - PARAMETER - AREA - - - - - - - - - - - -
      PA_Len = 2
      MAP-DELIMITER-IND
- - - - - - - - - - -HDR- - - - - - -[ Thu Nov 28 18:00:40 2002 ]
  MAPE-E Instance = 0
  MAPE-E Type = MAP_MSG_DLG_REQ (0000C7E2)
  MAPE-E Dialog_ID = 731
  MAPE-E Src  = 1D
  MAPE-E Dst  = 15
MAPE-E Rsp_req = 0  Class = 0  Status = 0  Err_info = 0  *Nxt = 0000
- - - - - - - - - - - PARAMETER - AREA - - - - - - - - - - - -
      PA_Len = 40
      MAP-OPEN-REQ
        MAPPN_dest_address(Q713)(1)
          L = 013
          Data:  Routing on GT, Global Title included(4),
                 Signalling Point Code (ITU) = 4-160-7 ( 9479)
                 Subsystem Number = HLR(6),
                 Global Title :
                    Translation Type = 0,
                    Numbering Plan = ISDN/Telephony(E164),
                 Nature Address Indicator = International number,
                    Address information = 85292395815
        MAPPN_orig_address(Q713)(3)
          L = 010
          Data:  Routing on GT, Global Title included(4),
                 No SPC in address
                 Subsystem Number = MSC(8),
                 Global Title :
                    Translation Type = 0,
                    Numbering Plan = ISDN/Telephony(E164),
                 Nature Address Indicator = International number,
                    Address information = 6596197979
        MAPPN_applic_context(11)
          L = 009
          Data: (Hex) 060704000001001402
                ShortMsgGatewayPackage_v1_or_v2 MAP V2
- - - - - - - - - -HDR - - - - - - -[ Thu Nov 28 18:00:40 2002 ]
```

```
          MAPE-E Instance = 0
          MAPE-E Type = MAP_MSG_SRV_REQ (0000C7E0)
          MAPE-E Dialog_ID = 731
          MAPE-E Src  = 1D
          MAPE-E Dst  = 15
         MAPE-E Rsp_req = 0  Class = 0  Status = 0  Err_info = 0  *Nxt = 0000
            - - - - - - - - - - PARAMETER - AREA - - - - - - - - - - -
             PA_Len = 25
             MAP-SEND-ROUTING-INFO-FOR-SM-REQ
               MAPPN_invoke_id(14)
                 L = 001
                 Data: 1
               MAPPN_msisdn(15)
                 L = 007
                 Data: Ext = No extension
                       Ton = International
                       Npi = ISDN
                       Address = 85292395815
               MAPPN_sm_rp_pri(16)
                 L = 001
                 Data: (1):High Priority
               MAPPN_sc_addr(17)
                 L = 006
                 Data: Ext = No extension
                       Ton = International
                       Npi = ISDN
                       Address = 6596197979
            - - - - - - - - - - - - - -HDR-[ Thu Nov 28 18:00:40 2002 ]


                    - HDR
          MAPE-E Instance = 0
          MAPE-E Type = MAP_MSG_DLG_REQ (0000C7E2)
          MAPE-E Dialog_ID = 731
          MAPE-E Src  = 1D
          MAPE-E Dst  = 15
         MAPE-E Rsp_req = 0  Class = 0  Status = 0  Err_info = 0  *Nxt = 0000
            - - - - - - - - - - PARAMETER - AREA - - - - - - - - - - -
             PA_Len = 2
             MAP-DELIMITER-REQ
            - - - - - - - - - - - - - - - - - - - - - - - - - - - - -
        *** Message received for inactive dialogue ***
        # ROUTER # =  MTU_SEND_ROUTING_INFO_REQ
                      12 732 +6596197979 +96657389862 9479 +96657389862
    +6596197979 0 1 0


         - - - - - - - - - - HDR- - - - - - - [ Thu Nov 28 18:00:40 2002 ]
          MAPE-R Instance = 0
```

```
        MAPE-R Type = MAP_MSG_DLG_IND (000087E3)
        MAPE-R Dialog_ID = 731
        MAPE-R Src  = 15
        MAPE-R Dst  = 1D
       MAPE-R Rsp_req = 0  Class = 0  Status = 0  Err_info = 0  *Nxt = 0000
        - - - - - - - - - PARAMETER - AREA - - - - - - - - - - -
            PA_Len = 16
            MAP-OPEN-CNF
              MAPPN_result(5)
                L = 001
                Data: (0):Accept
              MAPPN_applic_context(11)
                L = 009
                Data: (Hex) 060704000001001402
                        ShortMsgGatewayPackage_v1_or_v2 MAP V2
       - - - - - - HDR - - - - - - - - - - - - [ Thu Nov 28 18:00:40 2002 ]
        MAPE-R Instance = 0
        MAPE-R Type = MAP_MSG_SRV_IND (000087E1)
        MAPE-R Dialog_ID = 731
        MAPE-R Src  = 15
        MAPE-R Dst  = 1D
       MAPE-R Rsp_req = 0  Class = 0  Status = 0  Err_info = 0  *Nxt = 0000
        - - - - - - - - - - PARAMETER - AREA - - - - - - - - - - -
            PA_Len = 30
            MAP-SEND-ROUTING-INFO-FOR-SM-CNF
              MAPPN_invoke_id(14)
                L = 001
                Data: 1
              MAPPN_imsi(18)
                L = 008
                Data: Address = 454161002767389
              MAPPN_msc_num(19)
                L = 007
                Data: Ext = No extension
                      Ton = International
                      Npi = ISDN
                      Address = 33689000651
              MAPPN_lmsi(20)
                L = 004
                Data: Ext = Extension
                      Ton = Unknown
                      Npi = Unknown
                      Address = 003f78
        - - - - - - - - - - - - - - - - - - - - - - - - - - - - -
# ROUTER #   MTU_SEND_ROUTING_INFO_CNF :
              16 731 0 0 0 0 +454161002767389 +33689000651
```

```
- - - - - - - -HDR - - - - - - - - - - -[ Thu Nov 28 18:00:40 002 ]
   MAPE-R Instance = 0
   MAPE-R Type = MAP_MSG_DLG_IND (000087E3)
   MAPE-R Dialog_ID = 731
   MAPE-R Src  = 15
   MAPE-R Dst  = 1D
  MAPE-R Rsp_req = 0  Class = 0  Status = 0  Err_info = 0  *Nxt = 0000
    - - - - - - - - - - PARAMETER - AREA - - - - - - - - - - -
       PA_Len = 2
       MAP-CLOSE-IND
   - - - - - - - - - - - - - - - - - - - - - - - - - - - - - - -
 *** Message received for inactive dialogue ***
 # ROUTER # =  MTU_FORWARD_SHORT_MSG_MTA_REQ
               27 736 +6596197979 +6596197979 +33689000651  9479
+454161002767389   c0401276027fa91e   +0  +85292395815  0  0  0  1  254
20118202001400  0  3  0  18F4F29C0E6ABFC569761934BFA7E96374990C7A9BCD


                 - HDR                    [ Thu Nov 28 18:00:41 2002 ]
   MAPE-E Instance = 0
   MAPE-E Type = MAP_MSG_DLG_REQ (0000C7E2)
   MAPE-E Dialog_ID = 736
   MAPE-E Src  = 1D
   MAPE-E Dst  = 15
  MAPE-E Rsp_req = 0  Class = 0  Status = 0  Err_info = 0  *Nxt = 0000
    - - - - - - - - - - PARAMETER - AREA - - - - - - - - - - -
       PA_Len = 40
       MAP-OPEN-REQ
         MAPPN_dest_address(Q713)(1)
           L = 013
           Data:  Routing on GT, Global Title included(4),
                  Signalling Point Code (ITU) = 4-160-7 ( 9479)
                  Subsystem Number = MSC(8),
                  Global Title :
                     Translation Type = 0,
                     Numbering Plan = ISDN/Telephony(E164),
                  Nature Address Indicator = International number,
                     Address information = 33689000651
         MAPPN_orig_address(Q713)(3)
           L = 010
           Data:  Routing on GT, Global Title included(4),
                  No SPC in address
                  Subsystem Number = MSC(8),
                  Global Title :
                     Translation Type = 0,
                     Numbering Plan = ISDN/Telephony(E164),
                  Nature Address Indicator = International number,
                     Address information = 6596197979
```

```
              MAPPN_applic_context(11)
                 L = 009
                 Data: (Hex) 060704000001001501
                          ShortMsgRelayPackage_v1 MAP V1
- - - - - - - -HDR - - - - - - - - - - -[ Thu Nov 28 18:00:41 2002 ]
    MAPE-E Instance = 0
    MAPE-E Type = MAP_MSG_SRV_REQ (0000C7E0)
    MAPE-E Dialog_ID = 736
    MAPE-E Src  = 1D
    MAPE-E Dst  = 15
   MAPE-E Rsp_req = 0  Class = 0  Status = 0  Err_info = 0  *Nxt = 0000
       - - - - - - - - - - PARAMETER - AREA - - - - - - - - - - -
          PA_Len = 63
          MAP-FORWARD-SHORT-MESSAGE-REQ
            MAPPN_invoke_id(14)
               L = 001
               Data: 1
            MAPPN_sm_rp_da(23)
               L = 010
               Data: TA_IMSI
                     Address = 454161002767389
            MAPPN_sm_rp_oa(24)
               L = 008
               Data: TA_SC_ADR
                     Ext = No extension
                     Ton = International
                     Npi = ISDN
                     Address = 6596197979
            MAPPN_sm_rp_ui(25)
               L = 034
               Data: Message Type = SMS_DELIVER(SMS-MT)
                     TP_RP  = No request for reply path
                     TP_UDHI= No Header in TP-User-Data
                 TP_SRI = A status report will not be returned to the SME
                     TP_VPF = TP_VP field not present
                 TP_MMS = More messages are waiting for the MS in this SC
                     Originating mobile address              =
                        Type of number = Unknown
                        Numbering Plan = Unknown
                        Address =
TP-Protocol-Identifier = 00 (No Interworking: SME-to-SME protocol)
                     TP-Data-Coding-Scheme  = 00
                        Message Class = Default (Handset MEmory)
                     Alphabet     = Default alphabet(7 bit packed)
                 TP_Service_Centre_Time_Stamp = 02.11.28 20:00:41 00
                        TP-User-Data-length = 24
                        TP-User-Data =
```

```
                          test mobile switched off
    - - - - - - - HDR- - - - - - - - - - [ Thu Nov 28 18:00:41 2002 ]
   MAPE-E Instance = 0
   MAPE-E Type = MAP_MSG_DLG_REQ (0000C7E2)
   MAPE-E Dialog_ID = 736
   MAPE-E Src  = 1D
   MAPE-E Dst  = 15
  MAPE-E Rsp_req = 0  Class = 0  Status = 0  Err_info = 0  *Nxt = 0000
    - - - - - - - - - PARAMETER - AREA - - - - - - - - - -
       PA_Len = 2
       MAP-DELIMITER-REQ
    - - - - - - - - - - HDR - - - - - - - - [ Thu Nov 28 18:00:50 2002 ]
   MAPE-R Instance = 0
   MAPE-R Type = MAP_MSG_DLG_IND (000087E3)
   MAPE-R Dialog_ID = 736
   MAPE-R Src  = 15
   MAPE-R Dst  = 1D
  MAPE-R Rsp_req = 0  Class = 0  Status = 0  Err_info = 0  *Nxt = 0000
    - - - - - - - - - - PARAMETER - AREA - - - - - - - - - - -
       PA_Len = 16
       MAP-OPEN-CNF
         MAPPN_result(5)
           L = 001
           Data: (0):Accept
         MAPPN_applic_context(11)
           L = 009
           Data: (Hex) 060704000001001501
                   ShortMsgRelayPackage_v1 MAP V1
    - - - - - - - - - HDR - - - - - - [ Thu Nov 28 18:00:50 2002 ]
   MAPE-R Instance = 0
   MAPE-R Type = MAP_MSG_SRV_IND (000087E1)
   MAPE-R Dialog_ID = 736
   MAPE-R Src  = 15
   MAPE-R Dst  = 1D
  MAPE-R Rsp_req = 0  Class = 0  Status = 0  Err_info = 0  *Nxt = 0000
    - - - - - - - - - - PARAMETER - AREA - - - - - - - - - - -
       PA_Len = 5
       MAP-FORWARD-SHORT-MESSAGE-CNF
         MAPPN_invoke_id(14)
           L = 001
           Data: 1
    - - - - - - - - - - - - - - - - - - - - - - - - - - - - - -
# ROUTER #  MTU_FORWARD_SHORT_MSG_CNF :
             3 736 0 0 0 0


    - - - - - - - - - HDR - - - - - - - - - [ Thu Nov 28 18:00:50 2002 ]
   MAPE-R Instance = 0
```

```
MAPE-R Type = MAP_MSG_DLG_IND (000087E3)
MAPE-R Dialog_ID = 736
MAPE-R Src  = 15
MAPE-R Dst  = 1D
MAPE-R Rsp_req = 0  Class = 0  Status = 0  Err_info = 0  *Nxt = 0000
- - - - - - - - - - PARAMETER - AREA - - - - - - - - - - - -
     PA_Len = 2
     MAP-CLOSE-IND
- - - - - - - - - - - - - - -   - - - - - - - - - - - - - -
```

16.2   Example 2

Some MAP traces of nodes in Singapore (GT: +6596197979) and Hong Kong (GT: +85292347978) follow. Study the traces and then nswer these questions:

1. Where is the SMS initial request from?
2. Draw a diagram of the path used between routers.
3. Explain the behavior of the Singapore router: Why do you think it addressed the other Hong Kong node?
4. At what time did the sender get the confirmation?

```
- - - - - - HDR - - - - - - - - - - - [ Fri Nov 29 10:19:11 2002 ]
   MAPE-R Instance = 0
   MAPE-R Type = MAP_MSG_DLG_IND (000087E3)
   MAPE-R Dialog_ID = 33060
   MAPE-R Src  = 15
   MAPE-R Dst  = 1D
  MAPE-R Rsp_req = 0  Class = 0  Status = 0  Err_info = 0  *Nxt = 0000
   - - - - - - - - - PARAMETER - AREA - - - - - - - - - - -
        PA_Len = 42
        MAP-OPEN-IND
         MAPPN_dest_address(Q713)(1)
          L = 012
          Data:  Routing on SPC, Global Title included(4),
                 Signalling Point Code (ITU) = 4-002-4 ( 8212)
                 Subsystem Number = MSC(8),
                 Global Title :
                     Translation Type = 0,
                     Numbering Plan = ISDN/Telephony(E164),
                 Nature Address Indicator = International number,
                     Address information = 6596197979
         MAPPN_orig_address(Q713)(3)
          L = 013
          Data:  Routing on GT, Global Title included(4),
                 Signalling Point Code (ITU) = 4-160-7 ( 9479)
                 Subsystem Number = MSC(8),
```

```
                         Global Title :
                            Translation Type = 0,
                            Numbering Plan = ISDN/Telephony(E164),
                   Nature Address Indicator = International number,
                            Address information = 33899990000
             MAPPN_applic_context(11)
                L = 009
                Data: (Hex) 060704000001001902
                        ShortMsgRelayPackage_v2 MAP V2
- - - - - - - - - - - HDR - - - - - - - - [ Fri Nov 29 10:19:11 2002 ]
  MAPE-E Instance = 0
  MAPE-E Type = MAP_MSG_DLG_REQ (0000C7E2)
  MAPE-E Dialog_ID = 33060
  MAPE-E Src  = 1D
  MAPE-E Dst  = 15
MAPE-E Rsp_req = 0  Class = 0  Status = 0  Err_info = 0  *Nxt = 0000
  - - - - - - - - - - PARAMETER - AREA - - - - - - - - - - -
      PA_Len = 16
      MAP-OPEN-RSP
        MAPPN_result(5)
           L = 001
           Data: (0):Accept
        MAPPN_applic_context(11)
           L = 009
           Data: (Hex) 060704000001001902
                   ShortMsgRelayPackage_v2 MAP V2
- - - - - - - - - - - HDR - - - - - - [ Fri Nov 29 10:19:11 2002 ]
  MAPE-R Instance = 0
  MAPE-R Type = MAP_MSG_SRV_IND (000087E1)
  MAPE-R Dialog_ID = 33060
  MAPE-R Src  = 15
  MAPE-R Dst  = 1D
MAPE-R Rsp_req = 0  Class = 0  Status = 0  Err_info = 0  *Nxt = 0000
  - - - - - - - - - - PARAMETER - AREA - - - - - - - - - - -
      PA_Len = 85
      MAP-FORWARD-SHORT-MESSAGE-IND
        MAPPN_invoke_id(14)
           L = 001
           Data: 1
        MAPPN_sm_rp_da(23)
           L = 008
           Data: TA_SC_ADR
                   Ext = No extension
                   Ton = International
                   Npi = ISDN
                   Address = 6596197979
        MAPPN_sm_rp_oa(24)
```

```
                    L = 011
                    Data: TA_SC_ADR
                            Ext = No extension
                            Ton = Unknown
                            Npi = Unknown
                            Address = 404002cfc77fe9ea
              MAPPN_sm_rp_ui(25)
                    L = 055
                    Data: Message Type = SMS_SUBMIT(SMS-MO)
                            TP_RP  = Request for reply path
                            TP_UDHI= No Header in TP-User-Data
                            TP_SRR = Status Report not requested
                            TP_VPF = TP_VP field not present
                            TP_RD  = Accept SMS with same TP_MR
                            TP_Message reference = 26
                            Destination mobile address                =
                                Type of number = International
                                Numbering Plan = ISDN
                                Address = 60125165618
                        TP-Protocol-Identifier = 00 (No Interworking: SME-to-
                                SME protocol)
                            TP-Data-Coding-Scheme  =  00
                                Message Class = Default (Handset MEmory)
                        Alphabet     = Default alphabet(7 bit packed)
                            TP-User-Data-length = 47
                            TP-User-Data =
                        SMS quality check, sorry for the inconvenience.
    - - - - - - - - - - - - - - - - - - - - - - - - - - - - - - - -
  # ROUTER #  MTU_FORWARD_SHORT_MSG_MO_IND :
          4 33060 +33899990000 +60125165618 +0 404002cfc77fe9ea 1 0
 0 0 0 0 0  2FD3E61414AF87D9697A1E344697C76B1668FE96CBF320F35B0EA2A3
 CBA0B47BFC76DBCBEE74D93D2EBB00


                    - HDR                 [ Fri Nov 29 10:19:11 2002 ]
      MAPE-R Instance = 0
      MAPE-R Type = MAP_MSG_DLG_IND (000087E3)
      MAPE-R Dialog_ID = 33060
      MAPE-R Src  = 15
      MAPE-R Dst  = 1D
     MAPE-R Rsp_req = 0  Class = 0  Status = 0  Err_info = 0  *Nxt = 0000
      - - - - - - - - - - PARAMETER - AREA - - - - - - - - - - -
          PA_Len = 2
          MAP-DELIMITER-IND
    - - - - - - - - - - - - - - - - - - - - - - - - - - - - - - -
  # ROUTER #  MTU_FORWARD_SHORT_MSG_MO_IND :
```

```
              4 33060 +33899990000 +60125165618 +0 404002cfc77fe9ea 1 0
0  0  0  0  0  2FD3E61414AF87D9697A1E344697C76B1668FE96CBF320F35B0EA2A3C
BA0B47BFC76DBCBEE74D93D2EBB00
```

    # ROUTER # =   MTU_SEND_ROUTING_INFO_REQ
```
                  12 575 +6596197979 +60125165618 9479 +60125165618
+6596197979 0 2 0
```

                    - HDR   [Fri Nov 29 10:19:17 2002]
        MAPE-E Instance = 0
        MAPE-E Type = MAP_MSG_SRV_REQ (0000C7E0)
        MAPE-E Dialog_ID = 33060
        MAPE-E Src = 1D
        MAPE-E Dst = 15
        MAPE-E Rsp_req = 0 Class = 0 Status = 0 Err_info = 0 *Nxt = 0000
         - - - - - - - - - - PARAMETER - AREA - - - - - - - - - - -
          PA_Len = 51
          MAP-FORWARD-SHORT-MESSAGE-RSP
            MAPPN_invoke_id(14)
              L = 001
              Data: 1
            MAPPN_user_err(21)
              L = 001
              Data: (0):0k!
            MAPPN_deliv_fail_cse(31)
              L = 001
              Data:
            MAPPN_imsi(18)
              L = 008
              Data: Address = 502125323286468
            MAPPN_msc_num(19)
              L = 006
              Data: Ext = No extension
                    Ton = International
                    Npi = ISDN
                    Address = 6593659940
            MAPPN_sub_state(54)
              L = 001
              Data: (255):Unknown code received
            MAPPN_not_reach_rsn(56)
              L = 001
              Data: (255):Not applicable
            MAPPN_time_zone(128)
              L = 004
              Data: 8.00
            MAPPN_timing_adv(129)
              L = 001
              Data: (Hex) FF
```
```

```
                   MAPPN_age_loc_info(48)
                     L = 002
                     Data: age of location info 65535 minutes
                   MAPPN_fwing_options(43)
                     L = 001
                     Data: (Hex) FF
- - - - - - - -HDR - - - - - - - - -  - - - [ Fri Nov 29 10:19:17 2002 ]
MAPE-E Instance = 0
MAPE-E Type = MAP_MSG_DLG_REQ (0000C7E2)
MAPE-E Dialog_ID = 33060
MAPE-E Src  = 1D
MAPE-E Dst  = 15
MAPE-E Rsp_req = 0  Class = 0  Status = 0  Err_info = 0  *Nxt = 0000
  - - - - - - - - - - PARAMETER - AREA - - - - - - - - - - - -
      PA_Len = 5
      MAP-CLOSE-REQ
        MAPPN_release_method(7)
          L = 001
          Data: (0):Normal Release
- - - - - - - HDR - - - - - - - - - - - - [ Fri Nov 29 10:19:11 2002 ]
MAPE-E Instance = 0
MAPE-E Type = MAP_MSG_DLG_REQ (0000C7E2)
MAPE-E Dialog_ID = 575
MAPE-E Src  = 1D
MAPE-E Dst  = 15
MAPE-E Rsp_req = 0  Class = 0  Status = 0  Err_info = 0  *Nxt = 0000
  - - - - - - - - - - PARAMETER - AREA - - - - - - - - - - - -
      PA_Len = 40
      MAP-OPEN-REQ
        MAPPN_dest_address(Q713)(1)
          L = 013
          Data:  Routing on GT, Global Title included(4),
                 Signalling Point Code (ITU) = 4-160-7 ( 9479)
                 Subsystem Number = HLR(6),
                 Global Title :
                   Translation Type = 0,
                   Numbering Plan = ISDN/Telephony(E164),
                 Nature Address Indicator = International number,
                   Address information = 60125165618
        MAPPN_orig_address(Q713)(3)
          L = 010
          Data:  Routing on GT, Global Title included(4),
                 No SPC in address
                 Subsystem Number = MSC(8),
                 Global Title :
                   Translation Type = 0,
                   Numbering Plan = ISDN/Telephony(E164),
```

```
                    Nature Address Indicator = International number,
                      Address information = 6596197979
          MAPPN_applic_context(11)
             L = 009
             Data: (Hex) 060704000001001402
                   ShortMsgGatewayPackage_v1_or_v2 MAP V2
- - - - - - - HDR - - - - - - - - - [ Fri Nov 29 10:19:11 2002 ]
MAPE-E Instance = 0
MAPE-E Type = MAP_MSG_SRV_REQ (0000C7E0)
MAPE-E Dialog_ID = 575
MAPE-E Src  = 1D
MAPE-E Dst  = 15
MAPE-E Rsp_req = 0  Class = 0  Status = 0  Err_info = 0  *Nxt = 0000
- - - - - - - - - - PARAMETER - AREA - - - - - - - - - - -
       PA_Len = 25
       MAP-SEND-ROUTING-INFO-FOR-SM-REQ
         MAPPN_invoke_id(14)
            L = 001
            Data: 1
         MAPPN_msisdn(15)
            L = 007
            Data: Ext = No extension
                  Ton = International
                  Npi = ISDN
                  Address = 60125165618
         MAPPN_sm_rp_pri(16)
            L = 001
            Data: (1):High Priority
         MAPPN_sc_addr(17)
            L = 006
            Data: Ext = No extension
                  Ton = International
                  Npi = ISDN
                  Address = 6596197979
- - - - - - - - -HDR- - - - - - - -[ Fri Nov 29 10:19:11 2002 ]
MAPE-E Instance = 0
MAPE-E Type = MAP_MSG_DLG_REQ (0000C7E2)
MAPE-E Dialog_ID = 575
MAPE-E Src  = 1D
MAPE-E Dst  = 15
MAPE-E Rsp_req = 0  Class = 0  Status = 0  Err_info = 0  *Nxt = 0000
   - - - - - - - - - - PARAMETER - AREA - - - - - - - - - - -
       PA_Len = 2
       MAP-DELIMITER-REQ
- - - - - - - - - - - HDR - - - - - - - [ Fri Nov 29 10:19:11 2002 ]
MAPE-R Instance = 0
MAPE-R Type = MAP_MSG_DLG_IND (000087E3)
```

```
        MAPE-R Dialog_ID = 575
        MAPE-R Src  = 15
        MAPE-R Dst  = 1D
        MAPE-R Rsp_req = 0  Class = 0  Status = 0  Err_info = 0  *Nxt = 0000
          - - - - - - - - - PARAMETER - AREA - - - - - - - - - - -
            PA_Len = 16
            MAP-OPEN-CNF
              MAPPN_result(5)
                L = 001
                Data: (0):Accept
              MAPPN_applic_context(11)
                L = 009
                Data: (Hex) 060704000001001402
                        ShortMsgGatewayPackage_v1_or_v2 MAP V2
          - - - - - - - - - HDR - - - - - - - -[ Fri Nov 29 10:19:11 2002 ]
        MAPE-R Instance = 0
        MAPE-R Type = MAP_MSG_SRV_IND (000087E1)
        MAPE-R Dialog_ID = 575
        MAPE-R Src  = 15
        MAPE-R Dst  = 1D
        MAPE-R Rsp_req = 0  Class = 0  Status = 0  Err_info = 0  *Nxt = 0000
          - - - - - - - - - PARAMETER - AREA - - - - - - - - - - -
            PA_Len = 23
            MAP-SEND-ROUTING-INFO-FOR-SM-CNF
              MAPPN_invoke_id(14)
                L = 001
                Data: 1
              MAPPN_imsi(18)
                L = 008
                Data: Address = 502125323286468
              MAPPN_msc_num(19)
                L = 006
                Data: Ext = No extension
                      Ton = International
                      Npi = ISDN
                      Address = 6593659940
        - - - - - - - - - - - - - - - - - - - - - - - - - - - - - -
    # ROUTER #  MTU_SEND_ROUTING_INFO_CNF :
                16 575 0 0 0 0 +502125323286468 +6593659940


        - - - - - - - - - -HDR- - - - - - - - - [ Fri Nov 29 10:19:11 2002 ]
        MAPE-R Instance = 0
        MAPE-R Type = MAP_MSG_DLG_IND (000087E3)
        MAPE-R Dialog_ID = 575
        MAPE-R Src  = 15
        MAPE-R Dst  = 1D
        MAPE-R Rsp_req = 0  Class = 0  Status = 0  Err_info = 0  *Nxt = 0000
```

```
- - - - - - - - - - PARAMETER - AREA - - - - - - - - - - -
     PA_Len = 2
     MAP-CLOSE-IND
- - - - - - - - - - - - - - - - - - - - - - - - - - - - - - -
```

*** Message received for inactive dialogue ***
# ROUTER # = MTU_SEND_ROUTING_INFO_REQ
              12 582 +6596197979 +447904403378 9479 +447904403378
+6596197979 0 1 0


```
- - - - - - - - - - - HDR- - - - - - - - - -[ Fri Nov 29 10:19:12 2002 ]
   MAPE-E Instance = 0
   MAPE-E Type = MAP_MSG_DLG_REQ (0000C7E2)
   MAPE-E Dialog_ID = 583
   MAPE-E Src  = 1D
   MAPE-E Dst  = 15
   MAPE-E Rsp_req = 0  Class = 0  Status = 0  Err_info = 0  *Nxt = 0000
       - - - - - - - - - - PARAMETER - AREA - - - - - - - - - - -
        PA_Len = 40
        MAP-OPEN-REQ
          MAPPN_dest_address(Q713)(1)
            L = 013
            Data:  Routing on GT, Global Title included(4),
                   Signalling Point Code (ITU) = 4-160-7 ( 9479)
                   Subsystem Number = MSC(8),
                   Global Title :
                       Translation Type = 0,
                       Numbering Plan = ISDN/Telephony(E164),
                   Nature Address Indicator = International number,
                       Address information = 85292347978
          MAPPN_orig_address(Q713)(3)
            L = 010
            Data:  Routing on GT, Global Title included(4),
                   No SPC in address
                   Subsystem Number = MSC(8),
                   Global Title :
                       Translation Type = 0,
                       Numbering Plan = ISDN/Telephony(E164),
                   Nature Address Indicator = International number,
                       Address information = 6596197979
          MAPPN_applic_context(11)
            L = 009
            Data: (Hex) 060704000001001902
                   ShortMsgRelayPackage_v2 MAP V2
       - - - - - - - - - -HDR- - - - - - - - - - [ Fri Nov 29 10:19:12 2002 ]
   MAPE-E Instance = 0
   MAPE-E Type = MAP_MSG_SRV_REQ (0000C7E0)
   MAPE-E Dialog_ID = 583
```

```
MAPE-E Src  = 1D
MAPE-E Dst  = 15
MAPE-E Rsp_req = 0  Class = 0  Status = 0  Err_info = 0  *Nxt = 0000
- - - - - - - - - - PARAMETER - AREA - - - - - - - - - - -
PA_Len = 102
MAP-FORWARD-SHORT-MESSAGE-REQ
  MAPPN_invoke_id(14)
    L = 001
    Data: 1
  MAPPN_sm_rp_da(23)
    L = 019
    Data: TA_SC_ADR
          Ext = No extension
          Ton = International
          Npi = ISDN
          Address = 502125323286468 404002cfc77fe9ea
  MAPPN_sm_rp_oa(24)
    L = 008
    Data: TA_SC_ADR
          Ext = No extension
          Ton = International
          Npi = ISDN
          Address = 6593659940
  MAPPN_sm_rp_ui(25)
    L = 064
    Data: Message Type = SMS_DELIVER(SMS-MT)
          TP_RP  = Request for reply path
          TP_UDHI= No Header in TP-User-Data
       TP_SRI = A status report will not be returned to the SME
          TP_VPF = TP_VP field not present
    TP_MMS = No more messages are waiting for the MS in this SC
          Originating mobile address               =
             Type of number = Unknown
             Numbering Plan = Unknown
             Address =
TP-Protocol-Identifier = 00 (No Interworking: SME-to-SME protocol)
          TP-Data-Coding-Scheme  =  00
             Message Class = Default (Handset MEmory)
          Alphabet    = Default alphabet(7 bit packed)
       TP_Service_Centre_Time_Stamp = 02.11.29 18:19:12 00
             TP-User-Data-length = 47
             TP-User-Data =
          SMS quality check, sorry for the inconvenience.
          Additional Fields   -
          Destination Mobile Address =
             Type of number = International
             Numbering Plan = ISDN
```

```
                         Address = 60125165618
- - - - - - - - - -HDR - - - - - - - - - - [ Fri Nov 29 10:19:12 2002 ]
  MAPE-E Instance = 0
  MAPE-E Type = MAP_MSG_DLG_REQ (0000C7E2)
  MAPE-E Dialog_ID = 583
  MAPE-E Src  = 1D
  MAPE-E Dst  = 15
  MAPE-E Rsp_req = 0  Class = 0  Status = 0  Err_info = 0  *Nxt = 0000
      - - - - - - - - - - PARAMETER - AREA - - - - - - - - - -
        PA_Len = 2
        MAP-DELIMITER-REQ
- - - - - - - - - - -HDR- - - - - - - - - -[ Fri Nov 29 10:19:17 2002 ]
  MAPE-R Instance = 0
  MAPE-R Type = MAP_MSG_DLG_IND (000087E3)
  MAPE-R Dialog_ID = 583
  MAPE-R Src  = 15
  MAPE-R Dst  = 1D
  MAPE-R Rsp_req = 0  Class = 0  Status = 0  Err_info = 0  *Nxt = 0000
      - - - - - - - - - - PARAMETER - AREA - - - - - - - - - -
        PA_Len = 16
        MAP-OPEN-CNF
          MAPPN_result(5)
            L = 001
            Data: (0):Accept
          MAPPN_applic_context(11)
            L = 009
            Data: (Hex) 060704000001001902
                  ShortMsgRelayPackage_v2 MAP V2
- - - - - - - - - -HDR- - - - - - - - - - [ Fri Nov 29 10:19:17 2002 ]
  MAPE-R Instance = 0
  MAPE-R Type = MAP_MSG_SRV_IND (000087E1)
  MAPE-R Dialog_ID = 583
  MAPE-R Src  = 15
  MAPE-R Dst  = 1D
  MAPE-R Rsp_req = 0  Class = 0  Status = 0  Err_info = 0  *Nxt = 0000
      - - - - - - - - - - PARAMETER - AREA - - - - - - - - - -
        PA_Len = 48
        MAP-FORWARD-SHORT-MESSAGE-CNF
          MAPPN_invoke_id(14)
            L = 001
            Data: 1
          MAPPN_deliv_fail_cse(31)
            L = 001
            Data:
          MAPPN_imsi(18)
            L = 008
            Data: Address = 502125323286468
```

```
                    MAPPN_msc_num(19)
                      L = 006
                      Data: Ext = No extension
                            Ton = International
                            Npi = ISDN
                            Address = 6593659940
                    MAPPN_sub_state(54)
                      L = 001
                      Data: (255):Unknown code received
                    MAPPN_not_reach_rsn(56)
                      L = 001
                      Data: (255):Not applicable
                    MAPPN_time_zone(128)
                      L = 004
                      Data: 8.00
                    MAPPN_timing_adv(129)
                      L = 001
                      Data: (Hex) FF
                    MAPPN_age_loc_info(48)
                      L = 002
                      Data: age of location info 65535 minutes
                    MAPPN_fwing_options(43)
                      L = 001
                      Data: (Hex) FF
         - - - - - - - - - - - - - - - - - - - - - - - - - - - - -
     # ROUTER #   MTU_FORWARD_SHORT_MSG_CNF :
                    3 583 0 0 0 0 +502125323286468 +6593659940 ??? 255 255
 8.00 255 65535 ??? ??? ??? ??? 255
        - - - - - - - - -HDR- - - - - - - - - -[ Fri Nov 29 10:19:17 2002 ]
         MAPE-R Instance = 0
         MAPE-R Type = MAP_MSG_DLG_IND (000087E3)
         MAPE-R Dialog_ID = 583
         MAPE-R Src  = 15
         MAPE-R Dst  = 1D
        MAPE-R Rsp_req = 0  Class = 0  Status = 0  Err_info = 0  *Nxt = 0000
         - - - - - - - - - - PARAMETER - AREA - - - - - - - - - - -
             PA_Len = 2
             MAP-CLOSE-IND
         - - - - - - - - - - - - - - - - - - - - - - - - - - - - -
     # ROUTER # =   MTU_FORWARD_SHORT_MSG_MO_RSP
                    7 33060 0 0 0 0 +502125323286468 +6593659940 ??? -1 255
 8.00 255 65535 ??? ??? ??? ??? 255


        - - - - - - - - - - - - - - - - - - - - - - - - - - - - -
     HONG-KONG traces:
     - - - - - - - - - - - - - - - - - - - - - - - - - - - - -
```

```
- - - - - - - - - -HDR- - - - - - - - - [ Fri Nov 29 10:19:11 2002 ]
  MAPE-R Instance = 0
  MAPE-R Type = MAP_MSG_DLG_IND (000087E3)
  MAPE-R Dialog_ID = 33524
  MAPE-R Src  = 15
  MAPE-R Dst  = 1D
 MAPE-R Rsp_req = 0  Class = 0  Status = 0  Err_info = 0  *Nxt = 0000
  - - - - - - - - - - PARAMETER - AREA - - - - - - - - - - -
      PA_Len = 42
      MAP-OPEN-IND
       MAPPN_dest_address(Q713)(1)
         L = 013
         Data:   Routing on SPC, Global Title included(4),
                 Signalling Point Code (ITU) = 1-043-5 ( 2397)
                 Subsystem Number = MSC,
                 Global Title :
                    Translation Type = 0,
                    Numbering Plan = ISDN/Telephony(E164),
                 Nature Address Indicator = International number,
                    Address information = 85292347978
       MAPPN_orig_address(Q713)(3)
         L = 012
         Data:   Routing on GT, Global Title included(4),
                 Signalling Point Code (ITU) = 1-040-1 ( 2369)
                 Subsystem Number = MSC,
                 Global Title :
                    Translation Type = 0,
                    Numbering Plan = ISDN/Telephony(E164),
                 Nature Address Indicator = International number,
                    Address information = 6596197979
      MAPPN_applic_context(11)
         L = 009
         Data: (Hex) 060704000001001902
                 ShortMsgRelayPackage_v2 MAP V2
- - - - - - - - - - HDR- - - - - - - - -[ Fri Nov 29 10:19:11 2002 ]
  MAPE-E Instance = 0
  MAPE-E Type = MAP_MSG_DLG_REQ (0000C7E2)
  MAPE-E Dialog_ID = 33524
  MAPE-E Src  = 1D
  MAPE-E Dst  = 15
 MAPE-E Rsp_req = 0  Class = 0  Status = 0  Err_info = 0  *Nxt = 0000
  - - - - - - - - - - PARAMETER - AREA - - - - - - - - - -
      PA_Len = 16
      MAP-OPEN-RSP
       MAPPN_result(5)
         L = 001
         Data: (0):Accept
```

```
               MAPPN_applic_context(11)
                  L = 009
                  Data: (Hex) 060704000001001902
                          ShortMsgRelayPackage_v2 MAP V2
    - - - - - - - - - HDR- - - - - - - - - [ Fri Nov 29 10:19:11 2002 ]
      MAPE-R Instance = 0
      MAPE-R Type = MAP_MSG_SRV_IND (000087E1)
      MAPE-R Dialog_ID = 33524
      MAPE-R Src  = 15
      MAPE-R Dst  = 1D
     MAPE-R Rsp_req = 0  Class = 0  Status = 0  Err_info = 0  *Nxt = 0000
     - - - - - - - - - - PARAMETER - AREA - - - - - - - - - -
         PA_Len = 102
         MAP-FORWARD-SHORT-MESSAGE-IND
          MAPPN_invoke_id(14)
             L = 001
             Data: 1
          MAPPN_sm_rp_da(23)
             L = 019
             Data: TA_SC_ADR
                     Ext = No extension
                     Ton = International
                     Npi = ISDN
                     Address = 502125323286468 404002cfc77fe9ea
          MAPPN_sm_rp_oa(24)
             L = 008
             Data: TA_SC_ADR
                     Ext = No extension
                     Ton = International
                     Npi = ISDN
                     Address = 6593659940
          MAPPN_sm_rp_ui(25)
             L = 064
             Data: Message Type = SMS_DELIVER(SMS-MT)
                     TP_RP  = Request for reply path
                     TP_UDHI= No Header in TP-User-Data
                 TP_SRI = A status report will not be returned to the SME
                     TP_VPF = TP_VP field not present
             TP_MMS = No more messages are waiting for the MS in this SC
                     Originating mobile address              =
                         Type of number = Unknown
                         Numbering Plan = Unknown
                         Address =
    TP-Protocol-Identifier = 00 (No Interworking: SME-to-SME protocol)
                     TP-Data-Coding-Scheme  =  00
                       Message Class = Default (Handset MEmory)
                   Alphabet     = Default alphabet(7 bit packed)
```

```
                    TP_Service_Centre_Time_Stamp = 02.11.29 18:19:12 00
                          TP-User-Data-length = 47
                          TP-User-Data =
                       "SMS quality check, sorry for the inconvenience."
                        Additional Fields   -
                        Destination Mobile Address =
                            Type of number = International
                            Numbering Plan = ISDN
                            Address = 60125165618
       - - - - - - - - - - - - - - - - - - - - - - - - - - - -

    # ROUTER #   MTU_FORWARD_SHORT_MSG_MT_IND :
                     28  33524  +6596197979  +85292347978  +6593659940
  +502125323286468  +0  404002cfc77fe9ea  +60125165618  1  0  0  0  0  0  0
  2FD3E61414AF87D9697A1E344697C76B1668FE96CBF320F35B0EA2A3CBA0B47BFC76
  DBCBEE74D93D2EBB00


                      - HDR                   [ Fri Nov 29 10:19:11 2002 ]
        MAPE-R Instance = 0
        MAPE-R Type = MAP_MSG_DLG_IND (000087E3)
        MAPE-R Dialog_ID = 33524
        MAPE-R Src  = 15
        MAPE-R Dst  = 1D
        MAPE-R Rsp_req = 0  Class = 0  Status = 0  Err_info = 0  *Nxt = 0000
        - - - - - - - - - PARAMETER - AREA - - - - - - - - - -
            PA_Len = 2
            MAP-DELIMITER-IND
       - - - - - - - - - - - - - - -  - - - - - - - - - - - - - -

    # ROUTER #   MTU_FORWARD_SHORT_MSG_MT_IND :
                     28  33524  +6596197979  +85292347978  +6593659940
  +502125323286468  +0  404002cfc77fe9ea  +60125165618  1  0  0  0  0  0  0
  2FD3E61414AF87D9697A1E344697C76B1668FE96CBF320F35B0EA2A3CBA0B47BFC76
  DBCBEE74D93D2EBB00
    # ROUTER # =  MTU_FORWARD_SHORT_MSG_MTA_REQ
                     27  564  +85292347978  +85292347978  +6593659940 2369
  +502125323286468   404002cfc77fe9ea  +0  +60125165618  0  0  0  1  255
  20119281604200    0   3   0    2FD3E61414AF87D9697A1E344697C76B1668FE96
  CBF320F35B0EA2A3CBA0B47BFC76DBCBEE74D93D2EBB00


    - - - - - - - - - - HDR- - - - - - - - - [ Fri Nov 29 10:19:16 2002 ]
        MAPE-E Instance = 0
        MAPE-E Type = MAP_MSG_SRV_REQ (0000C7E0)
        MAPE-E Dialog_ID = 33524
        MAPE-E Src  = 1D
        MAPE-E Dst  = 15
        MAPE-E Rsp_req = 0  Class = 0  Status = 0  Err_info = 0  *Nxt = 0000
        - - - - - - - - - - PARAMETER - AREA - - - - - - - - - -
            PA_Len = 51
```

```
            MAP-FORWARD-SHORT-MESSAGE-RSP
              MAPPN_invoke_id(14)
                L = 001
                Data: 1
              MAPPN_user_err(21)
                L = 001
                Data: (0):Ok!
              MAPPN_deliv_fail_cse(31)
                L = 001
                Data:
              MAPPN_imsi(18)
                L = 008
                Data: Address = 502125323286468
              MAPPN_msc_num(19)
                L = 006
                Data: Ext = No extension
                      Ton = International
                      Npi = ISDN
                      Address = 6593659940
              MAPPN_sub_state(54)
                L = 001
                Data: (255):Unknown code received
              MAPPN_not_reach_rsn(56)
                L = 001
                Data: (255):Not applicable
              MAPPN_time_zone(128)
                L = 004
                Data: 8.00
              MAPPN_timing_adv(129)
                L = 001
                Data: (Hex) FF
              MAPPN_age_loc_info(48)
                L = 002
                Data: age of location info 65535 minutes
              MAPPN_fwing_options(43)
                L = 001
                Data: (Hex) FF
- - - - - - - - - - HDR- - - - - - - - - -[ Fri Nov 29 10:19:16 2002 ]
  MAPE-E Instance = 0
  MAPE-E Type = MAP_MSG_DLG_REQ (0000C7E2)
  MAPE-E Dialog_ID = 33524
  MAPE-E Src  = 1D
  MAPE-E Dst  = 15
MAPE-E Rsp_req = 0  Class = 0  Status = 0  Err_info = 0  *Nxt = 0000
  - - - - - - - - - - - PARAMETER - AREA - - - - - - - - - - - -
        PA_Len = 5
        MAP-CLOSE-REQ
```

```
                    MAPPN_release_method(7)
                       L = 001
                       Data: (0):Normal Release
          - - - - - - - - - HDR- - - - - - - - - [ Fri Nov 29 10:19:12 2002 ]
     MAPE-E Instance = 0
     MAPE-E Type = MAP_MSG_DLG_REQ (0000C7E2)
     MAPE-E Dialog_ID = 564
     MAPE-E Src  = 1D
     MAPE-E Dst  = 15
     MAPE-E Rsp_req = 0  Class = 0  Status = 0  Err_info = 0  *Nxt = 0000
        - - - - - - - - - - PARAMETER - AREA - - - - - - - - - - -
          PA_Len = 40
          MAP-OPEN-REQ
            MAPPN_dest_address(Q713)(1)
               L = 012
               Data: Routing on GT, Global Title included(4),
                     Signalling Point Code (ITU) = 1-040-1 ( 2369)
                     Subsystem Number = MSC,
                     Global Title :
                         Translation Type = 0,
                         Numbering Plan = ISDN/Telephony(E164),
                      Nature Address Indicator = International number,
                         Address information = 6593659940
            MAPPN_orig_address(Q713)(3)
               L = 011
               Data:  Routing on GT, Global Title included(4),
                     No SPC in address
                     Subsystem Number = MSC,
                     Global Title :
                         Translation Type = 0,
                         Numbering Plan = ISDN/Telephony(E164),
                      Nature Address Indicator = International number,
                         Address information = 85292347978
            MAPPN_applic_context(11)
               L = 009
               Data: (Hex) 060704000001001501
                     ShortMsgRelayPackage_v1 MAP V1
          - - - - - - - - - HDR- - - - - - - - - [ Fri Nov 29 10:19:12 2002 ]
     MAPE-E Instance = 0
     MAPE-E Type = MAP_MSG_SRV_REQ (0000C7E0)
     MAPE-E Dialog_ID = 564
     MAPE-E Src  = 1D
     MAPE-E Dst  = 15
     MAPE-E Rsp_req = 0  Class = 0  Status = 0  Err_info = 0  *Nxt = 0000
        - - - - - - - - - - PARAMETER - AREA - - - - - - - - - - -
          PA_Len = 85
          MAP-FORWARD-SHORT-MESSAGE-REQ
```

```
            MAPPN_invoke_id(14)
              L = 001
              Data: 1
            MAPPN_sm_rp_da(23)
              L = 010
              Data: TA_IMSI
                    Address = 502125323286468
            MAPPN_sm_rp_oa(24)
              L = 009
              Data: TA_SC_ADR
                    Ext = No extension
                    Ton = International
                    Npi = ISDN
                    Address = 85292347978
            MAPPN_sm_rp_ui(25)
              L = 055
              Data: Message Type = SMS_DELIVER(SMS-MT)
                    TP_RP  = No request for reply path
                    TP_UDHI= No Header in TP-User-Data
                  TP_SRI = A status report will not be returned to the SME
                    TP_VPF = TP_VP field not present
                  TP_MMS = No more messages are waiting for the MS in this
                    SC
                    Originating mobile address              =
                      Type of number = Unknown
                      Numbering Plan = Unknown
                      Address =
                  TP-Protocol-Identifier = 00 (No Interworking: SME-to-
                    SME protocol)
                    TP-Data-Coding-Scheme  =  00
                      Message Class = Default (Handset MEmory)
                   Alphabet     = Default alphabet(7 bit packed)
                  TP_Service_Centre_Time_Stamp = 02.11.29 18:06:24 00
                      TP-User-Data-length = 47
                      TP-User-Data =
                  SMS quality check, sorry for the inconvenience.
- - - - - - - - - - HDR- - - - - - - - - [ Fri Nov 29 10:19:12 2002 ]
  MAPE-E Instance = 0
  MAPE-E Type = MAP_MSG_DLG_REQ (0000C7E2)
  MAPE-E Dialog_ID = 564
  MAPE-E Src  = 1D
  MAPE-E Dst  = 15
  MAPE-E Rsp_req = 0  Class = 0  Status = 0  Err_info = 0  *Nxt = 0000
    - - - - - - - - - - PARAMETER - AREA - - - - - - - - - -
        PA_Len = 2
        MAP-DELIMITER-REQ
- - - - - - - - - - HDR- - - - - - - - - [ Fri Nov 29 10:19:16 2002 ]
```

```
              MAPE-R Instance = 0
              MAPE-R Type = MAP_MSG_DLG_IND (000087E3)
              MAPE-R Dialog_ID = 564
              MAPE-R Src  = 15
              MAPE-R Dst  = 1D
            MAPE-R Rsp_req = 0  Class = 0  Status = 0  Err_info = 0  *Nxt = 0000
            - - - - - - - - - - PARAMETER - AREA - - - - - - - - - - -
                  PA_Len = 16
                  MAP-OPEN-CNF
                    MAPPN_result(5)
                      L = 001
                      Data: (0):Accept
                    MAPPN_applic_context(11)
                      L = 009
                      Data: (Hex) 060704000001001501
                            ShortMsgRelayPackage_v1 MAP V1
        - - - - - - - - - - HDR- - - - - - - - - [ Fri Nov 29 10:19:16 2002 ]
              MAPE-R Instance = 0
              MAPE-R Type = MAP_MSG_SRV_IND (000087E1)
              MAPE-R Dialog_ID = 564
              MAPE-R Src  = 15
              MAPE-R Dst  = 1D
            MAPE-R Rsp_req = 0  Class = 0  Status = 0  Err_info = 0  *Nxt = 0000
            - - - - - - - - - - PARAMETER - AREA - - - - - - - - - - -
                  PA_Len = 5
                  MAP-FORWARD-SHORT-MESSAGE-CNF
                    MAPPN_invoke_id(14)
                      L = 001
                      Data: 1
            - - - - - - - - - - - - - - - - -   - - - - - - - - - - - - -
         # ROUTER #   MTU_FORWARD_SHORT_MSG_CNF :
                         3 564 0 0 0 0
        - - - - - - - - - - HDR- - - - - - - - - [ Fri Nov 29 10:19:16 2002 ]
              MAPE-R Instance = 0
              MAPE-R Type = MAP_MSG_DLG_IND (000087E3)
              MAPE-R Dialog_ID = 564
              MAPE-R Src  = 15
              MAPE-R Dst  = 1D
            MAPE-R Rsp_req = 0  Class = 0  Status = 0  Err_info = 0  *Nxt = 0000
            - - - - - - - - - - PARAMETER - AREA - - - - - - - - - - -
                  PA_Len = 2
                  MAP-CLOSE-IND
            - - - - - - - - - - - - - - - - - - - - - - - - - - - - - - -
         # ROUTER # =  MTU_FORWARD_SHORT_MSG_MO_RSP
                         7 33524 0 0 0 0 +502125323286468 +6593659940 ??? 255 255
        8.00 255 65535 ??? ??? ??? ??? 255
```

16.3   Example 3

Figure 16.1 shows how the routes in an SS7 network are set up:

GSM A is connected to IGP 1346.

GSM B is connected to IGP 4041.

GSM A wants to send an SCCP message to GSM node B.

Find the different translation rules and routing types that must be set up in the SS7 network to fulfill the desire to send an SCCP message from GSM A to GSM node B.

16.4   Example 4

In Figure 5.1 in Chapter 5, GSM A is sending an SMS to the subscriber +261323451234 of GSM B. GSM B wants to receive the SMS, but with the sending SMSC address of C (the SMS interworking network), which will be the proxy for the reception of all SMS-MT sent by GSM B's roaming partners as well as nonpartners:

B and C have an agreement (and an SCCP connection).

B and A have a roaming agreement.

C and A do not necessarily have an agreement.

B and C (for simplification of the explanation) have the same supplier for their international SCCP service. This supplier provides the basic service of translating to GSM B's HLRs addresses the "Called Party Addresses = MSISDN" used in the MAP_SEND ROUTING_INFO. In that case (if the GSM agrees), the translation table will point to a virtual HLR/MSC address that belongs to GSM B, but physically corresponds to equipment belonging to A.

Because voice roaming deals only with SCCP addresses that are individual network elements (HLR, VLR, MSC, SMSC), it is not impacted at all. The oriented graph model representation of Figure 16.2 corresponds to our problem:

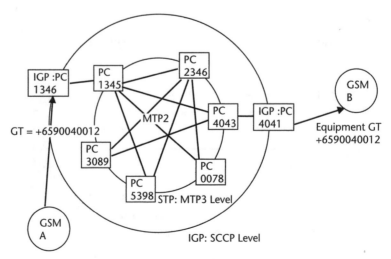

Figure 16.1   Setting up the routes in an SS7 network.

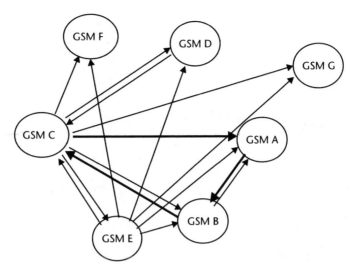

Figure 16.2   Rerouting of SMS when the IGP suppliers are different.

1. Design the solution when third-party SMS interworking network C and GSM B do not have the same IGP supplier.
2. Is there a (theoretical) way that the CNF from C (the SMS interworking network in Figure 16.2) goes back to A through B? Could your international carrier SCCP gateways do it?
3. In the setup of Figure 16.2, check whether or not A can (technically) send classical SMS-MT (confirmed) to C even though they do not have a roaming agreements. Who is likely to disagree of the various parties involved?

16.5   Example 5

What is wrong with the setup for the roaming in the GSM (in Africa) shown in Figure 16.3: the danger of using international point codes. A group of mobile operators uses a private STP to concentrate the international signaling traffic of its members (several operators in Africa, Asia, and so on). VSAT links exist between the gateway of these mobile operators and the private STP. *All have an international point code.*

The private STP operator has a direct connection with one of the main SS7 carriers, which offers SCCP service as well as MTP transit traffic. GSM Africa has set up a number of roaming agreements in the past weeks.

The private STP has received a big invoice (for the MSU traffic) sent to the IGP. It has set up an analyzer on the satellite link between the STP and GSM Africa, which shows that many SSTs have been sent from GSM Africa to various international point codes—without response. Answer these questions:

1. What is wrong (but works) with the setup for the roaming of GSM Africa?
2. What should the private STP operator do to prevent any such occurrence in the future? What should it do to switch from the present configuration to a better one without disrupting the roaming agreement of GSM Africa?

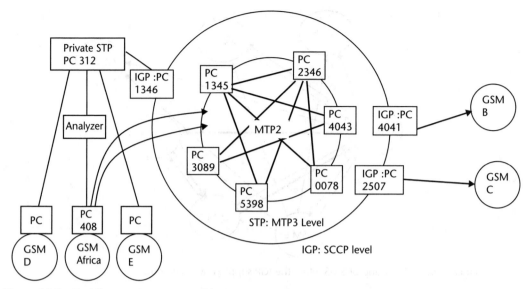

Figure 16.3 Find the roaming setup problem.

3. Why is there no response to the SST?
4. Should the SST on this satellite link be enabled?

16.6 Example 6

An IP trace of an SMPP dialogue between a client and an SMSC server follows. Study the trace and then answer these questions:

1. What is the destination address?
2. What is the *type of number* (TON) and *numbering plan indicator* (NPI)?
3. What is the result of the initial submission of the SMS?
4. Describe the link supervision protocol.

```
1    0.000000      thirdparty - customfr.custom.fr SMPP SMPP Unbind

0000   00 30 05 01 98 9a 00 03 47 4c 4b 00 08 00 45 00
.0......GLK...E.
0010   00 44 57 71 40 00 40 06 e7 13 d4 9b a9 7c d4 9b
.DWq@.@......|..
0020   a9 7b 94 a5 0a d7 16 00 4c 5e 15 d8 ce 27 80 18
.{......L^... ..
0030   16 d0 a9 68 00 00 01 01 08 0a 2d 2c 2d 8f 03 55
...h......-,-..U
0040   76 38 00 00 00 10 00 00 00 06 00 00 00 00 00 00
v8.............
0050   00 04                                          ..
```

    2   0.004972 customfr.custom.fr - thirdparty   SMPP SMPP Unbind -
resp: Ok

    0000   00 03 47 4c 4b 00 00 30 05 01 98 9a 08 00 45 00
..GLK..O......E.
    0010   00 44 16 b1 40 00 40 06 27 d4 d4 9b a9 7b d4 9b
.D..a.a. ....{..
    0020   a9 7c 0a d7 94 a5 15 d8 ce 27 16 00 4c 6e 80 18
.|........ ..Ln..
    0030   7d 78 b0 d6 00 00 01 01 08 0a 03 55 88 11 2d 2c
}x.........U..-,
    0040   2d 8f 00 00 00 10 80 00 00 06 00 00 00 00 00 00
-...............
    0050   00 04   ..

    3   0.005195   thirdparty - customfr.custom.fr TCP 38053   2775
[ACK] Seq=369118318 Ack=366530103 Win=5840 Len=0

    0000   00 30 05 01 98 9a 00 03 47 4c 4b 00 08 00 45 00
.0......GLK...E.
    0010   00 34 57 72 40 00 40 06 e7 22 d4 9b a9 7c d4 9b
.4Wr@.@.."...|..
    0020   a9 7b 94 a5 0a d7 16 00 4c 6e 15 d8 ce 37 80 10
.{......Ln...7..
    0030   16 d0 97 a1 00 00 01 01 08 0a 2d 2c 2d 8f 03 55
..........-,-..U
    0040   88 11   ..

    4   0.006240 customfr.custom.fr - thirdparty   TCP 2775   38053
[FIN, ACK] Seq=366530103 Ack=369118318 Win=32120 Len=0

    0000   00 03 47 4c 4b 00 00 30 05 01 98 9a 08 00 45 00
..GLK..O......E.
    0010   00 34 16 b4 40 00 40 06 27 e1 d4 9b a9 7b d4 9b
.4..a.a. ....{..
    0020   a9 7c 0a d7 94 a5 15 d8 ce 37 16 00 4c 6e 80 11
.|.......7..Ln..
    0030   7d 78 30 f8 00 00 01 01 08 0a 03 55 88 11 2d 2c
}x0........U..-,
    0040   2d 8f   -.

    5   0.039468   thirdparty - customfr.custom.fr TCP 38053   2775
[ACK] Seq=369118318 Ack=366530104 Win=5840 Len=0

    0000   00 30 05 01 98 9a 00 03 47 4c 4b 00 08 00 45 00
.0......GLK...E.

```
     0010    00  34  57  73  40  00  40  06  e7  21  d4  9b  a9  7c  d4  9b
.4Ws@.@..!...|..
     0020    a9  7b  94  a5  0a  d7  16  00  4c  6e  15  d8  ce  38  80  10
.{......Ln...8..
     0030    16  d0  97  9c  00  00  01  01  08  0a  2d  2c  2d  93  03  55
..........-,-..U
     0040    88  11                                                    ..
```

    6   0.230269   thirdparty - customfr.custom.fr TCP 38053   2775
[RST, ACK] Seq=369118318 Ack=366530104 Win=5840 Len=0

```
     0000    00  30  05  01  98  9a  00  03  47  4c  4b  00  08  00  45  00
.0......GLK...E.
     0010    00 34 57 74 40 00 40 06 e7 20 d4 9b a9 7c d4 9b   .4Wt@.@.. ...|..
     0020    a9  7b  94  a5  0a  d7  16  00  4c  6e  15  d8  ce  38  80  14
.{......Ln...8..
     0030    16  d0  97  85  00  00  01  01  08  0a  2d  2c  2d  a6  03  55
..........-,-..U
     0040    88  11                                                    ..
```

    7 10.553838   thirdparty - customfr.custom.fr TCP 38182   2775
[SYN] Seq=433049701 Ack=0 Win=5840 Len=0

```
     0000    00  30  05  01  98  9a  00  03  47  4c  4b  00  08  00  45  00
.0......GLK...E.
     0010    00 3c e1 cb 40 00 40 06 5c c1 d4 9b a9 7c d4 9b   ..@.@.\....|..
     0020    a9  7b  95  26  0a  d7  19  cf  d0  65  00  00  00  00  a0  02
.{.&.....e......
     0030    16  d0  4b  fb  00  00  02  04  05  b4  04  02  08  0a  2d  2c
..K...........-,
     0040  31 ae 00 00 00 00 01 03 03 00                1.........
```

    8 10.553885 customfr.custom.fr - thirdparty   TCP 2775   38182
[SYN, ACK] Seq=418940460 Ack=433049702 Win=32120 Len=0

```
     0000    00  03  47  4c  4b  00  00  30  05  01  98  9a  08  00  45  00
..GLK..0......E.
     0010    00 3c 18 fe 40 00 40 06 25 8f d4 9b a9 7b d4 9b   ..@.@.%....{..
     0020    a9  7c  0a  d7  95  26  18  f8  86  2c  19  cf  d0  66  a0  12
.|...&...,...f..
     0030    7d  78  b6  97  00  00  02  04  05  b4  04  02  08  0a  03  55
}x............U
     0040  8c 30 2d 2c 31 ae 01 03 03 00                .0-,1.....
```

    9 10.554096   thirdparty - customfr.custom.fr TCP 38182   2775
[ACK] Seq=433049702 Ack=418940461 Win=5840 Len=0

```
    0000    00  30  05  01  98  9a  00  03  47  4c  4b  00  08  00  45  00
.0......GLK...E.
    0010    00  34  e1  cc  40  00  40  06  5c  c8  d4  9b  a9  7c  d4  9b
.4..a.a.\....|..
    0020    a9  7b  95  26  0a  d7  19  cf  d0  66  18  f8  86  2d  80  10
.{.&.....f...-..
    0030    16  d0  4c  05  00  00  01  01  08  0a  2d  2c  31  ae  03  55
..L........-,1..U
    0040    8c  30                                                      .0
```

    10   10.554114   thirdparty - customfr.custom.fr TCP 38182 2775
[ACK] Seq=433049702 Ack=418940461 Win=5840 Len=0

```
    0000    00  30  05  01  98  9a  00  03  47  4c  4b  00  08  00  45  00
.0......GLK...E.
    0010    00  40  e1  cd  40  00  40  06  5c  bb  d4  9b  a9  7c  d4  9b
.@..a.a.\....|..
    0020    a9  7b  95  26  0a  d7  19  cf  d0  66  18  f8  86  2d  b0  10
.{.&.....f...-..
    0030    16  d0  d7  a3  00  00  01  01  08  0a  2d  2c  31  ae  03  55
..........-,1..U
    0040  8c 30 01 01 05 0a 18 f8 86 2c 18 f8 86 2d        .0.......,...-
```

    11   11.551347   thirdparty - customfr.custom.fr SMPP SMPP
Bind_transmitter

```
    0000    00  30  05  01  98  9a  00  03  47  4c  4b  00  08  00  45  00
.0......GLK...E.
    0010    00  5e  e1  ce  40  00  40  06  5c  9c  d4  9b  a9  7c  d4  9b
.^..a.a.\....|..
    0020    a9  7b  95  26  0a  d7  19  cf  d0  66  18  f8  86  2d  80  18
.{.&.....f...-..
    0030    16  d0  55  cb  00  00  01  01  08  0a  2d  2c  32  12  03  55
..U........-,2..U
    0040    8c  30  00  00  00  2a  00  00  00  02  00  00  00  00  00  00
.0...*..........
    0050    00  02  4e  69  6c  63  6f  6d  31  00  70  61  73  73  77  6f
..Custom1.passwo
    0060  72 64 00 45 53 4d 45 00 34 00 00 00            rd.ESME.4...
```

    12   11.551411 customfr.custom.fr - thirdparty   TCP 2775 38182
[ACK] Seq=418940461 Ack=433049744 Win=32120 Len=0

```
    0000    00  03  47  4c  4b  00  00  30  05  01  98  9a  08  00  45  00
..GLK..0......E.
    0010    00  34  19  3b  40  00  40  06  25  5a  d4  9b  a9  7b  d4  9b
.4.;@.@.%Z...{..
```

```
   0020    a9  7c  0a  d7  95  26  18  f8  86  2d  19  cf  d0  90  80  10
.|...&...-......
   0030    7d  78  e4  6a  00  00  01  01  08  0a  03  55  8c  94  2d  2c
}x.j.......U..-,
   0040    32  12                                                          2.
```

    13   11.630441 customfr.custom.fr - thirdparty   SMPP SMPP
Bind_transmitter - resp: "Ok"

```
   0000    00  03  47  4c  4b  00  00  30  05  01  98  9a  08  00  45  00
..GLK..0......E.
   0010    00  52  19  3f  40  00  40  06  25  38  d4  9b  a9  7b  d4  9b
.R.?@.@.%8...{..
   0020    a9  7c  0a  d7  95  26  18  f8  86  2d  19  cf  d0  90  80  18
.|...&...-......
   0030    7d  78  22  b8  00  00  01  01  08  0a  03  55  8c  9c  2d  2c
}x".......U..-,
   0040    32  12  00  00  00  1e  80  00  00  02  00  00  00  00  00  00
2..............
   0050    00  02  4e  49  4c  43  4f  4d  46  52  00  02  10  00  01  34
..CUSTOMFR.....4
```

    14   11.630615   thirdparty - customfr.custom.fr TCP 38182 2775
[ACK] Seq=433049744 Ack=418940491 Win=5840 Len=0

```
   0000    00  30  05  01  98  9a  00  03  47  4c  4b  00  08  00  45  00
.0......GLK...E.
   0010    00  34  e1  cf  40  00  40  06  5c  c5  d4  9b  a9  7c  d4  9b
.4..@.@.\....|..
   0020    a9  7b  95  26  0a  d7  19  cf  d0  90  18  f8  86  4b  80  10
.{.&.........K..
   0030    16  d0  4a  e6  00  00  01  01  08  0a  2d  2c  32  19  03  55
..J........-,2..U
   0040    8c  9c                                                          ..
```

    15   11.631323   thirdparty - customfr.custom.fr SMPP SMPP
Enquire_link

```
   0000    00  30  05  01  98  9a  00  03  47  4c  4b  00  08  00  45  00
.0......GLK...E.
   0010    00  44  e1  d0  40  00  40  06  5c  b4  d4  9b  a9  7c  d4  9b
.D..@.@.\....|..
   0020    a9  7b  95  26  0a  d7  19  cf  d0  90  18  f8  86  4b  80  18
.{.&.........K..
   0030    16  d0  4a  a5  00  00  01  01  08  0a  2d  2c  32  1a  03  55
..J........-,2..U
```

```
     0040    8c  9c  00  00  00  10  00  00  00  15  00  00  00  00  00  00
................
     0050    00  03                                                      ..
```

```
     16    11.633991  customfr.custom.fr - thirdparty         SMPP  SMPP
Enquire_link - resp: "0k"
```

```
     0000    00  03  47  4c  4b  00  00  30  05  01  98  9a  08  00  45  00
..GLK..0......E.
     0010    00  44  19  4c  40  00  40  06  25  39  d4  9b  a9  7b  d4  9b
.D.L@.@.%9...{..
     0020    a9  7c  0a  d7  95  26  18  f8  86  4b  19  cf  d0  a0  80  18
.|...&...K......
     0030    7d  78  63  ec  00  00  01  01  08  0a  03  55  8c  9c  2d  2c
}xc........U..-,
     0040    32  1a  00  00  00  10  80  00  00  15  00  00  00  00  00  00
2...............
     0050    00  03                                                      ..
```

```
     17    11.670896     thirdparty - customfr.custom.fr TCP 38182  2775
[ACK] Seq=433049760 Ack=418940507 Win=5840 Len=0
```

```
     0000    00  30  05  01  98  9a  00  03  47  4c  4b  00  08  00  45  00
.0......GLK...E.
     0010    00  34  e1  d1  40  00  40  06  5c  c3  d4  9b  a9  7c  d4  9b
.4..@.@.\....|..
     0020    a9  7b  95  26  0a  d7  19  cf  d0  a0  18  f8  86  5b  80  10
.{.&.........[..
     0030    16  d0  4a  c1  00  00  01  01  08  0a  2d  2c  32  1e  03  55
..J.......-,2..U
     0040    8c  9c                                                      ..
```

```
     18    71.558459     thirdparty - customfr.custom.fr SMPP  SMPP
Enquire_link
```

```
     0000    00  30  05  01  98  9a  00  03  47  4c  4b  00  08  00  45  00
.0......GLK...E.
     0010    00  44  e1  d2  40  00  40  06  5c  b2  d4  9b  a9  7c  d4  9b
.D..@.@.\....|..
     0020    a9  7b  95  26  0a  d7  19  cf  d0  a0  18  f8  86  5b  80  18
.{.&.........[..
     0030    16  d0  33  1c  00  00  01  01  08  0a  2d  2c  49  82  03  55
..3.......-,I..U
     0040    8c  9c  00  00  00  10  00  00  00  15  00  00  00  00  00  00
................
     0050    00  04                                                      ..
```

```
     19   71.563486 customfr.custom.fr - thirdparty      SMPP SMPP
Enquire_Link - resp: "Ok"


    0000    00 03 47 4c 4b 00 00 30 05 01 98 9a 08 00 45 00
..GLK..0......E.
    0010    00 44 25 0f 40 00 40 06 19 76 d4 9b a9 7b d4 9b
.D%.@.@..v...{..
    0020    a9 7c 0a d7 95 26 18 f8 86 5b 19 cf d0 b0 80 18
.|...&...[......
    0030    7d 78 34 fa 00 00 01 01 08 0a 03 55 a4 05 2d 2c
}x4........U..-,
    0040    49 82 00 00 00 10 80 00 00 15 00 00 00 00 00 00
I...............
    0050    00 04                                         ..


     20 71.563656      thirdparty - customfr.custom.fr TCP 38182  2775
[ACK] Seq=433049776 Ack=418940523 Win=5840 Len=0


    0000    00 30 05 01 98 9a 00 03 47 4c 4b 00 08 00 45 00
.0......GLK...E.
    0010    00 34 e1 d3 40 00 40 06 5c c1 d4 9b a9 7c d4 9b
.4..@.@.\....|..
    0020    a9 7b 95 26 0a d7 19 cf d0 b0 18 f8 86 6b 80 10
.{.&.........k..
    0030    16 d0 1b d4 00 00 01 01 08 0a 2d 2c 49 82 03 55
..........-,I..U
    0040    a4 05                                         ..


     21 131.605906         thirdparty - customfr.custom.fr SMPP SMPP
Enquire_Link


    0000    00 30 05 01 98 9a 00 03 47 4c 4b 00 08 00 45 00
.0......GLK...E.
    0010    00 44 e1 d4 40 00 40 06 5c b0 d4 9b a9 7c d4 9b
.D..@.@.\....|..
    0020    a9 7b 95 26 0a d7 19 cf d0 b0 18 f8 86 6b 80 18
.{.&.........k..
    0030    16 d0 04 1e 00 00 01 01 08 0a 2d 2c 60 f6 03 55
..........-, ..U
    0040    a4 05 00 00 00 10 00 00 00 15 00 00 00 00 00 00
...............
    0050    00 05                                         ..


     22 131.610378 customfr.custom.fr - thirdparty      SMPP SMPP
Enquire_Link - resp: Ok
```

```
    0000    00  03  47  4c  4b  00  00  30  05  01  98  9a  08  00  45  00
..GLK..0......E.
    0010    00  44  32  5b  40  00  40  06  0c  2a  d4  9b  a9  7b  d4  9b
.D2[a.a..*...{..
    0020    a9  7c  0a  d7  95  26  18  f8  86  6b  19  cf  d0  c0  80  18
.|...&...k......
    0030    7d  78  05  f1  00  00  01  01  08  0a  03  55  bb  79  2d  2c
}x.........U.y-,
    0040    60  f6  00  00  00  10  80  00  00  15  00  00  00  00  00  00
...............
    0050    00  05                                                          ..
```

    23 131.610547   thirdparty - customfr.custom.fr TCP 38182 2775
[ACK] Seq=433049792 Ack=418940539 Win=5840 Len=0

```
    0000    00  30  05  01  98  9a  00  03  47  4c  4b  00  08  00  45  00
.0......GLK...E.
    0010    00  34  e1  d5  40  00  40  06  5c  bf  d4  9b  a9  7c  d4  9b
.4..a.a.\....|..
    0020    a9  7b  95  26  0a  d7  19  cf  d0  c0  18  f8  86  7b  80  10
.{.&.........{..
    0030    16  d0  ec  cb  00  00  01  01  08  0a  2d  2c  60  f6  03  55
..........-, ..U
    0040    bb  79                                                          .y
```

    24 191.633367   thirdparty - customfr.custom.fr SMPP SMPP
Enquire_link

```
    0000    00  30  05  01  98  9a  00  03  47  4c  4b  00  08  00  45  00
.0......GLK...E.
    0010    00  44  e1  d6  40  00  40  06  5c  ae  d4  9b  a9  7c  d4  9b
.D..a.a.\....|..
    0020    a9  7b  95  26  0a  d7  19  cf  d0  c0  18  f8  86  7b  80  18
.{.&.........{..
    0030    16  d0  d5  16  00  00  01  01  08  0a  2d  2c  78  68  03  55
..........-,xh.U
    0040    bb  79  00  00  00  10  00  00  00  15  00  00  00  00  00  00
.y..............
    0050    00  06                                                          ..
```

    ....

    38 431.749398 customfr.custom.fr - thirdparty   SMPP SMPP
Enquire_link - resp: "0k"

```
    0000    00  03  47  4c  4b  00  00  30  05  01  98  9a  08  00  45  00
..GLK..0......E.
```

```
    0010    00  44  71  c9  40  00  40  06  cc  bb  d4  9b  a9  7b  d4  9b
.Dq.@.@......{..
    0020    a9  7c  0a  d7  95  26  18  f8  86  bb  19  cf  d1  10  80  18
.|...&..........
    0030    7d  78  1a  d7  00  00  01  01  08  0a  03  56  30  b5  2d  2c
}x.........V0.-,
    0040    d6  2e  00  00  00  10  80  00  00  15  00  00  00  00  00  00
..............
    0050    00  0a                                                          ..
```

    39 431.749580   thirdparty - customfr.custom.fr TCP 38182   2775
[ACK] Seq=433049872 Ack=418940619 Win=5840 Len=0

```
    0000    00  30  05  01  98  9a  00  03  47  4c  4b  00  08  00  45  00
.0......GLK...E.
    0010    00  34  e1  df  40  00  40  06  5c  b5  d4  9b  a9  7c  d4  9b
.4..@.@.\....|..
    0020    a9  7b  95  26  0a  d7  19  cf  d1  10  18  f8  86  cb  80  10
.{.&...........
    0030    16  d0  01  b5  00  00  01  01  08  0a  2d  2c  d6  30  03  56
..........-,.0.V
    0040    30  b5                                                          0.
```

    40 446.526112   thirdparty - customfr.custom.fr SMPP SMPP Submit_
sm

```
    0000    00  30  05  01  98  9a  00  03  47  4c  4b  00  08  00  45  00
.0......GLK...E.
    0010    00  6e  e1  e0  40  00  40  06  5c  7a  d4  9b  a9  7c  d4  9b
.n..@.@.\z...|..
    0020    a9  7b  95  26  0a  d7  19  cf  d1  10  18  f8  86  cb  80  18
.{.&...........
    0030    16  d0  22  24  00  00  01  01  08  0a  2d  2c  db  f6  03  56
.."$.......-,...V
    0040    30  b5  00  00  00  3a  00  00  00  04  00  00  00  00  00  00
0....:..........
    0050    2e  72  51  41  00  02  01  2b  30  00  00  01  33  33  36  30
.rQA...+0...3360
    0060    33  35  30  30  39  31  32  00  00  40  00  30  00  30  00  10
3500912..@.0.0..
    0070    00  f0  00  08  11  31  31  31  31  31  31  31          ....11111111
```

    41 446.538224 customfr.custom.fr - thirdparty   SMPP SMPP Submit_
sm - resp: "Ok"

```
    0000    00  03  47  4c  4b  00  00  30  05  01  98  9a  08  00  45  00
..GLK..0......E.
```

```
    0010    00  55  75  76  40  00  40  06  c8  fd  d4  9b  a9  7b  d4  9b
.Uuv@.@......{..
    0020    a9  7c  0a  d7  95  26  18  f8  86  cb  19  cf  d1  4a  80  18
.|...&.......J..
    0030    7d  78  d3  8c  00  00  01  01  08  0a  03  56  36  7c  2d  2c
}x.........V6|-,
    0040    db  f6  00  00  00  21  80  00  00  04  00  00  00  00  00  00
.....!..........
    0050    2e  72  36  30  30  30  64  30  37  31  38  66  37  66  63  39
.r6000d0718f7fc9
    0060    38  30  00                                                      80.
```

42 446.538421   thirdparty - customfr.custom.fr TCP 38182   2775
[ACK] Seq=433049930 Ack=418940652 Win=5840 Len=0

```
    0000    00  30  05  01  98  9a  00  03  47  4c  4b  00  08  00  45  00
.0......GLK...E.
    0010    00  34  e1  e1  40  00  40  06  5c  b3  d4  9b  a9  7c  d4  9b
.4..@.@.\....|..
    0020    a9  7b  95  26  0a  d7  19  cf  d1  4a  18  f8  86  ec  80  10
.{.&.....J......
    0030    16  d0  f5  cb  00  00  01  01  08  0a  2d  2c  db  f7  03  56
..........-,...V
    0040    36  7c                                                          6|
```

43 446.565786   thirdparty - customfr.custom.fr SMPP SMPP Query_sm

```
    0000    00  30  05  01  98  9a  00  03  47  4c  4b  00  08  00  45  00
.0......GLK...E.
    0010    00  58  e1  e2  40  00  40  06  5c  8e  d4  9b  a9  7c  d4  9b
.X..@.@.\....|..
    0020    a9  7b  95  26  0a  d7  19  cf  d1  4a  18  f8  86  ec  80  18
.{.&.....J......
    0030    16  d0  ba  0b  00  00  01  01  08  0a  2d  2c  db  fa  03  56
..........-,...V
    0040    36  7c  00  00  00  24  00  00  00  03  00  00  00  00  00  00
6|...$..........
    0050    2e  72  36  30  30  30  64  30  37  31  38  66  37  66  63  39
.r6000d0718f7fc9
    0060    38  30  00  00  00  00                                          80....
```

44 446.569087 customfr.custom.fr - thirdparty   SMPP SMPP Query_sm
- resp: "0k"

```
    0000    00  03  47  4c  4b  00  00  30  05  01  98  9a  08  00  45  00
..GLK..0......E.
```

```
    0010    01  10  75  7e  40  00  40  06  c8  3a  d4  9b  a9  7b  d4  9b
..u~@.@..:...{..
    0020    a9  7c  0a  d7  95  26  18  f8  86  ec  19  cf  d1  6e  80  18
.|...&........n..
    0030    7d  78  1c  33  00  00  01  01  08  0a  03  56  36  7f  2d  2c
}x.3.......V6.-,
    0040    db  fa  00  00  00  dc  80  00  00  03  00  00  00  00  00  00
................
    0050    2e  72  36  30  30  30  64  30  37  31  38  66  37  66  63  39
.r6000d0718f7fc9
    0060    38  30  00  00  01  44  14  00  00  01  44  14  01  00  01  00
80...D....D.....
    0070    14  02  00  01  00  14  03  00  01  00  14  04  00  04  3f  3f
..............??
    0080    3f  00  14  05  00  04  3f  3f  3f  00  14  06  00  04  3f  3f
?.....???.....??
    0090    3f  00  14  08  00  01  00  14  09  00  01  00  14  0a  00  01
?...............
    00a0    00  14  0b  00  01  00  14  0c  00  01  00  14  07  00  01  00
................
    00b0    14  0d  00  01  00  14  0e  00  01  00  14  0f  00  01  00  14
................
    00c0    10  00  01  00  14  11  00  04  3f  3f  3f  00  14  12  00  04
........???.....
    00d0    3f  3f  3f  00  14  13  00  04  3f  3f  3f  00  14  14  00  04
???.....???.....
    00e0    3f  3f  3f  00  14  15  00  04  3f  3f  3f  00  14  16  00  04
???.....???.....
    00f0    3f  3f  3f  00  14  17  00  04  3f  3f  3f  00  14  18  00  04
???.....???.....
    0100    3f  3f  3f  00  14  19  00  04  3f  3f  3f  00  14  1a  00  04
???.....???.....
    0110  3f 3f 3f 00 14 1b 00 01 00 14 1c 00 01 00          ???...........
```

16.7   Example 7: Connection of a GSM to a Third-Party SMS Network

Answer the following questions:

1. These are the full MTP/SCCP/TCAP/MAP traces for sending SMS by GSM
    operators in Saudi Arabia to a third-party SMS network in the United States.
    The destination number is Cingular (United States), which may be GSM or
    CDMA, hence the use of a third-party SMS network. Which one of the
    connection methods of Chapter 7 is used ?

2. What is the delay for the SRI_FOR_SM_CNF? What method is used: real
    HLR query or database lookup?

3. What is the delay between the FORWARD_SHORT_MESSAGE_MT_REQ and the CNF? Is real "paging" of the cellphone performed? What is the visited MSC GT?

4. What sort of complaints will the GSM network user of the third party receive from his or her own mobile customers?

```
- - - - - - - - - - - - - - - - - - - - - - - - - - - - - - - -
  Octet001   ANSI SS7                    Count=000001  Time=04/02/2004
14:50:50:562
- - - - - - - - - - - - - - - - - - - - - - - - - - - - - - - -
  11101111   BIB/BSN                    (239) 1/111
  11001001   FIB/FSN                    (201) 1/73
  ..111111   SU type/length             (63) MSU63
  00......   Spare                      0
- - - - - - - - - - - - - - - - - - - - - - - - - - - - - - - -
  Octet004   Service information octet
- - - - - - - - - - - - - - - - - - - - - - - - - - - - - - - -
  ....0011   Service indicator          (3) SCCP   Signaling Connection
                                            Control Part
  ..01....   Message priority           1
  10......   Network indicator          (2) N   National network
- - - - - - - - - - - - - - - - - - - - - - - - - - - - - - - -
  Octet005   Routing label
- - - - - - - - - - - - - - - - - - - - - - - - - - - - - - - -
  ........   DPC: Net-Clstr-Mbr         005-063-220
  ........   OPC: Net-Clstr-Mbr         001-044-120 lnp
  00010110   SLS                        22
- - - - - - - - - - - - - - - - - - - - - - - - - - - - - - - -
  Octet012   Message type
- - - - - - - - - - - - - - - - - - - - - - - - - - - - - - - -
  00001001   Message type               (9) UDT    Unitdata
- - - - - - - - - - - - - - - - - - - - - - - - - - - - - - - -
  Octet013   SCCP Protocol Class parameter
- - - - - - - - - - - - - - - - - - - - - - - - - - - - - - - -
  ....0000   Protocol class             (0) Class 0
  1000....   Message handling           (8) Return message on error
  00000011   Ptr - Called number        3
  00001100   Ptr - Calling #            12
  00010101   Pointer - Data             21
- - - - - - - - - - - - - - - - - - - - - - - - - - - - - - - -
  Octet017   SCCP Called Party Address parameter
- - - - - - - - - - - - - - - - - - - - - - - - - - - - - - - -
  00001001   Parameter length           9
  .......1   Subsystem # bit            (1) SSN present
  ......0.   Sgnl pt code bit           (0) SPC not present
  ..0010..   Global title ind           (2) Global title includes
                                            translation type only
```

```
     .0......   Routing bit          (0) Global title based routing (SCCP
                                         translation required)
     1.......   Natl Intnatl bit     (1) National address
     00000110   Subsystem number     (6) HLR Home Location Register
     00001010   Translation type     (10) Network Entity Addressing
     ........   Address signals      19257598435
   - - - - - - - - - - - - - - - - - - - - - - - - - - - - - - -
     Octet027   SCCP Calling Party Address parameter
   - - - - - - - - - - - - - - - - - - - - - - - - - - - - - - -
     00001001   Parameter length     9
     .......1   Subsystem # bit      (1) SSN present
     ......0.   Sgnl pt code bit     (0) SPC not present
     ..0010..   Global title ind     (2) Global title includes
                                         translation type only
     .0......   Routing bit          (0) Global title based routing (SCCP
                                         translation required)
     1.......   Natl Intnatl bit     (1) National address
     00001000   Subsystem number     (8) MSC    Mobile Switching Centre
     00001010   Translation type     (10) Network Entity Addressing
     ........   Address signals      966550309990
   - - - - - - - - - - - - - - - - - - - - - - - - - - - - - - -
     Octet037   SCCP Data parameter
   - - - - - - - - - - - - - - - - - - - - - - - - - - - - - - -
     00101001   Parameter length     41
     01100010   Tag                      (98) BGN Begin, constructor,
                                             application-wide
     00100111   Length               39
   - - - - - - - - - - - - - - - - - - - - - - - - - - - - - - -
     Octet040   OriginatingTransactionID
   - - - - - - - - - - - - - - - - - - - - - - - - - - - - - - -
     ...01000   Tag                      (8) OriginatingTransactionID
     010.....   Class and form       (2) Application-wide, primitive
     00000100   Length               4
     ........   Originating ID       0131BDF3
   - - - - - - - - - - - - - - - - - - - - - - - - - - - - - - -
     Octet046   ComponentPortion
   - - - - - - - - - - - - - - - - - - - - - - - - - - - - - - -
     ...01100   Tag                      (12) ComponentPortion
     011.....   Class and form           (3) Application-wide, constructor
     00011111   Length               31
   - - - - - - - - - - - - - - - - - - - - - - - - - - - - - - -
     Octet048   Invoke
   - - - - - - - - - - - - - - - - - - - - - - - - - - - - - - -
     ...00001   Tag                  (1) Invoke
     101.....   Class and form       (5) Context-specific, constructor
     00011101   Length               29
   - - - - - - - - - - - - - - - - - - - - - - - - - - - - - - -
```

```
Octet050   InvokeID
- - - - - - - - - - - - - - - - - - - - - - - - - - - - - -
...00010   Tag                       (2) InvokeID
000.....   Class and form            (0) Universal, primitive
00000001   Length                    1
00000001   InvokeID                  01
- - - - - - - - - - - - - - - - - - - - - - - - - - - - - -
Octet053   Operation Code
- - - - - - - - - - - - - - - - - - - - - - - - - - - - - -
...00010   Tag                       (2) Local
000.....   Class and form            (0) Universal, primitive
00000001   Length                    1
........   Operation Code            (45) SRIS SendRoutingInfoForSM
- - - - - - - - - - - - - - - - - - - - - - - - - - - - - -
Octet056   Parameter Sequence
- - - - - - - - - - - - - - - - - - - - - - - - - - - - - -
...10000   Tag                       (16) Parameter Sequence
001.....   Class and form            (1) Universal, constructor
00010101   Length                    21
- - - - - - - - - - - - - - - - - - - - - - - - - - - - - -
Octet058   MsIsdn
- - - - - - - - - - - - - - - - - - - - - - - - - - - - - -
...00000   Tag                       (0) MsIsdn
100.....   Class and form            (4) Context-specific, primitive
00000111   Length                    7
....0001   Numbering plan            (1) ISDN/Telephony Number Plan
                                         (REC E.164)
.001....   Nature of address         (1) international number
1.......   Extension Bit             (1) No extension
........   Address signals           19257598435
- - - - - - - - - - - - - - - - - - - - - - - - - - - - - -
Octet067   SM-RP-PRI
- - - - - - - - - - - - - - - - - - - - - - - - - - - - - -
...00001   Tag                       (1) Boolean
100.....   Class and form            (4) Context-specific, primitive
00000001   Length                    1
11111111   Contents                  (255) TRUE
- - - - - - - - - - - - - - - - - - - - - - - - - - - - - -
Octet070   Service centre address
- - - - - - - - - - - - - - - - - - - - - - - - - - - - - -
...00010   Tag                       (2) Service center address
100.....   Class and form            (4) Context-specific, primitive
00000111   Length                    7
....0001   Numbering plan            (1) ISDN/Telephony Number Plan
                                         (REC E.164)
.001....   Nature of address         (1) international number
1.......   Extension Bit             (1) No extension
```

```
........     Address signals          96655030999
- - - - - - - - - - - - - - - - - - - - - - - - - - - - - - - -
Checksum   CRC16................    0100011111011111 hex=47DF
- - - - - - - - - - - - - - - - - - - - - - - - - - - - - - - -
Octet001   ITU-T SS7                    Count=000001 Time=04/02/2004
14:50:50:853
                                          -
00111001   BIB/BSN                  (57)  0/57
11110101   FIB/FSN                  (245)  1/117
..111111   SU type/length           (63) MSU63
00......   Spare                 0
- - - - - - - - - - - - - - - - - - - - - - - - - - - - - - - -
Octet004   Service information octet
- - - - - - - - - - - - - - - - - - - - - - - - - - - - - - - -
....0011   Service indicator        (3) SCCP  Signaling Connection
                                              Control Part
..00....   Spare                 0
00......   Network indicator        (0) I  International Network
- - - - - - - - - - - - - - - - - - - - - - - - - - - - - - - -
Octet005   Routing label
- - - - - - - - - - - - - - - - - - - - - - - - - - - - - - - -
........   DPC                      4-040-6 RYH
........   OPC                      3-005-1 iVPC
0011....   SLS                      3
- - - - - - - - - - - - - - - - - - - - - - - - - - - - - - - -
Octet009   Message type
- - - - - - - - - - - - - - - - - - - - - - - - - - - - - - - -
00001001   Message type             (9) UDT    Unitdata
- - - - - - - - - - - - - - - - - - - - - - - - - - - - - - - -
Octet010   SCCP Protocol Class parameter
- - - - - - - - - - - - - - - - - - - - - - - - - - - - - - - -
....0000   Protocol class           (0) Class 0
0000....   Message handling         (0) No special options
00000011   Ptr - Called number    3
00001110   Ptr - Calling #        14
00011001   Pointer - Data         25
- - - - - - - - - - - - - - - - - - - - - - - - - - - - - - - -
Octet014   SCCP Called Party Address parameter
- - - - - - - - - - - - - - - - - - - - - - - - - - - - - - - -
00001011   Parameter length   11
.......0   Sgnl pt code bit     (0) SPC not present
......1.   Subsystem # bit      (1) SSN present
..0100..   Global title ind     (4) Global title incl. translation
              type, numbering plan, encoding scheme & addr nature
.0......   Routing bit          (0) Global title based routing
0.......   Reserved natl use  0
00001000   Subsystem number     (8) MSC    Mobile Switching Centre
```

```
00000000  Translation         (0) unknown
....0010  Encoding scheme     (2) BCD even
0001....  Numbering plan      (1) ISDN (Telephony) numbering plan
                                  (Rec. E.164/E.163)
.0000100  Nature of address    (4) International number
0.......  Spare               0
........  Address signals     966550309990
- - - - - - - - - - - - - - - - - - - - - - - - - - - - -
Octet026  SCCP Calling Party Address parameter
- - - - - - - - - - - - - - - - - - - - - - - - - - - - -
00001011  Parameter length    11
.......0  Sgnl pt code bit     (0) SPC not present
......1.  Subsystem # bit      (1) SSN present
..0100..  Global title ind     (4) Global title incl. translation
             type, numbering plan, encoding scheme & addr nature
.0......  Routing bit          (0) Global title based routing
0.......  Reserved natl use   0
00000110  Subsystem number     (6) HLR    Home Location Register
00000000  Translation         (0) unknown
....0010  Encoding scheme     (2) BCD even
0001....  Numbering plan      (1) ISDN (Telephony) numbering plan
                                  (Rec. E.164/E.163)
.0000100  Nature of address    (4) International number
0.......  Spare               0
........  Address signals     131231499300
- - - - - - - - - - - - - - - - - - - - - - - - - - - - -
Octet038  SCCP Data parameter
- - - - - - - - - - - - - - - - - - - - - - - - - - - - -
00101011  Parameter length    43
01100100  Tag                     (100) END End, constructor,
                                        application-wide
00101001  Length                  41
- - - - - - - - - - - - - - - - - - - - - - - - - - - - -
Octet041  DestinationTransactionID
- - - - - - - - - - - - - - - - - - - - - - - - - - - - -
...01001  Tag                  (9) DestinationTransactionID
010.....  Class and form       (2) Application-wide, primitive
00000100  Length               4
........  Destination ID       0131BDF3
- - - - - - - - - - - - - - - - - - - - - - - - - - - - -
Octet047  ComponentPortion
- - - - - - - - - - - - - - - - - - - - - - - - - - - - -
...01100  Tag                  (12) ComponentPortion
011.....  Class and form       (3) Application-wide, constructor
00100001  Length               33
- - - - - - - - - - - - - - - - - - - - - - - - - - - - -
Octet049  ReturnResult
```

```
- - - - - - - - - - - - - - - - - - - - - - - - - - - - - -
   ...00010   Tag                      (2) ReturnResult
   101.....   Class and form           (5) Context-specific, constructor
   00011111   Length                31
- - - - - - - - - - - - - - - - - - - - - - - - - - - - - -
   Octet051   InvokeID
- - - - - - - - - - - - - - - - - - - - - - - - - - - - - -
   ...00010   Tag                      (2) InvokeID
   000.....   Class and form           (0) Universal, primitive
   00000001   Length                1
   00000001   InvokeID              01
- - - - - - - - - - - - - - - - - - - - - - - - - - - - - -
   Octet054   Sequence
- - - - - - - - - - - - - - - - - - - - - - - - - - - - - -
   ...10000   Tag                      (16) Sequence
   001.....   Class and form           (1) Universal, constructor
   00011010   Length                26
- - - - - - - - - - - - - - - - - - - - - - - - - - - - - -
   Octet056   Operation Code
- - - - - - - - - - - - - - - - - - - - - - - - - - - - - -
   ...00010   Tag                      (2) Local
   000.....   Class and form           (0) Universal, primitive
   00000001   Length                1
   ........   Operation Code           (45) SRIS SendRoutingInfoForSM
- - - - - - - - - - - - - - - - - - - - - - - - - - - - - -
   Octet059   Parameter Sequence
- - - - - - - - - - - - - - - - - - - - - - - - - - - - - -
   ...10000   Tag                      (16) Parameter Sequence
   001.....   Class and form           (1) Universal, constructor
   00010101   Length                21
- - - - - - - - - - - - - - - - - - - - - - - - - - - - - -
   Octet061   IMSI
- - - - - - - - - - - - - - - - - - - - - - - - - - - - - -
   ...00100   Tag                      (4) Octetstring
   000.....   Class and form           (0) Universal, primitive
   00001000   Length                8
   ........   Address signals       310179257598435
   1111....   Filler                15
- - - - - - - - - - - - - - - - - - - - - - - - - - - - - -
   Octet071   Location info with LMSI
- - - - - - - - - - - - - - - - - - - - - - - - - - - - - -
   ...00000   Tag                      (0) Parameter
   101.....   Class and form           (5) Context-specific, constructor
   00001001   Length                9
- - - - - - - - - - - - - - - - - - - - - - - - - - - - - -
   Octet073   Network Node - Number
- - - - - - - - - - - - - - - - - - - - - - - - - - - - - -
```

```
...00001   Tag                   (1) Network Node - Number
100.....   Class and form        (4) Context-specific, primitive
00000111   Length              7
....0001   Numbering plan        (1) ISDN/Telephony Number Plan
                                     (REC E.164)
.001....   Nature of address     (1) international number
1.......   Extension Bit         (1) No extension
........   Address signals          13123149940
- - - - - - - - - - - - - - - - - - - - - - - - - - - - - -
Checksum   CRC16...............   1110001010111110 hex=E2BE
- - - - - - - - - - - - - - - - - - - - - - - - - - - - - -

- - - - - - - - - - - - - - - - - - - - - - - - - - - - - -
Octet001   ITU-T SS7      Count=000001 Time=04/02/2004 14:50:51:835
                                     -
11011101   BIB/BSN               (221)  1/93
10110010   FIB/FSN               (178)  1/50
..111111   SU type/length        (63) MSU63
00......   Spare                 0
- - - - - - - - - - - - - - - - - - - - - - - - - - - - - -
Octet004   Service information octet
- - - - - - - - - - - - - - - - - - - - - - - - - - - - - -
....0011   Service indicator     (3) SCCP  Signaling Connection
                                     Control Part
..00....   Spare               0
00......   Network indicator     (0) I  International Network
- - - - - - - - - - - - - - - - - - - - - - - - - - - - - -
Octet005   Routing label
- - - - - - - - - - - - - - - - - - - - - - - - - - - - - -
........   DPC                   3-005-1 iVPC
........   OPC                   4-040-6 RYH
1111....   SLS                   15
- - - - - - - - - - - - - - - - - - - - - - - - - - - - - -
Octet009   Message type
- - - - - - - - - - - - - - - - - - - - - - - - - - - - - -
00001001   Message type          (9) UDT   Unitdata
- - - - - - - - - - - - - - - - - - - - - - - - - - - - - -
Octet010   SCCP Protocol Class parameter
- - - - - - - - - - - - - - - - - - - - - - - - - - - - - -
....0000   Protocol class        (0) Class 0
1000....   Message handling      (8) Return  message on error
00000011   Ptr - Called number  3
00001110   Ptr - Calling #      14
00011001   Pointer - Data       25
- - - - - - - - - - - - - - - - - - - - - - - - - - - - - -
Octet014   SCCP Called Party Address parameter
- - - - - - - - - - - - - - - - - - - - - - - - - - - - - -
00001011   Parameter length        11
```

```
.......0   Sgnl pt code bit      (0) SPC not present
......1.   Subsystem # bit       (1) SSN present
..0100..   Global title ind      (4) Global title incl. translation
                    type, numbering plan, encoding scheme & addr nature
.0......   Routing bit           (0) Global title based routing
0.......   Reserved natl use     0
00001000   Subsystem number      (8) MSC    Mobile Switching Center
00000000   Translation           (0) unknown
....0001   Encoding scheme       (1) BCD odd
0001....   Numbering plan        (1) ISDN (Telephony) numbering plan
                                      (Rec. E.164/E.163)
.0000100   Nature of address     (4) International number
0.......   Spare                 0
........   Address signals       13123149940
0000....   Filler                0
- - - - - - - - - - - - - - - - - - - - - - - - - - - - - - - - -
Octet026   SCCP Calling Party Address parameter
- - - - - - - - - - - - - - - - - - - - - - - - - - - - - - - - -
00001011   Parameter length      11
.......0   Sgnl pt code bit      (0) SPC not present
......1.   Subsystem # bit       (1) SSN present
..0100..   Global title ind      (4) Global title incl. translation
                    type, numbering plan, encoding scheme & addr nature
.0......   Routing bit           (0) Global title based routing
0.......   Reserved natl use     0
00001000   Subsystem number      (8) MSC    Mobile Switching Centre
00000000   Translation           (0) unknown
....0001   Encoding scheme       (1) BCD odd
0001....   Numbering plan        (1) ISDN (Telephony) numbering plan
                                      (Rec. E.164/E.163)
.0000100   Nature of address     (4) International number
0.......   Spare                 0
........   Address signals       96655030999
0000....   Filler                0
- - - - - - - - - - - - - - - - - - - - - - - - - - - - - - - - -
Octet038   SCCP Data parameter
- - - - - - - - - - - - - - - - - - - - - - - - - - - - - - - - -
11001101   Parameter length      205
01100010   Tag                         (98) BGN Begin, constructor,
                                             application-wide
10000001   Length                129
........   Long form length      202
- - - - - - - - - - - - - - - - - - - - - - - - - - - - - - - - -
Octet042   OriginatingTransactionID
- - - - - - - - - - - - - - - - - - - - - - - - - - - - - - - - -
...01000   Tag                   (8) OriginatingTransactionID
010.....   Class and form        (2) Application-wide, primitive
```

```
00000100   Length                    4
........   Originating ID            0131BE6D
- - - - - - - - - - - - - - - - - - - - - - - - - - - - - -
Octet048   ComponentPortion
- - - - - - - - - - - - - - - - - - - - - - - - - - - - - -
...01100   Tag                 (12) ComponentPortion
011.....   Class and form       (3) Application-wide, constructor
10000001   Length                   129
........   Long form length 193
- - - - - - - - - - - - - - - - - - - - - - - - - - - - - -
Octet051   Invoke
- - - - - - - - - - - - - - - - - - - - - - - - - - - - - -
...00001   Tag                  (1) Invoke
101.....   Class and form       (5) Context-specific, constructor
10000001   Length                   129
........   Long form length 190
- - - - - - - - - - - - - - - - - - - - - - - - - - - - - -
Octet054   InvokeID
- - - - - - - - - - - - - - - - - - - - - - - - - - - - - -
...00010   Tag                  (2) InvokeID
000.....   Class and form       (0) Universal, primitive
00000001   Length                    1
00000001   InvokeID                 01
- - - - - - - - - - - - - - - - - - - - - - - - - - - - - -
Octet057   Operation Code
- - - - - - - - - - - - - - - - - - - - - - - - - - - - - -
...00010   Tag                  (2) Local
000.....   Class and form       (0) Universal, primitive
00000001   Length                    1
........   Operation Code      (46) MOFS MO-ForwardSM
- - - - - - - - - - - - - - - - - - - - - - - - - - - - - -
Octet060   Parameter Sequence
- - - - - - - - - - - - - - - - - - - - - - - - - - - - - -
...10000   Tag                 (16) Parameter Sequence
001.....   Class and form       (1) Universal, constructor
10000001   Length                   129
........   Long form length         181
- - - - - - - - - - - - - - - - - - - - - - - - - - - - - -
Octet063   IMSI
- - - - - - - - - - - - - - - - - - - - - - - - - - - - - -
...00000   Tag                  (0) IMSI
100.....   Class and form       (4) Context-specific, primitive
00001000   Length                    8
........   Address signals   310179257598435
- - - - - - - - - - - - - - - - - - - - - - - - - - - - - -
Octet073   Service centre address 0A (sm-RP-OA)
- - - - - - - - - - - - - - - - - - - - - - - - - - - - - -
```

```
...00100    Tag                    (4) Service centre address OA
                                       (sm-RP-OA)
100.....    Class and form         (4) Context-specific, primitive
00000111    Length                 7
....0001    Numbering plan         (1) ISDN/Telephony Number Plan
                                       (REC E.164)
.001....    Nature of address      (1) international number
1.......    Extension Bit          (1) No extension
........    Address signals        96655032999
```
- -
```
Octet082    Sm-RP-UI
```
- -
```
...00100    Tag                    (4) Octetstring
000.....    Class and form         (0) Universal, primitive
10000001    Length                 129
........    Long form length       159
......00    Message type ind.      (0) SMS-DELIVER
.....1..    More mesg. to send     (1) No more messages are waiting
                                       for the MS in this SC
...00...    Filler                 0
..0.....    Status report ind.     (0) A status report will not be
                                       returned to the SME
.1......    User data hdr. ind.    (1) The beginning of the TP-UD
            field contains a Header in addition to the short message
0.......    Reply path             (0) TP-Reply-Path parameter is not
                                       set in this SMS-SUBMIT/DELIVER
```
- -
```
Octet086    TP-Originating Address
```
- -
```
00001011    Length                 11
....0001    Numbering plan         (1) ISDN/Telephone numbering plan
                                       (E.164/E.163)
.001....    Type of number         (1) International number
1.......    Reserved               1
........    Address signals        96652121367
1111....    Filler                 15
```
- -
```
Octet094    TP-Protocol Identifier
```
- -
```
...00000    Telematic devices      (0) Implicit - device type is specific
            to this SC , or can be concluded on the basis of the address
..0.....    Telematic interwork.   (0) No interworking, but SME-to-SME
                                       protocol
00......    Interworking Id        (0) Assigns bits 0..5 as defined
                                       above
```
- -
```
Octet095    TP-Data coding scheme
```

```
- - - - - - - - - - - - - - - - - - - - - - - - - - - - - - -
  ....0000    Alphabet Ind           (0) Default alphabet
  0000....    SMS Coding Group       (0) Alphabet indication
- - - - - - - - - - - - - - - - - - - - - - - - - - - - - - -
  Octet096    TP-Service centre time stamp
- - - - - - - - - - - - - - - - - - - - - - - - - - - - - - -
  ........    Year                   04
  ........    Month                  04
  ........    Day                    02
  ........    Hour                   17
  ........    Minute                 57
  ........    Second                 33
  00100001    Time Zone              (33) GMT+8hr 15mins
- - - - - - - - - - - - - - - - - - - - - - - - - - - - - - -
  Octet103    TP-User data
- - - - - - - - - - - - - - - - - - - - - - - - - - - - - - -
  10011111    Length                 159
- - - - - - - - - - - - - - - - - - - - - - - - - - - - - - -
  Octet104    TP-User data header
- - - - - - - - - - - - - - - - - - - - - - - - - - - - - - -
  00000101    Length                 5
- - - - - - - - - - - - - - - - - - - - - - - - - - - - - - -
  Octet105    Information element
- - - - - - - - - - - - - - - - - - - - - - - - - - - - - - -
  00000000    Info element Id        (0) Concatenated short messages
  00000011    Length                 3
  ........    Contents               01 03 03
  ........    Fill bits              0
  ........    Contents               Undefined(the analyser hides the
                                     text!)
  ........    Spare                  0 0 0 0 0 0 0
- - - - - - - - - - - - - - - - - - - - - - - - - - - - - - -
  Checksum  CRC16................  1000100100111010 hex=893A
- - - - - - - - - - - - - - - - - - - - - - - - - - - - - - -
- - - - - - - - - - - - - - - - - - - - - - - - - - - - - - -
  Octet001    ITU-T SS7              Count=000001 Time=04/02/2004
14:50:52:168
- - - - - - - - - - - - - - - - - - - - - - - - - - - - - - -
  10110110    BIB/BSN                (182)  1/54
  11100010    FIB/FSN                (226)  1/98
  ..110010    SU type/length         (50) MSU50
  00......    Spare                  0
- - - - - - - - - - - - - - - - - - - - - - - - - - - - - - -
  Octet004    Service information octet
- - - - - - - - - - - - - - - - - - - - - - - - - - - - - - -
  ....0011    Service indicator      (3) SCCP Signaling Connection
                                         Control Part
```

```
..00....  Spare                   0
00......  Network indicator       (0) I  International Network
- - - - - - - - - - - - - - - - - - - - - - - - - - - - - - - -
Octet005  Routing label
- - - - - - - - - - - - - - - - - - - - - - - - - - - - - - - -
........  DPC                     4-040-6 RYH
........  OPC                     3-005-1 iVPC
0111....  SLS                     7
- - - - - - - - - - - - - - - - - - - - - - - - - - - - - - - -
Octet009  Message type
- - - - - - - - - - - - - - - - - - - - - - - - - - - - - - - -
00001001  Message type           (9) UDT   Unitdata
- - - - - - - - - - - - - - - - - - - - - - - - - - - - - - - -
Octet010  SCCP Protocol Class parameter
- - - - - - - - - - - - - - - - - - - - - - - - - - - - - - - -
....0000  Protocol class         (0) Class 0
0000....  Message handling       (0) No special options
00000011  Ptr - Called number    3
00001110  Ptr - Calling #        14
00011001  Pointer - Data         25
- - - - - - - - - - - - - - - - - - - - - - - - - - - - - - - -
Octet014  SCCP Called Party Address parameter
- - - - - - - - - - - - - - - - - - - - - - - - - - - - - - - -
00001011  Parameter length    11
.......0  Sgnl pt code bit       (0) SPC not present
......1.  Subsystem # bit        (1) SSN present
..0100..  Global title ind       (4) Global title incl. translation
                      type, numbering plan, encoding scheme & addr nature
.0......  Routing bit            (0) Global title based routing
0.......  Reserved natl use      0
00001000  Subsystem number       (8) MSC    Mobile Switching Center
00000000  Translation            (0) unknown
....0010  Encoding scheme        (2) BCD even
0001....  Numbering plan         (1) ISDN (Telephony) numbering plan
                      (Rec. E.164/E.163)
.0000100  Nature of address      (4) International number
0.......  Spare                  0
........  Address signals        966550309990
- - - - - - - - - - - - - - - - - - - - - - - - - - - - - - - -
Octet026  SCCP Calling Party Address parameter
- - - - - - - - - - - - - - - - - - - - - - - - - - - - - - - -
00001011  Parameter length    11
.......0  Sgnl pt code bit       (0) SPC not present
......1.  Subsystem # bit        (1) SSN present
..0100..  Global title ind       (4) Global title incl. translation
                      type, numbering plan, encoding scheme & addr nature
.0......  Routing bit            (0) Global title based routing
```

```
0.......  Reserved natl use      0
00001000  Subsystem number       (8) MSC    Mobile Switching Centre
00000000  Translation            (0) unknown
....0010  Encoding scheme        (2) BCD even
0001....  Numbering plan         (1) ISDN (Telephony) numbering plan
                                      (Rec. E.164/E.163)
.0000100  Nature of address      (4) International number
0.......  Spare                  0
........  Address signals        131231499400
- - - - - - - - - - - - - - - - - - - - - - - - - - - - - - - - -
Octet038  SCCP Data parameter
- - - - - - - - - - - - - - - - - - - - - - - - - - - - - - - - -
00001111  Parameter length       15
01100100  Tag                        (100) END End, constructor,
                                           application-wide
00001101  Length                 13
- - - - - - - - - - - - - - - - - - - - - - - - - - - - - - - - -
Octet041  DestinationTransactionID
- - - - - - - - - - - - - - - - - - - - - - - - - - - - - - - - -
...01001  Tag                        (9) DestinationTransactionID
010.....  Class and form             (2) Application-wide, primitive
00000100  Length                 4
........  Destination ID         0131BE6D
- - - - - - - - - - - - - - - - - - - - - - - - - - - - - - - - -
Octet047  ComponentPortion
- - - - - - - - - - - - - - - - - - - - - - - - - - - - - - - - -
...01100  Tag                    (12) ComponentPortion
011.....  Class and form         (3) Application-wide, constructor
00000101  Length                 5
- - - - - - - - - - - - - - - - - - - - - - - - - - - - - - - - -
Octet049  ReturnResult
- - - - - - - - - - - - - - - - - - - - - - - - - - - - - - - - -
...00010  Tag                    (2) ReturnResult
101.....  Class and form         (5) Context-specific, constructor
00000011  Length                 3
- - - - - - - - - - - - - - - - - - - - - - - - - - - - - - - - -
Octet051  InvokeID
- - - - - - - - - - - - - - - - - - - - - - - - - - - - - - - - -
...00010  Tag                        (2) InvokeID
000.....  Class and form             (0) Universal, primitive
00000001  Length                 1
00000001  InvokeID               01
- - - - - - - - - - - - - - - - - - - - - - - - - - - - - - - - -
     Checksum        CRC16................      1011011011101001
hex=B6E9
```

16.8   Example 8: SMS Interworking Between CDMA Networks

Answer the following questions:

1. In IS-41 what is the equivalent of the TCAP operations BEGIN and END?
2. What is the GSM SMS equivalent of this IS-41 operation?
3. What is the MSISDN of the cell-phone MSISDN destination number of the SMS (Globalstar USA uses CDMA)?

```
- - - - - - - - - - - - - - - - - - - - - - - - - - - - - - - -
    Octet001   ANSI SS7                  Count=000001 Time=10/20/2003
19:14:40:949
- - - - - - - - - - - - - - - - - - - - - - - - - - - - - - - -
    10011100   BIB/BSN                (156) 1/28
    11101110   FIB/FSN                (238) 1/110
    ..111111   SU type/length         (63) MSU63
    00......   Spare                  0
- - - - - - - - - - - - - - - - - - - - - - - - - - - - - - - -

    Octet004   Service information octet
- - - - - - - - - - - - - - - - - - - - - - - - - - - - - - - -
    ....0011   Service indicator      (3) SCCP  Signaling Connection
                                          Control Part
    ..00....   Message priority       0
    10......   Network indicator      (2) N  National network
- - - - - - - - - - - - - - - - - - - - - - - - - - - - - - - -

    Octet005   Routing label
- - - - - - - - - - - - - - - - - - - - - - - - - - - - - - - -
    ........   DPC: Net-Clstr-Mbr     001-044-230 nVPC
    ........   OPC: Net-Clstr-Mbr     005-021-204
    00011011   SLS                    27
- - - - - - - - - - - - - - - - - - - - - - - - - - - - - - - -
    Octet012   Message type
- - - - - - - - - - - - - - - - - - - - - - - - - - - - - - - -
    00001001   Message type           (9) UDT   Unitdata
- - - - - - - - - - - - - - - - - - - - - - - - - - - - - - - -
    Octet013   SCCP Protocol Class parameter
- - - - - - - - - - - - - - - - - - - - - - - - - - - - - - - -
    ....0000   Protocol class         (0) Class 0
    1000....   Message handling       (8) Return message on error
    00000011   Ptr - Called number    3
    00001100   Ptr - Calling #        12
    00010101   Pointer - Data         21
- - - - - - - - - - - - - - - - - - - - - - - - - - - - - - - -
    Octet017   SCCP Called Party Address parameter
- - - - - - - - - - - - - - - - - - - - - - - - - - - - - - - -
```

```
      00001001   Parameter length    9
      .......1   Subsystem # bit     (1) SSN present
      ......0.   Sgnl pt code bit    (0) SPC not present
      ..0010..   Global title ind    (2) Global title includes
                                         translation type only
      .0......   Routing bit         (0) Global title based routing
                                         (SCCP translation required)
  1.......   Natl Intnatl bit        (1) National address
      00000110   Subsystem number    (6) HLR      Home Location
                                         Register
      00001010   Translation type    (10) Network Entity Addressing
      ........   Address signals     905922121301
  - - - - - - - - - - - - - - - - - - - - - - - - - - - - - - - -
      Octet027   SCCP Calling Party Address parameter
  - - - - - - - - - - - - - - - - - - - - - - - - - - - - - - - -
      00001001   Parameter length    9
      .......1   Subsystem # bit     (1) SSN present
      ......0.   Sgnl pt code bit    (0) SPC not present
      ..0010..   Global title ind    (2) Global title includes
                                         translation type only
      .0......   Routing bit         (0) Global title based routing
                                         (SCCP translation required)
      1.......   Natl Intnatl bit    (1) National address
      00001011   Subsystem number    (11) SMS        Short Message
                                         Service
      00001010   Translation type    (10) Network Entity Addressing
      ........   Address signals     125437899970
  - - - - - - - - - - - - - - - - - - - - - - - - - - - - - - - -
      Octet037   SCCP Data parameter
  - - - - - - - - - - - - - - - - - - - - - - - - - - - - - - - -
      01001000 Parameter length    72
      11100010 Tag                 (226) QRY W Query with permission,
                                        constructor, private use
      01000110 Length              70
  - - - - - - - - - - - - - - - - - - - - - - - - - - - - - - - -
      Octet040   TCAP Transaction ID
  - - - - - - - - - - - - - - - - - - - - - - - - - - - - - - - -
      ...00111 Tag                     (7) TCAP Transaction ID
      110..... Class and form          (6) Private use, primitive
      00000100 Length                  4
      ........ Originating ID          041B01F1
  - - - - - - - - - - - - - - - - - - - - - - - - - - - - - - - -
      Octet046   TCAP Component Sequence
  - - - - - - - - - - - - - - - - - - - - - - - - - - - - - - - -
      ...01000 Tag                     (8) TCAP Component Sequence
      111..... Class and form      (7) Private use, constructor
      00111110 Length                  62
```

```
- - - - - - - - - - - - - - - - - - - - - - - - - - - - - - - -
   Octet048   Invoke component
- - - - - - - - - - - - - - - - - - - - - - - - - - - - - - - -
   ...01001   Tag                        (9) Invoke component
   111.....   Class and form             (7) Private use, constructor
   00111100   Length                     60
- - - - - - - - - - - - - - - - - - - - - - - - - - - - - - - -
   Octet050   TCAP Component ID
- - - - - - - - - - - - - - - - - - - - - - - - - - - - - - - -
   ...01111   Tag                        (15) TCAP Component ID
   110.....   Class and form             (6) Private use, primitive
   00000001   Length                     1
   11001111   Invoke ID                  CF
- - - - - - - - - - - - - - - - - - - - - - - - - - - - - - - -
   Octet053   Private TCAP Op Code
- - - - - - - - - - - - - - - - - - - - - - - - - - - - - - - -
   ...10001   Tag                        (17) Private TCAP Op Code
   110.....   Class and form             (6) Private use, primitive
   00000010   Length                     2
   .0001001   Operation Family           (9) IS-41
   0.......   Reply Expected             0
   00110101   Operation Specifier        (53) SMSDelptpt  SMS Delivery Point
To Point
- - - - - - - - - - - - - - - - - - - - - - - - - - - - - - - -
   Octet057   Parameter
- - - - - - - - - - - - - - - - - - - - - - - - - - - - - - - -
   ...10010   Tag                        (18) Parameter Set
   111.....   Class and form             (7) Private use, constructor
   00110011   Length                     51
- - - - - - - - - - - - - - - - - - - - - - - - - - - - - - - -
   Octet059   IS-41 Mobile Identification Number
- - - - - - - - - - - - - - - - - - - - - - - - - - - - - - - -
   ...01000   Tag                        (8) IS-41 Mobile Identification
                                             Number
   100.....   Class and form          (4) Context-specific, primitive
   00000101   Parameter length     5
   ........   Digit(s)             2542044662
- - - - - - - - - - - - - - - - - - - - - - - - - - - - - - - -
   Octet066   IS-41 Electronic Serial Number
- - - - - - - - - - - - - - - - - - - - - - - - - - - - - - - -
   ...01001   Tag                        (9) IS-41 Electronic Serial Number
   100.....   Class and form          (4) Context-specific, primitive
   00000100   Parameter length     4
   ........   Electronic serial numb 7401B82B
- - - - - - - - - - - - - - - - - - - - - - - - - - - - - - - -
   Octet072   IS-41 SMS_Bearer Data
- - - - - - - - - - - - - - - - - - - - - - - - - - - - - - - -
```

```
    ...11111  Tag                 (31) Extended tag
    100.....  Class and form       (4) Context-specific, primitive
    ........  Extended tag        (105) IS-41 SMS Bearer Data
    00011010  Length               26
    ........  Unsupported Format    00 03 10 AD 70 01 10 10 82 04 08 10
20 41 1F 66 FC 58 76 73 E9 87 90 0B 01 12
- - - - - - - - - - - - - - - - - - - - - - - - - - - - - - -
    Octet101  IS-41 SMS_Notification Indicator
- - - - - - - - - - - - - - - - - - - - - - - - - - - - - - -
    ...11111 Tag                  (31) Extended tag
    100..... Class and form        (4) Context-specific, primitive
    ........ Extended tag         (109) IS-41 SMS Notification
                                        Indicator
    00000001 Parameter length  1
    00000001 Notific. indicator  (1) Notify when available
- - - - - - - - - - - - - - - - - - - - - - - - - - - - - - -
    Octet105  IS-41 SMS_Teleservice Identifier
- - - - - - - - - - - - - - - - - - - - - - - - - - - - - - -
    ...11111  Tag                 (31) Extended tag
    100.....  Class and form       (4) Context-specific, primitive
    ........  Extended tag        (116) IS-41 SMS Teleservice
                                        Identifier
    00000010  Parameter length  2
    ........  Teleservice ID      (4099) CDMA Voice mail notification
- - - - - - - - - - - - - - - - - - - - - - - - - - - - - - -
    Checksum  CRC16.................  1000101110111100 hex=8BBC
- - - - - - - - - - - - - - - - - - - - - - - - - - - - - - -
- - - - - - - - - - - - - - - - - - - - - - - - - - - - - - -
    Octet001  ANSI SS7                 Count=000001 Time=10/20/2003
19:14:59:463
                                    -
    10111000  BIB/BSN              (184) 1/56
    10100010  FIB/FSN              (162) 1/34
    ..111000  SU type/length       (56) MSU56
    00......  Spare                0
- - - - - - - - - - - - - - - - - - - - - - - - - - - - - - -
    Octet004  Service information octet
- - - - - - - - - - - - - - - - - - - - - - - - - - - - - - -
    ....0011  Service indicator    (3) SCCP  Signaling Connection
                                        Control Part
    ..01....  Message priority   1
    10......  Network indicator    (2) N  National network
- - - - - - - - - - - - - - - - - - - - - - - - - - - - - - -
    Octet005  Routing label
- - - - - - - - - - - - - - - - - - - - - - - - - - - - - - -
    ........  DPC: Net-Clstr-Mbr   005-021-206
    ........  OPC: Net-Clstr-Mbr   001-044-230 nVPC
```

```
00010010  SLS                         18
- - - - - - - - - - - - - - - - - - - - - - - - - - - - - - - -
Octet012  Message type
- - - - - - - - - - - - - - - - - - - - - - - - - - - - - - - -
00001001  Message type              (9) UDT   Unitdata
- - - - - - - - - - - - - - - - - - - - - - - - - - - - - - - -
Octet013  SCCP Protocol Class parameter
- - - - - - - - - - - - - - - - - - - - - - - - - - - - - - - -
....0000  Protocol class            (0) Class 0
1000....  Message handling          (8) Return  message on error
00000011  Ptr - Called number       3
00001100  Ptr - Calling #           12
00010101  Pointer - Data            21
- - - - - - - - - - - - - - - - - - - - - - - - - - - - - - - -
Octet017  SCCP Called Party Address parameter
- - - - - - - - - - - - - - - - - - - - - - - - - - - - - - - -
00001001  Parameter length          9
.......1  Subsystem # bit           (1) SSN present
......0.  Sgnl pt code bit          (0) SPC not present
..0010..  Global title ind          (2) Global title includes
                                         translation type only
.0......  Routing bit               (0) Global title based routing
                                         (SCCP translation required)
1.......  Natl Intnatl bit          (1) National address
00000110  Subsystem number          (6) HLR   Home Location Register
00001010  Translation type          (10) Network Entity Addressing
........  Address signals           125437899970
- - - - - - - - - - - - - - - - - - - - - - - - - - - - - - - -
Octet027  SCCP Calling Party Address parameter
- - - - - - - - - - - - - - - - - - - - - - - - - - - - - - - -
00001001  Parameter length          9
.......1  Subsystem # bit           (1) SSN present
......0.  Sgnl pt code bit          (0) SPC not present
..0010..  Global title ind          (2) Global title includes
                                         translation type only
.0......  Routing bit               (0) Global title based routing
                                         (SCCP translation required)
1.......  Natl Intnatl bit          (1) National address
00000111  Subsystem number          (7) VLR    Visited Location
                                         Register
00001010  Translation type          (10) Network Entity Addressing
........  Address signals           905922121301
- - - - - - - - - - - - - - - - - - - - - - - - - - - - - - - -
Octet037  SCCP Data parameter
- - - - - - - - - - - - - - - - - - - - - - - - - - - - - - - -
00010110  Parameter length          22
```

```
11100100  Tag                    (228) RESP    Response, constructor,
                                        private use
00010100  Length                 20
- - - - - - - - - - - - - - - - - - - - - - - - - - - - - - -
Octet040  TCAP Transaction ID
- - - - - - - - - - - - - - - - - - - - - - - - - - - - - - -
...00111  Tag                    (7) TCAP Transaction ID
110.....  Class and form         (6) Private use, primitive
00000100  Length                 4
........  Responding ID          041B01F1
- - - - - - - - - - - - - - - - - - - - - - - - - - - - - - -
Octet046  TCAP Component Sequence
- - - - - - - - - - - - - - - - - - - - - - - - - - - - - - -
...01000  Tag                    (8) TCAP Component Sequence
111.....  Class and form         (7) Private use, constructor
00001100  Length                 12
- - - - - - - - - - - - - - - - - - - - - - - - - - - - - - -
Octet048  Return Result component
- - - - - - - - - - - - - - - - - - - - - - - - - - - - - - -
...01010  Tag                    (10) Return Result component
111.....  Class and form         (7) Private use, constructor
00001010  Length                 10
- - - - - - - - - - - - - - - - - - - - - - - - - - - - - - -
Octet050  TCAP Component ID
- - - - - - - - - - - - - - - - - - - - - - - - - - - - - - -
...01111  Tag                    (15) TCAP Component ID
110.....  Class and form         (6) Private use, primitive
00000001  Length                 1
11001111  Correlation ID         CF
- - - - - - - - - - - - - - - - - - - - - - - - - - - - - - -
Octet053  Parameter
- - - - - - - - - - - - - - - - - - - - - - - - - - - - - - -
...10010  Tag                    (18) Parameter Set
111.....  Class and form         (7) Private use, constructor
00000101  Length                 5
- - - - - - - - - - - - - - - - - - - - - - - - - - - - - - -
Octet055  IS-41 SMS_Cause Code
- - - - - - - - - - - - - - - - - - - - - - - - - - - - - - -
...11111  Tag                    (31) Extended tag
100.....  Class and form         (4) Context-specific, primitive
........  Extended tag           (153) IS-41 SMS_Cause Code
00000001  Parameter length       1
00100010  cause code             (34) No acknowledgment
- - - - - - - - - - - - - - - - - - - - - - - - - - - - - - -
Checksum  CRC16................  1011001001101000 hex=B268
```

Abbreviations and Acronyms

A3   Authentication algorithm A3

A38   Single algorithm performing the functions of A3 and A8

A5/1   Encryption algorithm A5/1

A5/2   Encryption algorithm A5/2

A5/X   Encryption algorithm A5/0–7

A8   Ciphering key generating algorithm A8

AA19   Standard GSM contract between two operators for the charging of the SMS-MT sent to their own subscribers by the other

AB   Access burst

AC   Access class (C0 to C15); application context

ACC   Automatic congestion control

ACCH   Associated control channel

ACK   Acknowledgment

ACM   Accumulated call meter (a zone of a SIM card); Address complete message (response to an ISUP call setup)

ACMmax   Maximum of the accumulated call meter

ACSE   Association control service element

ACU   Antenna combining unit

ADC   Administration center; analog-to-digital converter

ADN   Abbreviated dialing number

ADPCM   Adaptive differential pulse code modulation

AE   Application entity

AEC   Acoustic echo control

AEF   Additional elementary functions

AGCH   Access grant channel

AI   Action indicator

AMPS   Advanced Mobile Phone System; analog mobile radio system

ANSI   American National Standards Institute

AoC   Advice of charge

AoCC   Advice of charge charging supplementary service

AoCI   Advice of charge information supplementary service

AP   Application part

APLMN   Associated public land mobile network

AS   Affiliate server (SIGTRAN)

ASE   Application service element

ASN.1   Abstract Syntax Notation One

ARFCN   Absolute radio-frequency channel number

ARQ   Automatic request for retransmission

ASP   Application service provider (content provider for Internet services); application server (SIGTRAN)

ATT(flag)   Attach

AU   Access unit

AuC   Authentication center

AUT(H)   Authentication

BA   BCCH allocation

BAIC   Barring of all incoming calls supplementary service

BAOC   Barring of all outgoing calls supplementary service

BCC   BTS color code

BCCH   Broadcast control channel

BCD   Binary coded decimal

BCF   Base station control function

BCIE   Bearer capability information element

BER   Bit error rate

BFI   Bad frame indication

BI   All barring of incoming call supplementary services

BIB   Backward indicator bit

BIC-Roam   Barring of incoming calls when roaming outside the home PLMN country supplementary service

Bm   Full-rate traffic channel

BN   Bit number

BO   All barring of outgoing call supplementary services

BOIC   Barring of outgoing international calls supplementary service

BOIC-exHC   Barring of outgoing international calls except those directed to the home PLMN country supplementary service

BS   Basic service (group);
    Bearer service

BSG   Basic service group

BSC   Base station controller

BSIC   Base transceiver station identity code

BSIC-NCELL   BSIC of an adjacent cell

BSN   Backward sequence number

BSS   Base station system

BSSAP   Base station system application part

BSSAP-LE   BSSAP with location extension (for LBS)

BSSMAP   Base station system management application part

BSSOMAP   Base station system operation and maintenance application part

BTS   Base transceiver station

C   Conditional

CA   Cell allocation

CAI   Charge advice information

CAMEL   Customized application for mobile network

CB   Cell broadcast

CBC   Cell broadcast center

CBCH   Cell broadcast channel

CBMI   Cell broadcast message identifier

CC   Country code; Call control

CCBS   Completion of calls to busy subscriber supplementary service

CCCH   Common control channel

CCF   Conditional call forwarding

CCH   Control channel

CCITT   Comité Consultatif International Télégraphique et Téléphonique (The Consultative Committee on International Telegraphy and Telephony)

CCM   Current call meter

CCP   Capability/configuration parameter

CCPE   Control channel protocol entity

Cct   Circuit

CDMA   Code division multiple access

CDR   Call detailed record (billing record)

CDUR   Chargeable duration

CED   Called station identifier

CEIR   Central equipment identity register

CEND   End of charge point

CEPT   Conférence des Administrations Européennes des Postes et Telecommunications

CF   Conversion facility; all call forwarding services

CFB   Call forwarding on mobile subscriber busy supplementary service

CFNRc   Call forwarding on mobile subscriber "not reachable" supplementary service

CFNRy   Call forwarding on "no reply" supplementary service

CFU   Call forwarding unconditional supplementary service

CHP   Charging point

CHV   Card holder verification information

CI   Cell identity; CUG index

CIC   Circuit identification code

CIR   Carrier-to-interference ratio

CKSN   Ciphering key sequence number

CLI   Calling line identity

CLIP   Calling line identification presentation supplementary service

CLIR   Calling line identification restriction supplementary service

CM   Connection management

CMD   Command

CMM   Channel mode modify

CNF   Confirmation [answer to a REQ (request)]

CNG   Calling tone

COLI   Connected line identity

COLP   Connected line identification presentation supplementary service

COLR   Connected line identification restriction supplementary service

COM   Complete

CONNACK   Connect acknowledgment

C/R   Command/response field bit

CRC   Cyclic redundancy check (3 bit)

CRE   Call reestablishment procedure

CSPDN   Circuit-switched public data network

CT   Call transfer supplementary service; channel tester; channel type

CTR   Common technical regulation

CUG   Closed user group supplementary service

CW   Call waiting supplementary service

DA   Destination address

DAC   Digital-to-analog converter

DAMPS   Digital AMPS (the TDMA mobile radio system)

DB   Dummy burst

DCCH   Dedicated control channel

DCE   Data circuit terminating equipment

DCF   Data communication function

DCN   Data communication network

DCS1800   Digital Cellular System at 1,800 MHz

DET   Detach

DISC   Disconnect

DL   Data link (layer)

DLCI   Data link connection identifier

DLD   Data link discriminator

Dm   Control channel (ISDN terminology applied to mobile service)

DMR   Digital mobile radio

DNIC   Data network identifier

DNS   Domain name server

DP   Dial/dialed pulse; Destination point (of an IN service)

DPC   Destination point code

DRS Domain resolution server

DRX Discontinuous reception (mechanism)

DSE Data switching exchange

DSI Digital speech interpolation

DSS1 Digital Subscriber Signaling No. 1

DTAP Direct transfer application part

DTE Data terminal equipment

DTMF Dual-tone multiple-frequency (signaling)

DTX Discontinuous transmission (mechanism)

EA External alarms

EBSG Elementary basic service group

ECM Error correction mode (facsimile)

Ec/No Ratio of energy per modulating bit to the noise spectral density

ECT Explicit call transfer supplementary service

EEL Electric echo loss

EIA Electronic industries equipment

EIR Equipment identity register

EL Echo loss

EMC Electromagnetic compatibility

eMLPP Enhanced multilevel precedence and preemption service

EMMI Electrical man/machine interface

ENUM A protocol amplified in IETF RFC 2916 for fetching uniform resonance identifiers (URIs) given an E164 number

EPROM Erasable programmable read-only memory

ERP Ear reference point; equivalent radiated power

ERR Error

ESME External Short Message Entity (an ASP or ISP) connected by SMPP

ESN Electronic serial number

ETR ETSI technical report

ETS European Telecommunication Standard

ETSI European Telecommunications Standards Institute

E164 Format of the "ordinary" telephone numbers with a country code (CC) and a network destination code (NDC)

E212 Format of the IMSI telephone numbers with a mobile country code (MCC) and a mobile network code (MNC)

E214 Format of a destination address; a mix of E164 and E212

FA Full allocation; fax adapter

FAC Final assembly code

FACCH Fast associated control channel

FACCH/F Fast associated control channel/full rate

FACCH/H Fast associated control channel/half rate

FB   Frequency correction burst

FCCH   Frequency correction channel

FCS   Frame check sequence

FDM   Frequency division multiplex

FDN   Fixed dialing number

FEC   Forward error correction

FER   Frame erasure ratio

FH   Frequency hopping

FIB   Forward indicator bit

FISU   Fill-in signal units

FN   Frame number

FR   Full rate

FSG   Foreign subscriber gateway

FSN   Forward sequence number

ftn   Forwarded-to number

Gc   Protocol map interfacing a GGSN and an HLR

GCR   Group call register

GGSN   GPRS gateway support node; in a GPRS-equipped network, provides the interface between an operator's own IP network and the external IP network (GRX mostly)

GMLC   Gateway mobile location center

GMSC   Gateway mobile services switching center

GMSK   Gaussian minimum shift keying (modulation)

GPA   GSM PLMN area

GPRS   General packet radio service

GRX   The intranet IP network used by mobile operators to exchange GPRS data; is operated on a cooperative basis by the main international carriers

GSA   GSM system area

GSM   Global System for Mobile Communications

GSM MS   GSM mobile station

GSM PLMN   GSM public land mobile network

GT   Global title (E164 numbering address)

GTT   Global title translation

HANDO   Handover

HDLC   High-level data link control

HLC   High layer compatibility

HLR   Home location register

HOLD   Call hold supplementary service

HPLMN   Home PLMN

HPU   Hand-portable unit

HR   Half rate

HSN   Hopping sequence number

HU   Home units

I   Information frames (RLP)

IA   Incoming access (closed user group SS)

IA5   International Alphabet 5

IAM   Initial address message

IAP   Internet access provider; provides access to a modem or a permanent IP connection to the Internet, not necessarily a portal or content provider

IC   Interlock code (CUG SS)

ICB   Incoming calls barred (within the CUG)

ICC   Integrated circuit card

IC(pref)   Interlock code of the preferential CUG

ICM   In-call modification

ID   Identification/identity/identifier

IDN   Integrated digital network

IE   (Signaling) information element

IEC   International Electrotechnical Commission

IEI   Information element identifier

IETF   Internet Engineering Task Force

I-ETS   Interim European Telecommunications Standard

IGP   International gateway provider; provides SCCP access to the SS7 network

IMEI   International mobile station equipment identity

IMSI   International mobile subscriber identity

IN   Interrogating node

INAP   Intelligent network application part

InitialDP   CAMEL service to start an IN service

IP   Internet Protocol

IR21   International Roaming 21 document; provides a description of the detailed numbering plan as standardized by the GSM association

IRMIN   International roaming MIN (IS-41)

ISC   International switching center

ISDN   Integrated Services Digital Network

ISFS   Intelligent SMS filtering system

ISO   International Organization for Standardization

ISP   Internet service provider; provides content for Internet services

ISUP   ISDN User Part (of Signaling System No.7)

ITC   Information transfer capability

ITU   International Telecommunication Union

IVR   Interactive voice response

IWF   Interworking function

IWMSC   Interworking MSC

IWU   Interworking unit

k   Windows size

K   Constraint length of the convolutional code

Kc   Ciphering key

Ki   Individual subscriber authentication key

L1   Layer 1

L2ML   Layer 2 management link

L2R   Layer 2 relay

L2R BOP   L2R bit-oriented protocol

L2R COP   L2R character-oriented protocol

L3   Layer 3

LA   Location area

LAC   Location area code

LAI   Location area identity

LAN   Local area network

LAPB   Link Access Protocol Balanced

LAPDm   Link Access Protocol on the DM Channel

LBS   Location-based services

LCN   Local communication network

LCS   Location service

LCSC   LCS client

LCSS   LCS server

LE   Local exchange

LI   Length indicator; line identity

LLC   Low layer compatibility

Lm   Traffic channel with capacity lower than a Bm

LMSI   Local mobile station identity

LMU   Location measurement unit

LND   Last number dialed

LNP   Local number portability

LPLMN   Local PLMN

LR   Location register

LSSU   Link status signal units

LSTR   Listener sidetone rating

LTE   Local terminal emulator

LU   Local units; location update

LV   Length and value

M   Mandatory

MA   Mobile allocation

MACN   Mobile allocation channel number

MAF   Mobile additional function

MAH   Mobile access hunting supplementary service

MAI   Mobile allocation index

MAIO   Mobile allocation index offset

MAP   Mobile application part

MC   Message center (in the IS-41 network, equivalent to a GSM SMSC)

MCC   Mobile country code

MCI   Malicious call identification supplementary service

MD   Mediation device

MDL   (Mobile) management (entity); in the data link layer

ME   Maintenance entity; mobile equipment

MEF   Maintenance entity function

MF   Multiframe

MGT   Mobile global title

MHS   Message handling system

MIC   Mobile interface controller

MIN   Mobile identity number

MLC   Mobile location center

MLU   Mobile location units

MM   Man/machine; mobility management

MME   Mobile management entity

MMI   Man/machine interface

MMS   Multimedia messaging service

MM1   Protocols using IP standards (HTTP) to exchange MMS between the cell phone and the MMSC

MM4   In the MMS architecture, protocol to send MMS from one MMSC to another (interconnection); basically SMTP (e-mail)

MM5   In the MMS architecture, protocol to interrogate the HLRs

MM7   Protocols that use IP standards that allow the content provider to send MMS to an MMSC

MMS-IO   MMS interoperability

MNC   Mobile network code

MNP   Mobile number portability

MO   Mobile originated

MO-LR   Mobile originating location request

MoU   Memorandum of understanding

MPH   (Mobile) management (entity); in the physical layer (primitive)

MPTY   Multiple-party supplementary service

MRP   Mouth reference point

MS   Mobile station

MSC   Mobile services switching center; mobile switching center; includes anchor MSCs, which are mobile switching centers that are the first to assign a traffic channel to an MS; serving MSCs, which currently have the MS obtaining service at one of its cell sites; and tandem MSCs, which were previously the serving MSCs in the handoff chain

MSCID   MCS identity (a text string) in the IS-41 SMS protocol

MSCM   Mobile station class mark

MSCU   Mobile station control unit

MSISDN   Mobile station international ISDN number

MSRN   Mobile station roaming number

MSU   Message signal units

MT   Mobile terminated

MT (0,1,2)   Mobile termination

MT-LR   Mobile terminating location request

MTM   Mobile-to-mobile (call)

MTN   Maintenance regular message

MTP   Message transfer part

MTP2   MTP layer 2 (link control level)

MTP3   MTP layer 3 (network control sublevel; handles point codes)

MU   Markup

MUMS   Multiuser mobile station

MVNO   Mobile virtual network operator

MWD   Message waiting data (indication in an HLR)

M2UA   MTP2 user adaptation layer

M3UA   MTP3 user adaptation layer

N/W   Network

NAMPS   Narrow AMPS

NB   Normal burst

NBIN   Parameter in the hopping sequence

NCC   Network (PLMN) country code

NCELL   Neighboring (of current serving) cell

NCH   Notification channel

NDC   Network destination code

NDUB   Network determined user busy

NE   Network element

NEF   Network element function

NET   Norme Europeenne de Télécommunications

NF   Network function

NI   Network indicator

NIC   Network independent clocking

NI-LR   Network-induced location request

NM   Network management

NMC   Network management center

NMSI   National mobile station identification number

NPI   Number plan identifier

NPS   Network planning system

NSAP   Network service access point

NSS   Network vendors

NT   Network termination; Nontransparent

NTAAB   New Type Approval Advisory Board

NUA   Network user access

NUI   Network user identification

NUP   National user part (SS7)

O   Optional

OA   Outgoing access (CUG SS); Origin address

O&M   Operations and maintenance

OACSU   Off-the-air call setup

OCB   Outgoing calls barred within the CUG

OD   Optional for operators to implement for their aim

OLR   Overall loudness rating

OMC   Operations and maintenance center

OML   Operations and maintenance link

OPC   Originating point code

OS   Operating system

OSI   Open System Interconnection

OSI RM   OSI reference model

OSS   Originating supplementary services

OSSS   Originating SMS supplementary service

PABX   Private automatic branch exchange

PAD   Packet assembly/disassembly facility

PCH   Paging channel

PCM   Pulse code modulation

PD   Protocol discriminator; public data

PDN   Public data networks

PDU   Protocol data unit

PH   Packet handler; physical layer

PHI   Packet handler interface

PI   Presentation indicator

PICS   Protocol implementation conformance statement

PIN   Personal identification number

PIXT   Protocol implementation extra information for testing

PLMN   Public lands mobile network

PNE   Présentation des Normes Européennes

POI   Point of interconnection (with PSTN)

PP   Point-to-point

PPE   Primative procedure entity

Pref CUG   Preferential CUG

Ps   Location probability

PSPDN   Packet-switched public data network

PSTN   Public Switched Telephone Network

PUCT   Price-per-unit currency table

PW   Password

QA   Q (interface) adapter

QAF   Q adapter function

QoS   Quality of service

R   Value of reduction of the MS transmitted RF power relative to the maximum allowed output power of the highest power class of MS (A)

RA   Random mode request information field

RAB   Random access burst

RACH   Random access channel

RAND   Random number (used for authentication)

RBER   Residual bit error ratio

RDI   Restricted digital information

REC   Recommendation

REJ   Reject(ion)

REL   Release

REQ   Request

RF   Radio frequency

RFC, RFCH   Radio-frequency channel

RFN   Reduced TDMA frame number

RFU   Reserved for future use

RLP   Radio link protocol

RLR   Receiver loudness rating

rms   Root mean square (value)

RNTABLE   Table of 128 integers in the hopping sequence

ROSE   Remote operation service element

RPOA   Recognized private operating agency

RR   Radio resource

RSE   Radio system entity

RSL   Radio signaling link

RSZI   Regional subscription zone identity

RTE   Remote terminal emulator

RXLEV   Received signal level

RXQUAL   Received signal quality

S/W   Software

SABM   Set asynchronous balanced mode

SACCH   Slow associated control channel

SACCH/C4   Slow associated control channel/SDCCH/4

SACCH/C8   Slow associated control channel/SDCCH/8

SACCH/T   Slow associated control channel/traffic channel

SACCH/TF   Slow associated control channel/traffic channel full rate

SACCH/TH   Slow associated control channel/traffic channel half rate

SAP   Service access point

SAPI   Service access point indicator

SB   Synchronization burst

SC   Service center (used for SMS); service code

SCP   Service control point

SCCP   Signaling connection control part

SCF   Service control function

SCH   Synchronization channel

SCLC   SCCP connectionless control

SCMG   SCCP management

SCN   Subchannel number

SCOC   SCCP connection-oriented control

SCP   Service control point

SCRC   SCCP routing control

SCTP   Stram Control Transmission Protocol

SDCCH   Stand-alone dedicated control channel

SDL   Specification Description Language

SDP   Service data point

SDT   SDL development tool

SDU   Service data unit

SE   Support entity

SEF   Support entity function

SF   Status field

SFH   Slow-frequency hopping

SG   Signaling gateway (SIGTRAN)

SGSN   Support GPRS service node; in GSM 2.5G with GPRS, it has both circuit and IP interfaces and provides the GPRS service to a visiting cell phone; can deliver SMS-MT

SI   Screening indicator; service interworking; supplementary information

SID   Silence descriptor

SIGTRAN   Signal Transport Working Group; works on SS7/IP

SIF   Signaling information field

SIM   Subscriber identity module

SIO   Service information octet

SIP   Session Initiated Protocol (the VoIP protocol)

SLC   Signaling link code

SLPP   Subscriber LCS privacy profile

SLR   Send loudness rating

SLS   Signaling link selection

SLTA   Signaling link test message acknowledgment

SLTM   Signaling link test message (polling between adjacent point codes)

SM   Short message

SME   Short message entity

SMF   Service management function

SMG   Special mobile group

SMLC   Serving mobile location center

SMS   Short message service

SMSC   Short message service center

SMSCB   Short message service cell broadcast

SMSDPTP   SMS delivery point to point

SMSDBCKW   SMS delivery backward

SMSDFWD   SMS delivery forward

SMSSC   SMS service center

SMS-IO   SMS interoperability

SMS/PP   SMS/point-to-point

Smt   Short message terminal

SM-AL   Short message application layer

SM-TL   Short message transfer layer

SM-RL   Short message relay layer

SM-RP   Short message relay protocol

SN   Subscriber number

SNM   Signaling network management

SNR   Serial number

SOA   Suppress outgoing access (CUG SS)

SP   Service provider; signaling point; spare

SPC   Signaling point code

SPC   Suppress preferential CUG

SRES   Signed response (authentication)

SRI   SEND_ROUTING_INFO_FOR_SM

SRF   Service resource function

SS   Supplementary service; system simulator

SSC   Supplementary service control string

SSF   Subservice field

SSN   Subsystem number

SST   Subsystem test (polling between SCCP subsystems)

SSTA   Subsystem test acknowledgment

SS7   Signaling System No. 7

SSP   Service switching point

STMR   Sidetone masking rating

STP   Signaling transfer point

SU   Signal unit

SUA   SCCP user adaptation layer

SVN   Software version number

T   Timer; transparent; type only

TA   Terminal adapter; timing advance (between an MS and its serving BTS)

TAC   Type approval code

TAF   Terminal adaptation function

TBR   Technical basis for regulation

TC   Transaction capabilities

TCAP   Transaction capability application part

TCH   Traffic channel

TCH/F   A full-rate TCH

TCH/F2,4   A full-rate data TCH (2.4 Kbps)

TCH/F4,8   A full-rate date TCH (4.8 Kbps)

TCH/F9,6   A full-rate data TCH (9.6 Kbps)

TCH/FS   A full-rate speech TCH

TCH/H   A half-rate TCH

TCH/H2,4   A half-rate data TCH (2.4 Kbps)

TCH/H4,8   A half-rate data TCH (4.8 Kbps)

TCH/HS   A half-rate speech TCH

TCI   Transceiver control interface

TC-TR   Technical Committee Technical Report

TDMA   Time division multiple access

TE   Terminal equipment

TEI   Terminal endpoint identifier

TFA   Transfer allowed

TFP   Transfer prohibited

TI or TID   Transaction identifier (in the TCAP protocol)

TLV   Type, length, and value

TMN   Telecommunications management network

TMSI   Temporary mobile subscriber identity

TN   Time slot number

TOA   Time of arrival

TON   Type of number

TP   Transfer Protocol (in the MAP protocol)

TRX   Transceiver

TS   Time slot; technical specification; teleservice

TSC   Training sequence code

TSDI   Transceiver speech and data interface

TSS   Terminating supplementary services

TTCN   Tree and tabular combined notation

TUA   TCAP user adaptation layer

TUP   Telephone user part (SS7)

TV   Type and value

TXPWR   Transmit power; Tx power level in the MS_TXPWR_REQUEST and MS_TXPWR_CONF parameters

UA   User adaptation

UDI   Unrestricted digital information

UDT   Unit data message (of SCCP)

UDUB   User determined user busy

UI   Unnumbered information (frame)

UIC   Union Internationale des Chemins de Fer

UP   User part

UPCMI   Uniform PCM Interface (13-bit)

UPD   Up to date

USSD   Unstructured supplementary service data

UUS   User-to-user signaling supplementary service

VAD   Voice activity detection

VAP   Videotex access point

VBS   Voice broadcast service

VGCS   Voice group call service

VLR   Visitor location register

VMS   Voice mail system

VMSC   Visited MSC (recommend not to be used)

VPLMN   Visited PLMN

VPN   Virtual private network

VoIP   Voice-over-IP protocol

VSC   Videotex service center

V(SD)   Send state variable

VTX host   Components dedicated to Videotex service

WAN   Wide-area network

WAP   Wireless application protocol

WDP   WAP data protocol

WLL   Wireless local loop

WLNP   Wireless local number portability

WPA   Wrong password attempts (counter)

WS   Work station

WSP   WAP session protocol

WTP WAP transport protocol
XID Exchange identifier
ZC Zone code

About the Authors

Arnaud Henry-Labordère is a professor of operations research at Ecole Nationale des Ponts et Chaussées, Paris, and is also a member of the PRISM Laboratory, CNRS, at the University of Versailles. He was the chairman-founder of FERMA, a VMS and VAS platforms manufacturer, and then the chairman-founder of Nilcom, which has implemented a global SMS interworking network.

Vincent Jonack is currently a telecommunications system architect at Coframi and is the founder of Victoria Telecom, a consulting company specializing in telecommunications security architecture. Previously he worked for SAGEM, CNET, and Nilcom.

Index

3G UMTS networks, 202–3

A

Abstract Syntax Notation (ASN.1), 52
Access provider name (APN), 158
Accumulated call meter (ACM), 188
Address translation, 127–34
   GT, in GMSC, 127–30
   GT, in SMSC, 130–31
   GT, not possible, 135
Advanced Mobile Phone System (AMPS), 57
Advice_of_Charge (AoC), 187, 188–89
Alert mechanism, 11–13
ALERT_SERVICE_CENTER/CNF message,
   12
American National Standards Institute (ANSI),
   37
American Numbering Plan Association
   (ANPA), 60
Analyzer traces, 185–87
ANY_TIME_INTERROGATION_REQ/CNF
   message, 199
Application service providers (ASPs), *xvi*
Association control service element (ACSE), 30

B

Barring SMS-MO, 104
Barring SMS-MT, 101–8
   filtering at SCCP, 101–2
   filtering based on content, 105–8
   at GMSCs SCCP level, 102
   HLR, 103
   importance, 101
   intelligent, 104–8
   MAP, 103–4
   at MSC level, 103
   origin address-based, 104–5
   origin address type, 103
Base station system application part (BSSAP),
   200

Best flow problem, 165–72
   centralized network traffic regulation
      principle, 171–72
   continuous concave price function, 167
   global optimum algorithm, 171
   income model, 166
   mathematical optimization model, 168–71
   network model, 167–68
   noncontinuous price function, 166–67
Billing
   records and methods, 23–26
   source, 23
   virtual SMSCs, coherence, 111
Bind_transceiver PDU, 138, 141
Bind_transceiver_resp PDU, 139, 142

C

Call detailed records (CDRs), 23
   SMS-MO, 25
   SMS-MT, 26
CAMEL services
   CONTINUE, 182
   details, 182–83
   examples, 184
   existing, using, 183
   INITIAL_DP, 181, 182, 184
   specificity of, 183–84
   See also Customized application for mobile
      network (CAMEL)
CANCEL_LOCATION_REQ/CNF message,
   23
Cell ID method, 198–200
Cellular messaging teleservice (CMT), 78
Charge advice information (CAI), 188, 189
Circuit identification code (CIC), 35
CLOSE, 17
Code division multiple access (CDMA), 2
   addressing HLRs in, 83–84
   IMSI in, 60
   networks, SMS interworking between,
      294–300

Concatenated short messages, 9–11
  illustrated, 10
  maximum number of, 11
  reference number, 11
Continuous concave price function, 167
Conversion unit
  as GMSC of SMSC, 132–33
  as HLRs, 131–32
  private, 131–33
Customized application for mobile network
    (CAMEL), xv
  analyzer traces, 185–87
  Application Part (CAP), 177
  gateways, 184–85
  phase 1, 177
  primitives for prepaid service, 182–83
  service details, 182–83
  service specificity, 183–84
  for SMS prepaid services, 178–79
  transactions, 185–87

D

Destination networks
  fixed-line SMS interconnection, 96–97
  identification of, 96–98
  MMS and fixed-line SMS interconnection,
    97–98
  MMS interconnection, 96
Djsktra algorithm, 165
Domain name server (DNS), 152
Domain resolution servers (DRSs), 97
  agreement amount, 97
  architecture for telephone number mapping,
    98
Dual-tone multiple-frequency (DTMF), 181

E

Entropy
  average, 222–23
  global, 223
  numbering plan, 222–23
Examples (worked-out), 233–300
  example 1, 233–50
  example 2, 250–67
  example 3, 268
  example 4, 268–69
  example 5, 269–70
  example 6, 270–80
  example 7, 280–93
  example 8, 294–300
Extended cell ID method, 200

External short message entities (ESMEs), 137

F

Filtering
  based on content, 105–8
  functioning of, 107–8
  intelligent SMS system in relay mode, 106–7
  outgoing SMS, 108
  principle, 106–7
  at SCCP, 101–2
  See also Barring SMS-MT
Fixed-line SMS interconnection, 96–98
Foreign subscriber gateway (FSG) architecture,
    219–20
  illustrated, 219
  number setup and, 219
FORWARD_SHORT_MESSAGE_REQ/CNF
    message, 8
Frank-Wolfe algorithm, 171
Full-in signal units (FISUs), 32

G

Gateway mobile switching centers (GMSCs)
  configuration for third-party routing,
    127–34
  defined, 1
  MAP barring by, 103–4
Gateways
  CAMEL, 184–85
  SCCP, 134–35
  super-routing, 120–21
General packet radio service (GPRS), xv
  gateway (GGSN), 23
  procedures, 23
Global optimum algorithm, 171
Global System for Mobile Communications.
    See GSM
Global title (GT)
  address translation in GMSC, 127–30
  address translation in SMSC, 130–31
Global title translation (GTT), 37
GSM
  connection to third-party SMS network,
    280–93
  interworking between IS-41 and, 75–83
  IS-41 interworking through SMPP, 143–44
  mapping, to IS-136-710, 78
  mapping, to IS-637, 76–78
  modems, 215–16
  specifications of user information, 75
  standards, xiii

GSM networks
   architecture, 1–3
   IS-41 analogy, 72–75
   mobility management, 58
   networks use, 1
   SMS standard procedures in, 1–28
GSM-to-IS-41 protocol conversion, 94–95
   in sending SMSC, 95
   transparent text SMS-MT service for, 94–95
GT address translation (GMSC), 127–30
   functioning of, 127–28
   method, 129–30
   private conversion unit use, 131–33
   reasons for, 128–29
   table, 130
GT address translation (SMSC), 130–31
   layout, 131
   private conversion unit use, 131–33
   process, 130–31

H
Half-SCCP roaming, 112–14
   failure, 113
   failure solution, 113–14
   illustrated, 113
   See also Roaming
Home location registers (HLRs), 29
   addressing, in TDMA/CDMA networks,
      73–84
   barring, 103
   direct interrogation of, 126–27
   foreign network, 15–16
   interrogation of, 6–7
   multiple spanning of, 222
   numbering plan computation, 223–24
   number of, 16
   spanning, 222
Home public lands mobile network (HPLMN),
      5
Hyperbola
   computing, from TOA difference, 204
   equation, 203–4
   notation, 203–4
   rotated/translated, 204
Hyperbolic n-triangulation, 203–6
   best localization estimate, 204–5
   exact solution, 205–6
   illustrated, 201

I
Information elements (IEs), 69–70

defined, 69
   nonspecific, 70
   SMS-specific, 69–70
INFORM_SERVICE_CENTER_REQ message,
      13
INSERT_SUBSCRIBER_DATA profile, 22
INSERT_SUBSCRIBER_DATA_REQ/CNF
      message, 21
Intelligent barring, 104–8
   filtering based on content, 105–8
   origin address-based, 104–5
   See also Barring SMS-MT
Intelligent network application part (INAP), 30
   defined, 177
   implementation through, 30
Intelligent peripherals (IPs), 177
Intelligent SCCP routing, 133–34
Interactive voice response (IVR), 187
Interconnections, 96–98
   fixed-line SMS, 96–97
   MMS, 96
   MMS and fixed-line SMS, 97–98
   SMSC to third-party network, 118–20
International Forum on ANSI-41 Standards
      Technology (IFAST), 60
International gateway providers (IGPs)
   intelligent SCCP routing by, 133–34
   SCCP, 118–19
International Mobile Subscriber Identity
      (IMSI)
   card, 4
   in CDMA/TDMA networks, 60
International PC addressing, 119–20
International roaming MIN (IRM), 60
International Telecommunication Union-
Telecommunications Standardization Sector
      (ITU-T), 29
   defined, 29
   point codes, 37
Internet service providers (ISPs), 137
IS-41, xiii, 2
   analogy with GSM network, 72–75
   functional entities, 57
   GSM interworking through SMPP, 143–44
   interworking with GSM, 75–83
   numbering for SMS delivery, 83
   SMS procedures, 66–68
   SMS protocol description, 68–70
   SMS services, 66
IS-41 networks, 57
   defined, 57
   mobility management, 58

IS-41 networks (continued)
  SMS interworking issues, 61–63
  SMS-MO implementation, 61–63
  SMS-MO procedure, 58
  SMS-MT implementation, 63
  SMS-MT procedure, 59
  SMS services implementation, 61–63
  SMS standard procedures, 57–84
  voice call, 57–58
IS-41_SMS_DELIVERY_FORWARD_REQ/
    CNF message, 65
IS-41_SMS_DELIVERY_POINT_TO_POINT_
    REQ/CNF message, 64
IS-41_SMS_NOTIFICATION_REQ/CNF
    message, 65
IS-41 SMS router, 70–75
  data model, 71
  as SMS router, 72
  specification, 71–72
  See also Routers
IS-136-170, 78
IS-637
  defined, 76
  mapping GSM to, 76–78
  WMT, 76–78

L

Lagrange bound, 211
Latin multiplication algorithm
  defined, 162
  enumerating loopless paths with, 161–65
Least cost paths, 165
Least trouble paths, 165
Level N MNP, 87–90
  common gateway carrier, 87–89
  direct voice and signaling links, 87
  SMS reception cannot be maintained, 89
  SMS reception maintained, 89–90
  See also Mobile number portability (MNP)
Linearized tangential problem, 171
Link status signal units (LSSUs), 32, 33
  defined, 33
  signaling link functions, 34
Load tests, 26–28
  configuration, 26
  results/performance model, 26–28
Local number portability (LNP), 86
Location-based services (LBS), 197–213
  3G UMTS networks, 202–3
  cell ID method, 198–200
  extended cell ID method, 200
  hyperbolic n-triangulation, 203–6

  methods, 198–201
  MLUs and BSSAP-LE, 200–201
  mobile measured power level, 201–2
  mobile-originated, 197–98
  MSC location method, 198
  validity, 197
Loopless paths, 161–65

M

M2UA layer, 52
MAP_OPEN_IND message, 193
MAP_PROCESS_UNSTRUCTURED_SS_
    REQUEST_CNF message, 193
MAP_PROCESS_UNSTRUCTURED_SS_
    REQUEST_REQ message, 192
MAP_UNSTRUCTURED_SS_NOTIFY_
    REQ/CNF message, 194
MAP_UNSTRUCTURED_SS_REQUEST_
    REQ/CNF messages, 193
Mathematical optimization model, 168–71
  incidence matrix edge to paths, 169
  incidence matrix source x destination to
      paths, 169–71
  linear case, 171
  process, 168–69
Maximum value of call meter (ACMmax), 188
Message signal units (MSUs), 32, 33
  defined, 33
  formats, 36
Message submission operations, 139–43
Message transfer part (MTP), 29, 30–37
  layer 1, 31
  layer 2, 31–34
  layer 3, 34–37
  management messages, 36
Message waiting data (MWD), 12
MM1 protocol
  defined, 147
  over M-IMAP, 149–50
  over WAP, 147–49
MM1_Submit.REQ
  example, 152
  mapping to RFC2822 header, 153
MM1_Submit.RES, 153
MM4_Forward.REQ, 154
MM4 protocol, 150–51
MM7 protocol
  command example, 157
  defined, 151
  transaction flow, 156
MM7_Submit.REQ, 157
MM7_Submit.RES, 158

MMS
   cell phone profile setup, 156–60
   data access profile, 157–59
   deferred retrieval, 150
   defined, xiii
   immediate retrieval, 150
   interworking, 145–60
   interworking architectures with third party,
      151–56
   MM1 over M-IMAP, 149–50
   MM1 over WAP, 147–49
   MM4, 150–51
   MM7, 151
   MMSC profile, 159–60
   network architecture, 147
   PDUs, building, 148–49
   profile illustration, 159
   protocol interfaces, 148
   relay/server, 145
   sending, 149
   sending/receiving model, 145–47
   standard protocols, 147–51
   transaction flows, 149, 151
   user agent, 146
   user databases, 145
   uses, 3
   VAS application, 146–47
MMS interconnection, 96
   with fixed-line SMS interconnection, 97–98
   importance, 96
Mobile directory numbers (MDNs), 83
Mobile identity number (MIN), 59
   assignment, 59
   international roaming (IRM), 60
Mobile-initiated USSD services, 191–94
Mobile Internet Message Access Protocol
      (M-IMAP)
   defined, 149
   MM1 over, 149–50
Mobile location centers (MLCs), *xv*
Mobile location units (MLUs), 200
Mobile network codes (MNCs), 26
Mobile number portability (MNP), *xiv*
   defined, 85, 86
   handled by entry international SCCP
      gateway, 90–91
   handled by individual operators, 87–90
   handled by SMS interworking network, 92
   implementations, 86–91
   SMS routing strategies, 91–92
   for SMS with GSM and IS-41 operators,
      92–96

unregulated country process, 91
Mobile-originated LBS, 197–98
   functions, 197–98
   principle, 198
   See also Location-based services
Mobile station international ISDN number
      (MSISDN), xiv
Mobile subscriber roaming number (MSRN),
      22
Mobile switching centers (MSCs), xv
   barring at, 103
   location method, 198
   number of, 231
   search problem, 224–25
Mobile virtual network operators (MVNOs),
      195
Mobility Application Part (MAP) protocol, 2
   ASN.1 specification, 46
   barring by GMSC, 103–4
   dialog models, 16–18
   operation invocation, 47
   parts, 45
   primitives, 18
   Provider layer, 45–46
   segmented messages flowchart, 48
   traces, 233, 250
   User layer, 46
Mobility procedures, 19–23
   defined, 19
   network equipment model, 23
   telephone call to mobile, 22–23
   update location, 20–22
MT3UA layer, 52
Multimedia messaging. *See* MMS
Multimedia messaging center (MMSC)
   manufacturers, xvi
   profile, 159–60
Multimedia messaging service environment
      (MMSE), 145

N

Network access system (NAS), 157–58
Network destination code (NDC), xiv
Network equipment model
   illustrated, 16
   for mobility procedure, 23
   summary, 16
Noncontinuous price function, 166–67
N-triangulation, 203
Numbering plan indicator (NPI), 270
Numbering plans
   after one try, 224

Numbering plans (continued)
    after three tries, 224
    after two tries, 224
    average entropy, 222–23
    computation purpose, 221
    entropy, as quality indicator, 222–23
    HLR, computing, 223–24
Numbers
    setup, 219
    short-code, 218–19
    SMS-MO premium, 215–20

O

Office code number (OCN), 96
OPEN, parameters, 17–18
Optimal routing algorithms, 161–76
Origin address (OA), 103
    barring, 104–5
    type, barring at MSC level, 103
OSI layer, 30
Over the air (OTA) provisioning, 8

P

Paths
    least cost, 165
    least trouble, 165
    loopless, 161–65
    shortest, 165
    valuations, 176
PDUs
    bind_transceiver, 138, 141
    bind_transceiver_resp, 139, 142
    list of, 138–39
    MMS, building, 148–49
    submit_sm, 139–41, 142
    submit_sm_resp, 141–43
Premium number services, 215–20
Prepaid customers, 178–84
    CAMEL services for, 179–84
    credit reloading for, 179
    SMS payment from, 178
    voice call charging for, 180
Prepaid SMS
    with AoC-enabled networks, 188–89
    with CAMEL, 178–87
    with service nodes (SNs), 187–88
Protocol data units. *See* PDUs
PROVIDE_ROAMING_NUMBER_REQ/
    CNF message, 24
PROVIDE_SUBSCRIBER_INFO_REQ/
    CNF message, 199

PROVIDE_SUBSCRIBER_LOCATION_
    REQ/CNF message, 203
Public lands mobile network (PLMN), 29
Pulse code modulation (PCM), 31

R

READY_FOR_SM_REQ/CNF message, 11
Receiver (RX) sessions
    defined, 137
    example, 140
Recursive algorithm, 211–13
Relay mode, 123–25
    defined, 123–24
    illustrated, 124
Relay SCCP roaming, 116–17
    illustrated, 118
    implementing, 116–17
    international PC addressing vs., 119–20
Remote operation service element (ROSE), 30
Reply-path
    function, 4
    setting up, 8–9
REPORT_SM_DELIVERY_STATUS_REQ/
    CNF message, 12
Retry mechanism, 11–13
Roaming, 4–5
    agreement implementation, 85–86
    agreements, restricting, 117–18
    half-SCCP, 112–13
    relay SCCP, 116–17, 118
    setup problem, finding, 270
Routers
    IS-41, 70–75
    SS7, 172–76
Routing
    design, 48–49
    indicator field, 39
    intelligent SCCP, 133–34
    label, 36–37
    SCCP, 39–42
    SMS strategies, 91–92, 93
    third-party, 127–36
    voice call, to announcement machine,
        181–82

S

SCCP routing, 39–42
    intelligent, by IGP, 133–34
    third-party, 134–36
Searches average problem, 228–31
    asymptotic bound of M_N, 230–31

case *N*=2 MSCs, 228–29
case *N*=3 MSCs, 229–30
upper bound estimate, 230
SEND_INFO_FOR_MO_SMS_REQ/CNF
    message, 6
SEND_ROUTING_INFO_FOR_GPRS_
    REQ/CNF message, 25
SEND_ROUTING_INFO_FOR_SM_
    REQ/CNF message, 6, 7
SEND_ROUTING_INFO_REQ/CNF message,
    24
Service center, changing, 5
Service control function (SCF), 177
Service control points (SCPs), 30
Service indicator (SI), 34
    defined, 34
    field, 35
Service information octet (SIO), 34
Service management function (SMF), 177
Service nodes (SNs), 187–88
    defined, 187
    prepaid SMS with, 187–88
Service-oriented design, 50
Service resource function (SRF), 177
Service switching function (SSF), 177
Service switching points (SSPs), 30
Session management operations, 138–39
Short code numbers, 218–19
Shortest paths, 165
Short message application layer (SM-AL), 72
Short Message Relay Protocol (SM-RP), 3
    DA (destination address), 4
    originating address (OA), 4
    UI, 4
Short messages
    concatenated, 9–11
    maximum number of, 11
    sequence number, 11
Short message service. *See* SMS
Short message transfer layer (SM-TL), 72
Signaling
    data link level, 31
    link functions, 31–34
    message handling, 35
    network functions, 34–37
    network management, 35
    transmission between GSM and IS-41
        network, 136
Signaling connection control part (SCCP),
    37–42
    addresses, 18–19
    addressing scheme, 38

barring SMS-MT at, 102
calling party address, 42
connectionless control (SCLC), 38, 39
connectionless transport service, 43
connection-oriented control (SCOC), 38
defined, 29
envelope, 18
filtering service at, 101–2
gateway, 134–35
IGP, 118–19
layer architecture, 38–39
management (SCMG), 38, 39
message format, 38
relay roaming, 116–17, 118
routing control (SCRC), 38
SS7 connection setup, 37
tasks, 37
See also SCCP routing
Signaling link code (SLC), 35
Signaling link selection (SLS), 35
Signaling System No. 7. *See* SS7
Signaling transfer point (STP), 30, 85
    packet switches, 31
    private, 269
Signal Transport (SIGTRAN), 51
Simplified Mail Transfer Protocol (SMTP), 3
SIM toolkit application, 202
SMPP protocol, 137–38
    ancillary submission operations commands,
        139
    commands, 138
    defined, 137, 138
    dialogue, 270
    example session, 138–39
    GSM IS-41 interworking with, 143–44
    message delivery operations commands, 139
    message submission operations commands,
        139
    operations, 138–39, 143
    receiver (RX), 137
    session management operations commands,
        138
    sessions, 137
    transceiver (TRX), 137
    transmitter (TX), 137
SMS
    billing records and methods, 23–26
    defined, *xiii*
    delivery failures, 11–13
    fixed-line interconnection, 96–98
    global service network, 53

SMS (continued)
  interworking between CDMA networks, 294–300
  interworking operating margin, xv
  IS-41 procedure for, 63–75
  IS-41 protocol description, 68–70
  network equipment model, 16
  payment from prepaid customers, 178
  prepaid, 177–89
  price as function of offered throughput, 166–67
  price function with minimum offered capacity, 167–68
  rerouting, 269
  routing strategies, 91–92, 93
  standard procedures (GSM networks), 1–28
  standard procedures (IS-41 networks), 57–84
  USSD advantages, 191
SMS centers (SMSCs), xiii
  configuration for third-party routing, 127–34
  load tests, 26–28
  next-generation, 54
  originating address control, 5–6
  sending commands to, 14–15
  virtual, 109–14
SMS_COMMAND type, 15
SMS_DELIVER CMT service, 79
SMS delivery
  IS-41 numbering for, 83
  from IS-41 SME to MAP SME, 78–82
  from MAP SME to IS-41 SME, 82–83
SMS_DELIVERY_ACK CMT service, 80
SMS_DELIVERY_BACKWARD_REQ/CNF message, 61
SMS_DELIVERY_POINT_TO_POINT Invoke operation, 73–74
SMS_DELIVERY_POINT_TO_POINT_REQ/ CNF message, 62
SMS_DELIVERY_POINT_TO_POINT Return Result operation, 74
SMS_MANUAL_ACK CMT service, 80
SMS message terminated (SMS-MT), xiv
  CDRs, 26
  foreign network HLRs for, 15–16
  forwarding, 8
  GSM to IS-41 destination, 92–95
  implementation (GSM), 6–14
  implementation (IS-41), 63
  inbound, barring, 101–8
  IS-41 network to GSM destination, 95–96

  test configuration, 26, 27
SMS mobile originated (SMS-MO), xiv
  barring, 104
  CDRs, 25
  implementation (GSM), 3–6
  implementation (IS-41), 61–63
  mobile operator connection for, 123–36
  numbers, 218
  premium number business, 215–16
  with real SIM card, 218
  teleservice provisioning control, 6
SMS networks, 127
  detailed modeling of, 172–76
  full graph model, 175
  graph model, 162
  margin, maximizing, 161
  third-party, GSM connection, 280–93
SMS_NOTIFICATION Invoke operation, 75
SMS_NOTIFICATION Return Result operation, 75
SMS_REQUEST Invoke operation, 74
SMS_REQUEST_REQ/CNF message, 63
SMS_REQUEST Return Result operation, 74–75
SMS services
  functional entities, 66
  global, 54
  implementation in GSM networks, 2, 3–16
  implementation in IS-41 networks, 61–63
  IS-41, 64–68
SMS_STATUS_REPORT type, 14
SMS_SUBMIT CMT service, 79
SM_SUBMIT message, 5
Specification and Description Language (SDL), 52
SS7, 2
  connection setup with SCCP, 37
  data links, 50
  fault-tolerant system, 50
  interworking with, 52
  routes setup, 268
  signaling network, 31
  signal units, 32–34
  specification, 29
SS7-based-over-IP (SS7oIP), 51
SS7 routers, 172–76
  graph model illustration, 172
  modeling, 172–73
  network hosting, 173
  virtual, modeling, 173–76
STATUS_REPORT, 127

Stream Control Transmission Protocol (SCTP), 51
Sturm's theorem, 208–10
   defined, 210
   definition variation, 209
   quotient Sturm sequence, 210
   sequence theorem, 209–10
SUA layer, 52
Submit_sm PDU, 139–41, 142
Submit_sm_resp PDU, 141–43
Subscriber identity module (SIM) cards, 1
Subservice field (SSF), 34
Subsystem numbers (SSNs), 37
Super-routing gateway, 120–21
Support GPRS service node (SGSN), 7

T
Temporary mobile subscriber identity (TMSI), 202
Terminating supplementary services (TSSs), 67
Theory of resultants, 206
Third-party routing, 127–36
   GMSC for, 127–34
   SCCP, 134–36
   SMSC configuration for, 127–34
Time division multiple access (TDMA), 2
   addressing HLRs in, 83–84
   IMSI in, 60
Time of arrival (TOA), 200
   difference, 200–201
   in hyperbola computation, 204
   measurements, 200–201, 202
Transaction capability application part (TCAP), 16, 30, 42–45
   architecture, 43–44
   defined, 42–43
   features, 43
   implementation, 43
   operation invocation, 44–45, 47
   transaction identity (TID), 113
Transceiver (TRX) sessions
   defined, 137
   example, 141
Transit agreements
   implementation, 114–20
   implementation optimization, 118
   international point code use, 118–20
   virtual SMSC has all roaming agreements and, 114–18
Transmission Control Protocol/Internet Protocol (TCP/IP), 51
Transmitter (TX) sessions

   defined, 137
   example, 140
Transparent mode, 125–26
   defined, 125
   design, 126
   illustrated, 125
TUA layer, 52
Type of number (TON), 270

U
Unstructured supplementary service data (USSD), *xv*, 191–96
   advantages, 191
   application programming, 194
   call-back application, 195–96
   defined, 191
   example service, 194–95
   flexibility, 196
   free call-back service, 196
   functioning of, 191–94
   mobile-initiated services, 191–94
   mobile-originated sessions, 196
   in scratch card recharge, 195
Update location procedure, 20–22
UPDATE_LOCATION_REQ/CNF message, 21

V
Value-added services (VAS) applications, 146–47
   provider (VASP), 151
   types, 146–47
Virtual HLR/MSC approach, 123–27
   direct interrogation, 126–27
   relay mode, 123–25
   SMS interworking network, 127
   transparent mode, 125–26
Virtual private networks (VPNs), 29, 177
Virtual roaming subscriber architecture, 216–18
Virtual SMSCs, 109–14
   all roaming agreements, 114–18
   architecture, 109–10
   architecture illustration, 110
   billing coherence, 111
   business model, 109
   detailed implementation, 112–14
   GT, 111–12
   half-SCCP roaming, 112–13
   multiple, 120–21
   payment issues, 110–11

Virtual SMSCs (continued)
 principle, 109–12
 See also SMS centers (SMSCs)
Visitor location registers (VLRs), 29
Voice mail system (VMS), *xiii*

W

Wireless local number portability (WLNP), 86
Wireless messaging teleservice (WMT), 76
 mapping GSM SM-TL and IS-637, 76–78

SM-teleservice layer, 76
Worked-out examples, 233–300
 example 1, 233–50
 example 2, 250–67
 example 3, 268
 example 4, 268–69
 example 5, 269–70
 example 6, 270–80
 example 7, 280–93
 example 8, 294–300

Handbook of Land-Mobile Radio System Coverage, Garry C. Hess

Handbook of Mobile Radio Networks, Sami Tabbane

High-Speed Wireless ATM and LANs, Benny Bing

Interference Analysis and Reduction for Wireless Systems, Peter Stavroulakis

Introduction to 3G Mobile Communications, Second Edition, Juha Korhonen

Introduction to Digital Professional Mobile Radio, Hans-Peter A. Ketterling

Introduction to GPS: The Global Positioning System, Ahmed El-Rabbany

An Introduction to GSM, Siegmund M. Redl, Matthias K. Weber, and Malcolm W. Oliphant

Introduction to Mobile Communications Engineering, José M. Hernando and F. Pérez-Fontán

Introduction to Radio Propagation for Fixed and Mobile Communications, John Doble

Introduction to Wireless Local Loop, Second Edition: Broadband and Narrowband Systems, William Webb

IS-136 TDMA Technology, Economics, and Services, Lawrence Harte, Adrian Smith, and Charles A. Jacobs

Location Management and Routing in Mobile Wireless Networks, Amitava Mukherjee, Somprakash Bandyopadhyay, and Debashis Saha

Mobile Data Communications Systems, Peter Wong and David Britland

Mobile IP Technology for M-Business, Mark Norris

Mobile Satellite Communications, Shingo Ohmori, Hiromitsu Wakana, and Seiichiro Kawase

Mobile Telecommunications Standards: GSM, UMTS, TETRA, and ERMES, Rudi Bekkers

Mobile Telecommunications: Standards, Regulation, and Applications, Rudi Bekkers and Jan Smits

Multiantenna Digital Radio Transmission, Massimiliano "Max" Martone

Multipath Phenomena in Cellular Networks, Nathan Blaunstein and Jørgen Bach Andersen

Multiuser Detection in CDMA Mobile Terminals, Piero Castoldi

Personal Wireless Communication with DECT and PWT, John Phillips and Gerard Mac Namee

Practical Wireless Data Modem Design, Jonathon Y. C. Cheah

Prime Codes with Applications to CDMA Optical and Wireless Networks, Guu-Chang Yang and Wing C. Kwong

QoS in Integrated 3G Networks, Robert Lloyd-Evans

Radio Engineering for Wireless Communication and Sensor Applications,
Antti V. Räisänen and Arto Lehto

Radio Propagation in Cellular Networks, Nathan Blaunstein

Radio Resource Management for Wireless Networks, Jens Zander and
Seong-Lyun Kim

RDS: The Radio Data System, Dietmar Kopitz and Bev Marks

Resource Allocation in Hierarchical Cellular Systems, Lauro Ortigoza-Guerrero
and A. Hamid Aghvami

RF and Microwave Circuit Design for Wireless Communications,
Lawrence E. Larson, editor

Sample Rate Conversion in Software Configurable Radios, Tim Hentschel

Signal Processing Applications in CDMA Communications, Hui Liu

SMS and MMS Interworking in Mobile Networks, Arnaud Henry-Labordère and
Vincent Jonack

Software Defined Radio for 3G, Paul Burns

Spread Spectrum CDMA Systems for Wireless Communications, Savo G. Glisic and
Branka Vucetic

Third Generation Wireless Systems, Volume 1: Post-Shannon Signal Architectures,
George M. Calhoun

Traffic Analysis and Design of Wireless IP Networks, Toni Janevski

Transmission Systems Design Handbook for Wireless Networks, Harvey Lehpamer

UMTS and Mobile Computing, Alexander Joseph Huber and Josef Franz Huber

Understanding Cellular Radio, William Webb

Understanding Digital PCS: The TDMA Standard, Cameron Kelly Coursey

Understanding GPS: Principles and Applications, Elliott D. Kaplan, editor

Understanding WAP: Wireless Applications, Devices, and Services,
Marcel van der Heijden and Marcus Taylor, editors

Universal Wireless Personal Communications, Ramjee Prasad

WCDMA: Towards IP Mobility and Mobile Internet, Tero Ojanperä and
Ramjee Prasad, editors

Wireless Communications in Developing Countries: Cellular and Satellite Systems,
Rachael E. Schwartz

Wireless Intelligent Networking, Gerry Christensen, Paul G. Florack, and
Robert Duncan

Wireless LAN Standards and Applications, Asunción Santamaría and Francisco J. López-Hernández, editors

Wireless Technician's Handbook, Second Edition, Andrew Miceli

For further information on these and other Artech House titles, including previously considered out-of-print books now available through our In-Print-Forever® (IPF®) program, contact:

Artech House	Artech House
685 Canton Street	46 Gillingham Street
Norwood, MA 02062	London SW1V 1AH UK
Phone: 781-769-9750	Phone: +44 (0)20 7596-8750
Fax: 781-769-6334	Fax: +44 (0)20 7630-0166
e-mail: artech@artechhouse.com	e-mail: artech-uk@artechhouse.com

Find us on the World Wide Web at:
www.artechhouse.com